Theory and Practice of Animal Taxonomy and Biodiversity

Theory and Practice of Animal Taxonomy and Biodiversity

Eighth Edition

VC Kapoor

PhD, DSc, FNAAS
Ex-Professor and Head of Zoology, and
Ex-Coordinator of Research
College of Basic Sciences and Humanities
Punjab Agricultural University, Ludhiana, India

Oxford & IBH Publishing Co. Pvt. Ltd.

New Delhi

(*A Unit of* CBS Publishers & Distributors Pvt Ltd)

CBS

CBS Publishers & Distributors Pvt Ltd

New Delhi • Bengaluru • Chennai • Kochi • Kolkata • Lucknow • Mumbai
Hyderabad • Jharkhand • Nagpur • Patna • Pune • Uttarakhand

Theory and Practice of Animal Taxonomy and Biodiversity

Eighth Edition

ISBN-13: 978-81-204-1799-1
ISBN-10: 81-204-1799-2

© 2017, 2013, 2008, 2001, 1998, 1994, 1988, 1983; VC Kapoor
Reprint 2017, 2018, 2019, 2020, 2022, **2024**

OXFORD & IBH

New Delhi
(A Unit of CBS Publishers & Distributors Pvt Ltd)

Published by **Satish Kumar Jain** and produced by **Varun Jain** for

CBS Publishers & Distributors Pvt Ltd
4819/XI Prahlad Street, 24 Ansari Road, Daryaganj, New Delhi 110 002, India.
Ph: 011-23289259, 23266861   Website: www.cbspd.com
                                   e-mail: delhi@cbspd.com

Corporate Office: 204 FIE, Industrial Area, Patparganj, Delhi 110 092
Ph: 011-4934 4934   Fax: 011-4934 4935
                          e-mail: publishing@cbspd.com; publicity@cbspd.com

Branches

- **Bengaluru:** Seema House 2975, 17th Cross, KR Road, Banasankari 2nd Stage, Bengaluru 560 070, Karnataka, India
  Ph: +91-80-26771678/79   Fax: +91-80-26771680   e-mail: bangalore@cbspd.com
- **Chennai:** 7, Subbaraya Street, Shenoy Nagar, Chennai 600 030, Tamil Nadu, India
  Ph: +91-44-26680620, 26681266   Fax: +91-44-42032115   e-mail: chennai@cbspd.com
- **Kochi:** 42/1325, 1326, Power House Road, Opp KSEB, Power House, Ernakulum Kochi 682 018, Kerala, India
  Ph: +91-484-4059061-65,67   Fax: +91-484-4059065   e-mail: kochi@cbspd.com
- **Kolkata:** 147, Hind Ceramics Compound, 1st Floor, Nilgunj Road, Belghoria, Kolkata-700056, West Bengal, India
  Ph: +033-25633055, 033-25633056   e-mail: kolkata@cbspd.com
- **Lucknow:** Basement, Khushnuma Complex, 7 Meerabai Marg (Behind Jawahar Bhawan), Lucknow-226001, UP, India
  Ph: +0522-4000032   e-mail: tiwari.lucknow@cbspd.com
- **Mumbai:** PWD Shed, Gala no 25/26, Ramchandra Bhatt Marg, Next to JJ Hospital Gate no. 2, Opp. Union Bank of India, Noorbaug, Mumbai-400009, Maharashtra, India
  Ph: 022-66661880/89   e-mail: mumbai@cbspd.com

Representatives

- Hyderabad   0-9885175004
- Patna   0-9334159340
- Jharkhand   0-9811541605
- Pune   0-9664372571
- Nagpur   0-8692091830
- Uttarakhand   0-9716462459

Printed at Chaman Enterprises, Daryaganj, New Delhi, India

To My Grandchildren
KARTIKAY
and
KHEVNA

Preface to Eighth Edition

It gives me great pleasure in bringing out the eighth edition of this book. I always start working on every new edition, right from the day I receive the first copy of the previous one. The seventh edition was out in 2013 and a lot of new information has been added since then. It is always my privilege to provide the latest information to the readers so that they are not deprived of the present day ever changing scenario.

I have revised all the chapters, more specifically the chapters on Biodiversity and Nomenclature. The International Code of Zoological Nomenclature is likely to be published in 2018 since "Electronic Publications" have now been recognized as valid. I am confident that the present edition will be very useful to the readers.

I shall be grateful to the readers in sending me suggestions and comments so that the same are taken care of in the future editions.

V.C. KAPOOR

Contents

Preface to Eighth Edition   vii

Chapter 1. **Introduction**   1

    Stages in Taxonomy   3
    Importance of Taxonomy   4
    Problems of Taxonomists   11
    Aims and Tasks of a Taxonomist   12
    Taxonomy as a Profession   13

Chapter 2. **Taxonomy and Biodiversity**   19

    Taxonomy   19
    Taxonomic Organisations in Support of
        Biodiversity   22
    Biodiversity   24
    Importance of Biodiversity   26
    Benefits of Biodiversity   27
    Conservation of Biodiversity   34
    Threat to Biodiversity   35
    Ozone and Biodiversity   36
    Degradation of Ecosystems and
        Biodiversity Loss   41
    Gain in Biodiversity   43
    Marine Biodevrsity   44
    Biodiversity Awareness   46
    New Techniques and Biodiversity Preservation   50
    Role of United Nations in Biodiversity   50
    Support to Biodiversity   52
    National Biodiversity Authority   53

Agricultural Biodiversity   53
Carbon Credits   55

Chapter 3. Rise of Taxonomy   **63**

Chapter 4. Newer Trends in Taxonomy   **71**

Morphological Approach   71
Immature Stages and Embryological Approach   76
Ecological Approach   78
Behavioural Approach   79
Cytological Approach   81
Biochemical Approach   89
Numerical Taxonomy   97
Differential Systematics   100
Conclusion   101

Chapter 5. Zoological Classification   **103**

Kinds of Classifications   104
Phyletic Lineages   110
Components of Classification   113
Linnaean Hierarchy   114

Chapter 6. Concept of Species   **117**

Typological Species Concept   119
Nominalistic Species Concept   119
Biological Species Concept   120
Evolutionary Species Concept   123
Phylogenetic Species Concept   124
Recognition Species Concept   124
Ecological Species Concept   125
Genotypic Cluster Species Concept   125
Cohesion Species Concept   126
Genic Species Concept   126
Differential Fitness Species Concept   127
Conclusions   127
How Many Species?   129
Other Kinds of Species   130
Polytypic Species   134
Subspecies   135
Other Infraspecific Groups   139
Superspecies   141

Chapter 7. **Taxonomic Collection—Identification—**
Description and Publication   **134**

Species Registry   148
Collecting Ways   151
Data of Collection   166
Preservation of Collected Material   166
Curating   167
Preparation of Specimens   167
Relaxing Insects   168
Mounting   169
Storage   175
Cataloguing of Specimens   176
Arrangement of the Specimens   177
Maintaining Quality of Collection   177
Identification   178
Methods of Identification   179
Problems Encountered in Identification   196
Sending Insects and other Arthropods by Mail   197
Description   202
Subjects of Description   203
Taxonomic Characters   203
Future of Descriptive Taxonomy   205
Taxonomic Publications   205
Preparation of Taxonomic Publications   210
Taxonomic Paper   213

Chapter 8. **Reference Works in Taxonomy**   **215**

Zoological Record   215
Literature Earlier to Zoological Record   218
Biological Abstracts   219
Dissertation Abstracts   219
Nomenclator Zoologicus   220
The Century of Dictionary   220
Directories   220
Entomology Abstracts   221
Helminthology Abstracts   221
Protozoological Abstracts   221
Review of Agricultural Entomology   221
Review of Medical and Veterinary Entomology   222
Agrindex   222
Bibliography of Agriculture with Subject Index   223

Publications of International Commission on
Zoological Nomenclature   223
Guides to Journals   225
Books   226
Latin and Greek Terminology in Taxonomy   227
Latin Abbreviations   227
Latin Words   229
Linnaean Signs   229
Taxonomy and Biodiversity on the Web   229

Chapter 9. Zoological Nomenclature   235

Origin of the Code   236
Biocode   239
Phylocode   240
International Code of Zoological Nomenclature 242
Rules of Nomenclature   243
Electronic Publication   271

References   273

Glossary   307

Index   325

1

Introduction

The word taxonomy is derived from the Greek words *taxis* (= arrangement) and *nomos* (= law). It was first coined by A.P. de Candolle, a Professor of Montpellier University in France, in his Botany treatise in 1813, as a French word 'Taxonomie,' evidently formed on the analogue of astronomie, economie, agronomie and other similar words. Later in 1819, he spelled it as 'Taxeonomie'. The Greek scholars criticised this spelling as according to them, the correct Greek spelling should have been 'Taxinomie'. Under these circumstances, the correct word presently in use should be spelt as 'Taxinomy' instead of 'Taxonomy'. But the present day taxonomists prefer the already established term 'Taxonomy' as it is in use now for over 190 years. Taxonomy is the science of classifying living things into groups based on their similarities. This science of naming and classifying organisms is the original bioinformatics and a fundamental basis for biology.

The terms taxonomy, classification and systematics are often used interchangeably because they apply to same concept of arranging things in their 'correct relationship'. Systematics stems from the Latinized Greek word 'Systema' applied to the systems of classification developed by Linnaeus in the 4th edition of his historical book, *Systema Naturae* in 1735. It is sometimes incorrectly used in place of taxonomy. Taxonomy is actually the study of the principles and practices of classification and as such it is only a part of systematics.

Various workers have tried to separate these terms into distinct compartments. According to Mason (1950), "taxonomy is the synthesis of all the facts about (organisms), into a concept arid expression of the interrelationships of (organisms)", Heslop-Harrison (1953) explains it as the "study of the principles and practices of classification, in particular the methods, the principles, and even in part the results of biological classification". Simpson (1961) defines it as the theoretical study of classification, including its bases, principles, procedures and rules. Davis and Heywood (1963) consider taxonomy as a way of arranging and interpreting information. Biackweider (1967) explains it as the day-to-day practice of handling different kinds of organisms, it includes collection and identification of specimens, the publication of data, the study of literature, and the analysis of variations shown by the specimens, it is a science of placing biological form in order (Johnson, 1974).

Christoffersen (1995) defined taxonomy as the practice of recognising, naming, and ordering taxa into a system of words consistent with any kind of relationships among taxa that the investigator has discovered in nature. Taxonomy thus becomes the most basic activity in biology, dealing exclusively with the discovery, ordering and communication of patterns of biological taxa.

The classification is the ordering of animals into groups (or sets) on the basis of their relationships, that is, of associations by contiguity, similarity or both (Simpson, 1961). Or it is the arrangement of the individuals into groups (taxa) and the groups info a system in which the data about the kinds determine their position in the system and thereafter are reflected by the position (Biackweider, 1967).

The relationship of taxonomy to systematics is somewhat like that of theoretical physics to the whole field of physics. Taxonomy includes classification and nomenclature, but leans heavily on systematics for its concepts. Systematics includes both taxonomy and evolution. Simpson (1961) explains systematics as that scientific study which deals with kinds and diversity of organisms and any or all relationships among them Or it is that science which includes both taxonomy and classification, and all the other aspects of dealing with kinds of organisms and the data accumulated about them (Biackweider, 1967). It is thus concerned with organising biological knowledge; taxonomy, with constructing the frame-work for the organisation; and classification is a hierarchy of names of taxa representing the components of a group of plants or animals, with (included or referred to) characters sufficient to diagnose and differentiate the taxa or it is nothing but a list of taxa names indented to indicate category levels. Christoffersen (1995) defined systematics as the theory, principles and practice of identifying (discovering) systems, i.e., of ordering the diversity of organisms (parts) into more general systems of taxa (wholes) according to the most general casual processes. According to Wagele (2005) the terms taxonomy and systematics could be synonyms but in practice both function differently. Systematics searches for a phylogenetic system but not necessarily to acquire special knowledge on the distinction, validity of proper names and the number of species. In this way systematics works mainly on the phytogeny of supra-specific taxa but can identify new species. Dubois (2005) found even taxonomy and nomenclature different from each other in their function. Taxonomy recognizes classificatory units or taxa where as nomenclature gives unique scientific name to each of these units. Taxonomy is a scientific discipline and nomenclature is a technique.

In simple terms, actually there are two parts of systematics. The first part, *taxonomy*, is concerned with describing and naming the many kinds of organisms that exist today, those that have been extinct for many, even millions of years and also those that are becoming extinct. The second part of systematics, *evolution*, is concerned with understanding just how all of these kinds of animals arose in the first place and what processes are at work today to maintain or change them. There are two distinct taxonomic systems in operation among professional zoologists today. The traditional taxonomy is still largely adopted by field workers, conservationists and animal husbandry people The **cladistic** taxonomy is mostly supported by evolutionary' biologists. In cladistic taxonomy, an attempt is made to group species in accordance with their evolutionary history. Thus, the original purpose of taxonomy was the recognition, categorization and identification of organisms.

STAGES IN TAXONOMY

It is now well-known that taxonomy of a given group passes through several stages. These stages are referred to as alpha (analytical phase), beta (synthetic phase) and gamma (biological phase) taxonomy. Alpha taxonomy is the level at which the species are characterised and named; beta taxonomy refers to the arrangement of the species into a natural system of lower and higher categories: and gamma taxonomy is the analysis of intraspecific variations and evolutionary studies, i.e., study of speciation. But in actual practice it is rather difficult to dissociate them because these overlap and intergrade. There are only a few groups of animals (some vertebrates, especially the birds, and a few insect orders like Lepidoptera) where the taxonomy has reached up to gamma level otherwise in almost majority of the groups the works are still at alpha and beta levels.

IMPORTANCE OF TAXONOMY

It is very important to know the living organisms around us. Each species occurs in nature in many different forms like sexes, larvae, nymphs, pupae, seasonal forms, morphs and other phena. Taxonomy is thus the language for communication. Over one and a half million species of animals have been described and it is estimated that about 4 to 40 million or more species still await discovery. Imagine when all these animals did not have a proper name- it would have created total confusion and anarchy. If a research worker does not have a reliable name of the animal, he is working, it will be impossible to communicate and publish the work. It is, therefore, necessary to put such a large number of species into definite groups so that the extent of their harmful and edible properties are established. There is thus subtle relationship between the characterisation of an organism by a systematist and finding solution of a particular problem by a scientist. The assignment of a name to an organism provides the only key to all the information available about that species and its relatives. Careful and accurate identification and classification are of vital importance. Sailor (1969) rightly pointed out that in all the steps necessary to environmental research, from accumulation of data through evaluation of results, storage of information, retrieval of Information, and use of the information in action programmes, the taxonomist acts as a necessary catalyst who allows the control of pests through manipulation of habitats as well as in the management of our environment in the cheapest and most successful way.

The contributions of systematics to biology are both direct and indirect. In some cases total dependence on systematics is not ruled out. To know specifically the areas, both in theoretical and applied biology, benefited by systematics, it is necessary to discuss its role with appropriate examples.

Theoretical Biology

Many of us are unaware of its role in theoretical biology. Systematics has played a very important role in laying the foundations of some important fields of biology. It is responsible in making conceptual contributions like **population thinking.** It is also responsible in solving the problems **of multiplication** of **species.** One of its greatest contributions lies in the understanding of the **structure of species** and of the evolutionary role of **peripheral populations.** It was only the taxonomist who reaffirmed the role of **natural selection** as evolutionary factor in contrast to **mutation theory** of Mendel. **Mimicry** and other evolutionary areas have also been clearly understood through taxonomy. It has also played a useful role in the development of **behavioural science.** It is a key to the study of **ecology** as no ecological survey or work can be undertaken unless all the species of ecological importance are identified.

Applied Biology

Taxonomy has played quite a wide role in the important fields of applied biology. The applied biologists, too, are heavily relying on it for laying accurate and foolproof experiments and getting quick useful results. It is now well-known that chemical control of pests with insecticides can satisfy only a short term need. The appearance of insecticide resistant strains of pests, and the problem of hazardous residues in food, water, air, etc., have led to the conclusion that the best and perhaps the only way to save our crops from insects and others is to organise an integrated pest management programme. Such a programme is aimed at integrating chemical methods with the use of **resistant plant varieties, predators and parasites, pheromones, hormones and lethal genes.** All these methods (except the use of insecticides) are highly specific and can only succeed if the identity of the pest or pests is accurately determined. A wrong identification may upset the entire control strategy. To know the useful role of systematics in specific areas of applied biology, it is pertinent to discuss them separately.

Agriculture and Forestry

Presently we are faced with the acute problem of saving our crops and trees from the attack of various kinds of pests, it is essential to know the correct names of such pests before their proper control or eradication. Every species has its own niche in nature and differs from its related species in food preference, breeding season, tolerance to various stimuli, resistance to predators, competitors, pathogens, etc., and all these are essential for an applied worker before applying control measures. All this information can be easily obtained by screening the literature if the identity of the pest is known.

It is also necessary to collect samples of animals occurring on a particular crop when the crop shows early signs of injury or disease. In many cases the crop continues to be damaged even after spraying with suitable recommended pesticides. This may be either due to the entry of some new pest(s) or appearance of some resistant strain of a pest.

It is also sometimes very important to have local observation of the destruction of the crop. When the injury is obvious and the leaves are being eaten, it is often easy to discover the insects like caterpillars or leaf-feeding beetles; some may even defoliate the plants. On the other hand there are many leaf-eating insects that feed at night and hide away in the daytime. Other insects suck juices of plants, e.g., plant bugs and aphids. Besides, there are many other insects that bore into plants or trees, e.g., wood-boring beetles, and their larvae, some moth caterpillars, etc. There are also maggots of certain flies, the fruit flies, that are found in fruits and vegetables. This information is important before controlling any pest with the use of chemicals. On getting the correct identity of the pest species, it becomes easier to collect information about its habits which is vital for its effective control. For example even with leaf-eating insects, specialised ways of control have to be applied taking into consideration whether it feeds on the upper or lower surface of the leaves. Similarly, many of the plant diseases are caused by certain vectors. The correct identification of a particular vector is vital for bringing the disease under control for killing its vector.

Biological Control

As the use of insecticides is declining and replaced by specific methods of biological control, the use of accurate Identification of the pest and its natural enemies is becoming increasingly important. Natural enemies of pests can often be introduced for biological control to the enormous advantage of agriculture, forestry, etc. When successful, the biological control is much more economical than chemical control because it need not be repeated and has no injurious side effects. The controlling agent (parasite, predator, pathogenic organism) is always there in the environment and able to multiply as needed to reduce the members of the pest species. Specific controlling agents are often difficult to distinguish from closely related and nearly identical forms that feed on different hosts. Failure to discriminate between populations of different hosts, and usually other different attributes as well, can result in expensive confusion and even introduction of wrong control agents. For example, in the early 1940s a parasite *Archytas incertus* (Macq.) was introduced from Uruguay and Argentina into the USA for the control of armyworm. Its import was discontinued when the taxonomists reported that the species was already common in southern United States. Similarly, a gall fruit fly, *Procecidochares utilis* Stone was imported from New Zealand Into India for the control of crofton weed, *Agerartina trapezoideum* (Sprengel) in Darjeeling District. This fly has since been well established there and has also migrated to Nepal where, too, it has become quite well established on this weed. This fly was not imported in Nepal because of its timely discovery there. There are many examples of successful biological control programmes in many parts of the world. In all these cases the correct identification of the parasites (including weed destroyers) and their hosts has been the key factor.

The systematists are presently greatly involved in designing and implementing the biological control programmes of pests and diseases most effectively. Since it is now well-known that knowledge of an organism's relationship with others permits us to develop means of control more easily, total dependence on correct identification of both the parasite and the host is not ruled out.

Public Health

Taxonomy plays a great role in public health programmes. There are a number of diseases which are spread by many arthropods which are disease-specific. So our control strategy should be planned in such a way that only the target species is attacked. This is possible only on getting the correct identity of that species. For example, some species of *Anopheles* mosquitoes are responsible for transmitting malaria and others not. A good example of such a case is the occurrence of malaria in epidemic form in Europe. The mosquito, *A. maculipennis* Meigen, a supposed vector of malaria, was found prevalent throughout Europe. Large sums of money were wasted in controlling this mosquito, although malaria was not prevalent all ever Europe in spite of the presence of this mosquito. The taxonomists Hackett (1937) and Bates (1940) were consulted who took samples of the mosquito from all over Europe. They discovered *A. maculipennis* complex comprising several sibling species of which few were responsible for transmitting malaria. The control measures were then applied only to the target species and malaria was brought under control. In this way a lot of money and manpower were saved. Thus, correct identification ensures a maximum of effective control at minimum cost.

Quarantine

Many new pests and diseases of plants, animals and human beings have already entered many countries and still others are following suit. Their spread from one country to the other is through transportation of various crops, ornamental plants, and mainly through the agency of human-beings. Respective governments have established quarantine laboratories at aerodromes, ports, etc., to check the spread of new pests and diseases. Even when the American astronauts came back from the moon, they were kept in quarantine laboratory for a few days to avoid the risk of introducing new pests or diseases from that heavenly body. Taxonomists play a vital role here in providing correct and prompt identification of the pest or disease. Unless this is done, no further action is possible in such matters. The most common example of such measures was the rule till quite recently of compulsory inoculation for cholera for every passenger travelling from one country to another. Indian bird pest, Rose-Ringed Parakeet has been found to attack vineyards across Europe. The European Environment Agency has found the parrots introduced from India-intentionally released or accidentally escaped- creating havoc across the continent resulting in expected reduced wine production.

Wildlife Management

Presently *great* attention is being paid to conserve and propagate wildlife. Many programmes have been initiated all over the world to teach the people the importance of fauna and flora for human welfare. The indiscriminate killing of animals and felling of trees have already resulted in great disturbance in the natural environment. Many species of animals have become extinct and still many others are following suit. Taxonomists help the environmental protection protectors to identify all such animals which are endangered by man's activities. This job becomes more challenging in view of widespread support in favour of preservation and protection of our biodiversity. See more detail in chapter 2.

Mineral Prospecting

The sequestering of rocks and geological events in an area is basic to any search for fossil fuels and mineral deposits. Some igneous rocks may be dated by radioactive decay, but sedimentary rocks can be dated only by their enclosed fauna and flora. The palaeontologists play a major role in the identification of the fauna arid flora and thus give clear picture of the correct sequence of geological events. Such works have been of great value in the success of oil industry in America. Since the complex faulting and folding of rocks pose great problems in the correct identification of the sequence of rocks, it is only through the fossils that their true sequence is determined.

National Defence

Information concerning disease vectors and parasites is an obvious application of systematic to national defence. During World War II, Japanese paper balloons carrying fire bombs created havoc in the forests of the American northwest. Eventually a balloon was recovered with several bags of ballast sand attached. The sand contained a large number of shells of micro-organisms, which, after careful identification and interpretation, were found to represent both cold and warm water worms which occur in a small area off one of the mainland islands of Japan. Subsequent bombing of this beach area destroyed the balloon launching sites. The use of such biological means in the wars are economical and require less efforts in their operation. In these cases, both in making the bombs and their destruction, the correct identification of the organisms has been the first step. Tills is not the isolated example and the need for taxonomy in national defence still continues. Moreover, the identification of potential disease vectors is vital to the health of both; military and civilian populations all over the world.

Environmental Problems

Systematists have also played a useful role in tackling the various environmental problems. Pollutants such as certain chemical pesticide residues may persist in the environment or even concentrate in certain plants or animals. Tracing the movement of these pollutants to determine their effects requires identification of the species within the food chains. It is also now dear that a biological approach, to be successful, requires a thorough understanding of the taxonomic relationships between pest species and component species of the ecosystem. The taxonomists-dominated survey team of Hoffmann and others (1949) gave extremely useful information on the effects of aeroplane application of DDT on forest invertebrates in the USA.

Presently water pollution is also considered as one of the foremost national problem. Since each species has its own requirements for oxygen, nitrogen, and other organic and inorganic compounds in water, certain algae, aquatic Insects, helminth parasites of fish, and some microscopic organisms are reliable indicators of the degree and nature of pollution. The identification of species present in a particular location provides a rapid and inexpensive monitoring system for detecting pollution.

Soil Fertility

Many animals play an important role in increasing the fertility of the soil. The soil is tunnelled in such a way that it becomes, more aerated and is also enriched by their secretions and dead bodies. It is necessary to know such animals for their proper management in agriculture.

In Commerce

Products like honey, silk, lac, and dyes are provided by insects. Besides, there are many other animals (including insects) which are directly eaten as food or provide us other useful commercial products. Taxonomists can play an important role in increasing and improving the quality of these products by manipulating the useful species. The introduction of any useful species is possible only through correct identification. Many such species have been established in India and other parts of the world only through intelligent introduction based on sound systematics. It is only through correct identification that pertinent information about the biology of the species (to be introduced) is collected from the literature. The Italian honey bee, *Apis meilifera* and the fish, European carp, *Cyprinus carpio* are two well-known examples of successful introduction in India and which was possible only through correct identification.

PROBLEMS OF TAXONOMISTS

There are many problems which the taxonomists encounter daily but the most important ones are discussed below.

To Characterise Species

The biological species are complex populations or groups of populations which cannot be measured precisely. These can only be sampled and the samples show all the limits or all the variability of the whole population. It is, therefore, agreeable that the biological species which environmental biologists see and study cannot be treated with mathematical precision.

To Arrange Species in Hierarchy of Higher Categories

The higher categories are always arbitrary although new numerical and mathematical methods are developed to describe them, measure the gap between them, and arrange them in many ways.

AIMS AND TASKS OF A TAXONOMIST

The primary aim of a taxonomist must be the construction of classes of living things about which scientifically useful inductive generalisations can be made. Many workers (Blackweider and Boyden. 1952; Michener, 1963; Ehrlich, 1964; Blackweider. 1967; Mayr. 1969; Darlington, 1969) have enumerated various aims and tasks of a taxonomist. For the sake of convenience to readers, these are summarised below.

1. To catalogue the diversity of life on earth and to preserve large samples, both of extant and extinct organisms, drawn from the diversity in various sorts of collection.

2. To differentiate the various kinds of organisms and to point out their characteristics (both qualitatively and quantitatively) through descriptions, keys, illustrations, etc.

3. To provide names for each kind of organism, so that all concerned can know what they are talking about and so that information can be recorded, stored and retrieved when needed. In other words the goal of the taxonomy is to create a common language so that we know what organisms we are talking about.

4. To develop a set of principles in regard to the choice and relative importance of characters with the ultimate aim of arranging species in hierarchy of higher categories.

5. To estimate genetic and phylogenetic relationships among organisms.

6. To contribute towards the understanding of evolutionary process.

7. To integrate the data from all fields of biology, like behaviour, genetics, physiology, etc., and to detect and then summarise significant patterns possibly with the help of modern electronic computers.

8. To document and preserve specimens to provide a useful reservoir of data.

9. To help in clarifying the place of systematic or taxonomy in general biology by revising their aims and priorities, realistically restructuring the efforts in applied taxonomy and reaffirming faith in taxonomy.

TAXONOMY AS A PROFESSION

Taxonomy requires as much wisdom and intelligence as any other field of biology. But the progress in taxonomy is slow and steady and is without the brilliant discoveries which sometimes come quickly in other fields. Due to this it has never been an attractive profession. We are equally unfavourably placed in terms of scientific capabilities of identifying, working with and adding value to biodiversity resources. Every year more than 100,000 students get a bachelor's degree in one of the Life Sciences. But only a very small fraction of these get an exposure to India's living wealth. Practically none of them are able to name more than 5 to 10 species of plants or animals put together. This is because they are rarely encouraged to look at living creatures abounding around them, their training being confined to identifying a few dead specimens or dissecting a still living cockroach or a dead pigeon. Given such a training programme few teachers of biology know much of the living wealth of India either when this country is recognised as one of the eight mega centres of plant biodiversity having 45,000 species of plants, i.e., 12% of the global floristic wealth. Similarly, India has a large animal genetic diversity with proven potential as is evident from the availability of 26 breeds of cattle, 8 of buffalo. 40 of sheep, 20 of goat, 4 of camel, several breeds of horses, pigs, poultry and 2200 fish and shell fish species and a very large number of other animal species, economically important or unimportant but playing a greater role in keeping firm ecological balance. Furthermore, it receives step-motherly treatment from the financial agencies as well; whenever any economy is to be thought of, the axe first falls on it. Above all the students, too, find the glamorous fields like molecular biology, biotechnology, etc., more attractive and with better employment prospects. In this way the taxonomists remain the most unfortunate beneficiaries. This is very unfortunate because the biological science would prosper better if the systematic fields were more actively cultivated.

Many taxonomists feel that the lower status allotted to their subject is due to man's own interests which are changing with the time. Originally the cataloguing of the fauna and flora was the exclusive interest of biologists. They were impressed by the variety and grandeur of nature, and undertook to describe and name the organisms. Then followed successively the rise of other fields like comparative anatomy, physiology and genetics, and biotechnology is the latest of these. Thus the tide moved way from taxonomy, and finally even went against it. Nevertheless expensive contributions made by taxonomists cannot be forgotten. It was they who have furnished the bricks on which the whole structure of organic evolution has been constructed.

The arrangement of taxa and establishment of the similarities between them (or Taxonomy) are essential ingredients for all other studies in Biological Diversity. It is not single specialization that stands on its own, but is a lateral study encompassing the full sweep of life whereas the rest of biology is a vertical study of biological organization within a very few species. It is therefore necessary that both taxonomy and rest of biology should be combined to achieve a full study of Biological Diversity. Taxonomy lays the foundations for the Tree of Life and is prerequisite for ecology and conservation. Taxonomy is an exploration of the still largely unknown Biodiversity of Life. There is still a great deal of fauna and flora to be discovered. Molecular Taxonomy cannot be successful without initial identification through traditional taxonomic methods.

Presently the picture is very different and fast changing in favour of taxonomy. A taxonomist is now no more a caretaker of his collection but a well trained naturalist. He goes out in the fields and supplements his studies with more and more information he gathers from field observations as well as from other branches of science like cell biology, physiology and molecular biology. The ferment introduced into systematics in recent years by biochemists, ethologists, statisticians, and theorists has given a new vitality to it. Sophisticated techniques and equipments, together with the essential role of systematics in increasingly important environmental studies, have made it an exciting and challenging science. Today his taxonomic character does not mean only aspects of morphology but also includes others like physiology, biochemistry, genetics, etc., to establish the relationships among different species. This difficult task of integrating the data from other fields of biology is being aided by modern electronic computers. The use of electronic devices in the analysis of sounds of frogs, cicadas and grasshoppers has helped in the breaking up of complexes of species. The taxonomic research which was at one time considered secondary in importance to other rapidly developing branches of zoology, has now been regarded as basic to these as well as other fields of biology. Even then only those who are willing to make greater than usual sacrifices, and who are sustained by an unquenchable interest in the subject, can expect to make a success in taxonomic research.

Taxonomists are presently employed by universities, research institutes, museums, central and state government agencies, industries and zoos. A well trained taxonomist is well qualified to teach a course in zoology as he has a broad background in morphology, physiology, genetics and ecology, which others, due to lack of time, are not able to acquire. In all types of zoological researches the availability of taxonomic specialists is a guarantee for the identification and homogeneity of the working material; inversely a shortage of such specialists leads to very serious consequences. Still there are not enough openings for the taxonomists. Sometimes a person trained in taxonomy has to look for other interests. When at one stage there are not enough taxonomists available, this diversion makes the situation more acute It not only results in loss of a few trained persons but also affects the morale of those who are being trained, or still others who are thinking of entering this profession. Furthermore, when continuity in a field is broken, progress not only stops, but the amount of available knowledge also actually declines or temporarily halts until at some later date someone else with waste of time and energy, takes up the group and learns it again.

Even the Convention of Biological Diversity has acknow-ledged the existence of this 'Taxonomic Impediment' which is affecting the sound management of biodiversity. The shortage of trained taxonomists and curators are showing clear sign of negativity in the protection and conservation of plants and animals, especially the endangered ones. The inability to identify species is a major component of the taxonomic impediment. Still there are millions of undescribed species and there are very few taxonomists to do this job, especially for biodiversity-rich but economically poorer countries. Identification of large and charismatic vertebrates, like birds and mammals, is quite well done but majority of the organisms belonging to arthropods and lower plants, fungi and microorganisms are yet to be discovered. Taxonomy is suffering from lack of funding. Even when a well trained and experienced taxonomist retires, he is not replaced due to shortage of funds as well as non-availability of suitable replacement. Money has to be made available, especially when thousands of species are threatened with imminent extinction (Wheeler, 2004; Wheeler et al., 2004). Presently, there are only 7000-11000 taxonomists worldwide, only few of them are in developing countries which contain most of the world's biodiversity. Moreover, correct identification of organisms is fundamental for the sound assessment and conservation of biodiversity. Species description is seen as an old-fashioned way of doing research. In any case the taxonomists, too, need to improve their image as still various scientists believe that they are artificially increasing the number of species by raising subspecies to species levels, resulting in **"Taxonomic inflation."** Frequent synonymies, homonymies, new combinations are further adding confusion to diversity studies. To overcome such confusion it would be better for the taxonomists to face species problem only. Taxonomists should take thier opportunity of **"BIODIVERSITY BOON"** in their favour and strengthen the taxonomic research to work earnestly for not only discovering biodiversity but also in their conservation.

Until the early 1990's, many taxonomic services were provided free of charge by expert institutes in Europe and elsewhere. To-day, developing countries typically lack the ability to pay for external taxonomic services and have inadequate and poorly resourced local taxonomic capacity, in many important organism groups, trained taxonomists are totally lacking in most of these regions. At present when many species are yet to be identified and described, lesser number of taxonomists are hampering this important job. Surprisingly, most of the taxonomists are located in countries in North which are relatively poor in their biodiversity. Biodiversity is quite rich in developing countries of the tropics and so a large number of taxonomists are needed in these countries. Recently, Rodman and Cody (2003), in USA, proposed National Science Foundation's Partnerships for enhancing expertise in taxonomy (PEET) as a model to overcome the present taxonomic impediment. This programme has substantial budget to fund projects, enabling intensive training, targeting poorly known groups of organisms for revisionary works or monographic research. Many trainees from under this programme have secured employment in the USA and abroad in academics, museums or government agencies on positions relevant to systematics.

Synthesis of Systematic Resources (SYNTHESYS) is an initiative launched by **Consortium of European Taxonomic Facilities,** it functions in two ways:

i), here the European researchers get access to the collections comprising more than half of the world's natural history specimens, world class libraries, facilities for microscopy, physical, chemical and molecular analysis and experiences hosts and trainers at 20 European institutions: ii), this pad is related to networking activities focused on creating a single museum service, an integrated European resource bringing together the collections of the major institutions in Europe.

The **PEET** and **SYSTHESYS,** together with others like **PBI (Planetary Biodiversity Initiative)** of the US National Science Foundation, The **EDIT programme (European Distributed Institute of Taxonomy), UK-NERC funded CATE (Creating a Taxonomic e-science)** project could serve as role models for future organizations promoting biodiversity studies.

2

Taxonomy and Biodiversity

TAXONOMY

The importance of taxonomy has already been discussed in the previous chapter. Since, in the present days, biodiversity is the most talked about subject, the role of taxonomy in the conservation of biodiversity is necessary to be discussed separately as well. Taxonomy and biodiversity are interdependent and closely knitted. Biodiversity is simply a very special, unique and wonderful feature of the living world. It needs to be understood and then conserved by humans, both from practical and ethical point of view. Taxonomy is the science of identifying (incl. naming), describing and classifying organisms Global biodiversity (or The Majesty of Life) is being lost due to human activities at unbelievable fast rate. In view of this taxonomy takes a centre stage to save our biodiversity. How can we develop protected areas when we do not know what is to be protected? How can developing countries reap the benefit of the use of biodiversity when we do not know the biological diversity which is to be used? How can we identify and find harmful invasive species if we cannot distinguish them from native species? Taxonomy plays a major role here to provide basic understanding of various components of biodiversity which finally help in strong decision-making in conservation and sustainable use.

Biodiversity is a concept and has species as its keystone. The scientific names of animal species are crucial to effective global communication about biodiversity. Without broad agreement on the name of a disease-bearing microbe, vital food species, or threatened animal, we cannot even begin to combat, exploit or conserve them; i.e. in scientific language, humans are *Homo sapiens,* honey bees are *Apis mellifera* and cockroaches are *Blatta orientalis* and this is true ail over the world. The universal acceptance and adoption of a system for naming organisms is an incredible achievement of mankind. In the past 250 years of research the taxonomists have described nearly 1.8 million species of animals, plants and microbes, it is estimated that this number may go up to 40 million or more species alive today. Thus, the species alive today are only a very small percentage (5–9%) of the billions of species which have lived on this Earth since first evolved over 3.5×109 years ago. Over 75% of described species belong to Arthropods. Presently, more than billion specimens have been gathered over the last 250 years and most of them are housed in Natural History Museums all over the world. How can such a vast number of specimens/species be arranged, categorized and classified? This is done through "TAXONOMY". Taxonomy provides a framework for understanding organic diversity. Taxonomists first group organisms with common characteristics into species. Although the populations of species are things that can interbreed, taxonomists study morphology and genetics to work out which organisms are the same species. Taxonomy, thus gives us a vivid picture of existing organic diversity of the Earth, provides us much of the information permitting a reconstruction of the phylogeny of life; reveals numerous interesting evolutionary phenomena and supplies classifications which are of great explanatory value in most of biological studies and paleontology.

Taxonomy is a vital component of biodiversity management. Taxonomic skills and products are thus essential for implementing sustainable development (see page 50 also), including conservation of biodiversity and food security. Besides identifying organisms in an area, taxonomy also documents their wide distribution. Moreover, systematic data in the form of collection data are most important as they serve as the only direct evidence of species distributions (Funk and Richardson, 2002). This helps scientists to select which areas are needed to be conserved. The description of new animal phyla, Loricifera in 1993, Cycliophora in 1995 and Mycrognathozoa in 2000 and a new insect order, Maniophasmatodea (Klass et al.,

2002) together with tropical arthropods alone could number over ten million species is a clear example to show how little is known regarding the magnitude of global species richness (Simonetti, 1997). Thus, it is right if one says that '**Taxonomy**' and '**Biodiversity**' are two sides of the same coin and neither of them is complete without the other.

Thus, taxonomy is indispensable for conserving and sharing the benefits of biodiversity. Presently a little less than two million species of an estimated 5–40 million species have been described. Knowing species and their distribution is the central point to formulate measures for their protection and also for new opportunities for life sciences to realize benefits from biodiversity. Taxonomy is now crucial to the management of biodiversity, public health, agriculture and many other aspects of life and society. It is a science not only for sustainable development but also for sustainable developed science itself.

The world presently lives under greatest mass extinction since the dinosaurs perished 65 million years ago. Most of the loss of the biodiversity is caused by human activities by intensifying habitat loss, introduction of exotic species through trade and travel and climatic change. The survival of our planet is dependant on a fine balancing act where every organism has its role to play. The disturbance in this balance will have far reaching consequences for other organisms and ultimately the planet itself. At least 40% of the world's economy and 80% of needs of the poor are earned from biological resources. Besides, the richer the diversity of life, the greater the opportunity for medical discoveries and economic development and adaptive responses to such new challenges as climatic change. The number of species of plants, animals and microorganisms, the enormous diversity in genes in these species, the different ecosystems of the planet, such as deserts, rainforests and coral reefs are all parts of biological diversity of Earth. By finding out all about biodiversity in order to protect genes, species, habitats and ecosystems, we would be protecting our own future as well as that of our own planet.

Thus, taxonomists play major role in the identification of taxa requiring conservation action; taxonomic understanding of species on red list; reserve site selection criteria based on taxon richness and endemism; identifying agents for use in biological control, invasive species management and disease control; taxonomy of keystone species for ecosystem services, etc. Taxonomy helps us to formulate methods for conserving biodiversity. These methods are very effective when working in an area where the animals are not well known. By studying the major groups of known animals in an area, taxonomists can find out more about the environment and its protection. Thus, taxonomy provides us with basic understanding about the components of biodiversity which is necessary for effective decision-making about conservation and sustainable use. Without taxonomists the natural world cannot be documented. Taxonomy keeps pace with the present as well as draws the wealth of knowledge accumulated throughout its history. Taxonomy in the twenty-first century will be completely different than its past. Taxonomists do not forget their past, but improve it by adopting more and more creative ways for better understanding of the living world around us.

TAXONOMIC ORGANISATIONS IN SUPPORT OF BIODIVERSITY

Presently, the importance of taxonomy in the conservation of our environment is so largely felt that more and more effective ways are adopted to strengthen it the world over. A number of international organizations are being established for coordinating this effort the world over. A few such important organizations are discussed below:

i) **BioNET International:** It is an international non-profit organization dedicated for promoting taxonomy, especially in the biodiversity rich but economically poorer countries of the world. It was established at the United Kingdom sponsored international workshop held in Natural History Museum, London in 1993. It provides forum for collaboration that is equally open to all taxonomists and to other users of taxonomy. In partnership, locally and internationally, it contributes to raising awareness of importance of taxonomy to society building and sharing of capacity and meeting taxonomic needs via innovative tools and approaches.

ii) **Global Taxonomic Initiative (GTI):** It was created under the United Nation's Convention of Biodiversity

(CBD) to provide taxonomic information and expertise needed to support implementation of the objectives of the Convention of Biodiversity. GTI is specifically intended to build a capacity to generate, collect, disseminate taxonomic information on thematic and cross-cutting issues. Taxonomy, in the context of GTI, is also fundamental to work on other CBD issues including invasive alien species, indicators of biodiversity and bio-safety. All the countries signatory to Convention of Biodiversity (CBD) are bound by its programme and nominate a National Focal Point for the GTI in their respective countries. The National Focal Point (NFP) acts as a central point of contact for national centers of taxonomic expertise and as a link for information sharing to other countries through their respective focal points. Its main goal is to reduce the rate of loss of biodiversity.

The working of GTI is coordinated at world level through fourteen loops—**Andinonet** (Andean Countries); **Aseanet** (South-East Asia); **Carinet** (Carribbean); **Eafrinet** (East Africa); **Easianet** (East Asia); **Euroloop** (Europe); **Nafrinet** (Northern Africa); **Pacinet** (South Pacific); **Safrinet** (Southern Africa); **Wafrinet** (West Africa); **Latinet** (South America, S. Cone); **Mesoamerinet** (Mesoamerica); **Nameriloop** (North America) and **Sacnet** (South Asia).

iii) **Integrated Taxonomic Information System (ITIS):** This is formed in partnership of Federal agencies of USA (like USDA, Smithsonian Institution, Environment Protection Agency, Nature Serve, etc.) to meet their mutual needs for scientifically credible taxonomic information not only for Nation's biota but also of the World, it provides authoritative taxonomic information on plants, animals, fungi and microbes of North America and the World.

iv) **Fauna and Flora International (FFI):** It was established in 1903 as a non-profit organization with registered offices both in the United Kingdom as well as USA. It is World's longest established international conservation body. It is known for its source based approach for sustainable conservation work, providing solutions that simultaneously help wildlife, humans and the environment.

v) **Virtual Museum of Natural History (VMNH):** It is developed with the aim to act as a primary access point for any researcher interested in obtaining accurate information about any animal taxon, down to species level. The primary search and link mechanism is the **VMNH's International Curator Project.** VMNH was incorporated as a non-profit organization in Washington, USA in 1998. The programme of this project is to develop interactive keys to animal groups on a worldwide scale.

BIODIVERSITY

Biodiversity is derived from two Latin words-*bios* means *life* and *diversitas* means *variety*. Biodiversity is thus defined as the 'FULL VARIETY OF LIFE ON EARTH'. in other words it refers to the totality of different kinds of living species, their forms, levels and combinations on Earth. Biodiversity is presently the most talked about term the world over. It was coined as a 'contraction' of 'biological diversity' in 1985. But one can also find, in traces, study of aspects of biodiversity as far as **Aristotle, the father of biological classification.** A symposium held in 1986 on BIODIVERSITY and the subsequent publication of its proceedings (Norton, B.G., ed.,1986) and follow-up book on BIODIVERSITY edited by Wilson in 1988 and again his another book on BIODIVERSITY in 1992 gave it much popularity. Now it will be difficult to count as to how very often 'biodiversity' is used every day by scientists, conservationists, economists, politicians, etc.

Biological diversity or biodiversity is a term which is used to describe the variety of life on Earth-means wide variety of ecosystems and living organisms; it is thus the foundation of life on Earth. Biodiversity is a measure of the health of ecosystem, biome, or an entire planet. It is extremely complex, dynamic and varied as compared to any other feature of the Earth. Its innumerable plants, animals and microbes physically and chemically unite the atmosphere (i.e. mixture of gases around the Earth), geosphere (i.e. solid part of the Earth) and hydrosphere (i.e. the Earth's water, ice and water vapour) into one environmental system in which millions of species, including human beings, exist.

Biosphere is presently threatened in two ways. One is the continuous degradation of the physical environment due to ozone depletion, global warming by greenhouse gases, etc. and second is the depletion of biodiversity. A clean environment is a question of life and death. Generally three kinds of biodiversities exist—**habitat diversity, genetic diversity and species diversity. Habitat diversity** relates to the variety of habitats, biotic communities and the ecological differences, together with the tremendous diversity present within ecosystems in terms of habitat differences and the variety of ecological processes. **Genetic diversity** refers to the variation of genes within species, i.e. genetic variation between distinct populations of the same species as well as within a population. The large differences in the amount and distribution of genetic variation is attributed in part to the enormous variety and complexity of habitats and the different ways the organisms obtain their living. **Species diversity** refers to the variety of species, i.e. species richness, species abundance and taxonomic or phylogenetic diversity. The species richness count the number of species in a defined area; species abundance means the relative numbers among species (such a sample may contain several very common species, a few less common and numerous rare species); and in taxonomic or phylogenetic diversity, one considers the genetic relationships between different groups of species.

Earth is relatively still unexplored planet in terms of its fauna and flora. The most important driving force behind the current interest in biodiversity is to know how many different life forms exist today on this planet. New species are being continuously added on this planet. Presently, it is believed that the number of species on earth (both below as well as above ground) may be anywhere between 40 to 100 million. Although we spend a fortune in exploring the galaxy and other parts of the Universe yet we have scarcely thought of exploring the biodiversity on our planet on which our very existence is ultimately dependent. Moreover, the warm and humid tropical regions of the earth, between the tropic of Cancer and the tropic of Capricorn are rich in diversity of plant and animal life. This is called 'MEGADIVERSITY'. India is one of the 17 "mega diverse" countries. It is composed of diversity of ecological habitats like forests, grasslands, wetlands, coastal and marine ecosystems and desert ecosystems. Presently, little more than 70% of the country has been surveyed discovering more than 45,000 plant species and 90,000 animal species (including about 60,000 insect species, more than 2500 fish species, 242 amphibian species. 465 reptile species and about 400 mammal species).

Out of the present biodiversity of the planet, more than half is concentrated in a few countries like Brazil, Colombia, Equador. Peru, Mexico, Zaire, Madagascar, Australia, China, India, Indonesia and Malaysia. With only 2.5% of the land area, India already accounts for 7–8% of the global recorded species in its ten biogeographically regions. Man, being intelligent and at apex of all food chains, is able to manipulate, control and remold biodiversity as per his needs and desires. This has greatly interfered with the natural process and resulted in biodiversity decline. During the last 3.5 billion years, evolution has witnessed the rise and fall of many species. Natural changes in Earth's climate change all life, transforming ecosystems from the lush, tropical forest of the dinosaur's reign to the frozen tundras. Our planet thus gradually changes over time. Only the species best able to adapt these changes will survive to the next century, only to face new challenges to their survival.

IMPORTANCE OF BIODIVERSITY

Thus, biodiversity is of prime importance in relation to existence of mankind. All the components of nature remain in a perfect balance, not only interwoven but also interdependent. The disturbance to any type of component can threaten the whole life support system of which human-beings are also part. At the ecosystem level, biodiversity not only provides conditions but also drives the processes that sustain the economy and our very survival as a species. The fast depletion of the greener pastures of our planet, leading to destruction of the natural habitats of the wildlife is causing great concern to our conservationists. The activities of deforestation, illegal wildlife trade, destruction of forest areas for human purposes have prompted conservationists (working through non-governmental organizations) to see conservation of wildlife and their habitats as major issue in India and other parts of the world. Presently, there is great awareness about the importance of biodiversity in our daily life and many non-government organizations (in addition to government role) have come forward to gather information by involving our countrymen for the protection and conservation of our biodiversity as a whole. In India, one such organization, ATREE (Ashoka Trust for Research in Ecology and Environment) was established in Bangalore in 1996. Its function is to promote environmental conservation and sustainable development by generating rigorous interdisciplinary knowledge that engages actively with academia, policy makers, practitioners, activists, students and wider public audiences. SAFE (South Asian Forum for Environment), in Kolkata, India, helps local communities to achieve sustainable development, especially in areas with rich biodiversity. Its support for **"biorights"** serves as good financial tool for the people. Similarly, RANWA (Research and Action in National Wealth Administration), in Pune, India, is another non-profit organization which works for conservation of wild life and ecosystems. There are more such non-government organizations like Centre for Environment Education, Centre for Science and Environment, etc., dealing at local, states, regional levels and national level doing good work for the protection and conservation of fauna and flora besides the ecosystems as a whole. The environment is no longer seen as a peripheral concern, but a route to better, balanced and sustainable economic growth (Times of India, 2011, Sept. 11, p. 13). Another important body is 'T.C. Narendra Trust for Animal Taxonomy established in 2006 in Calicut, Kerala, India. It is non-profit organization intended to make strong support of taxonomy to biodiversity.

BENEFITS OF BIODIVERSITY

There are many benefits and services provided by biodiversity which are summarized below:

   i) **Maintenance of soil fertility** – It is done mainly through the activities of the underground microbes and animal species.

   ii) **Maintenance of water quality** – For example molluscs remove nutrients from the water, helping to prevent nutrient enrichment and attendant problems, such as eutrophication arising from fertilizer run-off, etc.

iii) **Natural pest control** – i.e. using friendly organisms against harmful organisms destroying crops, spreading diseases, etc in natural way as both are part of the same ecosystem

iv) **Detoxification and decomposition of wastes** – some 130 million metric tons of wastes is processed every year by earth's decomposing organisms.

v) **Pollination and crop production** – Various animal species like bees, butterflies, bats, birds help many flowering plants to reproduce through transportation of pollen; at least one-third food for humanity is based on such kind of reproduction. Pollinators, including some 20,000 species of wild bees , contribute to the growth of fruit, vegetables and many nuts, as well as flowering plants. Plants that depend on pollination make up 35% of global production volume with a value of $577 billion a year. Many pollinator species are threatened by extinction including some 16% vertebrates like birds and bats. In view of many pollinator species are facing extinction, there will be shortage of food production due to negative effect of subdued pollination of agricultural produce.

vi) **Climate stabilization** – Ecosystems have direct or indirect role in stabilization of regional or local weather patterns. The release of moisture in the atmosphere by rainforests causes regular rainstorms, limiting water loss from the region which helps to control the surface temperature in cold climates. The rainforests thus contribute both the soil formation as well as regulate the climate through photosynthesis. Forests also act as insulators and windbreaks which then help in mitigating the impacts of freezing temperatures.

vii) **Prevention of natural disasters** – Forests and grasslands prevent land erosion, loss of nutrient and landslides through the action of roots which strongly bind together the soil particles. Coral reefs and mangrove swamps protect the land surrounded by them by reducing the effect of erosion. Animals are also being used to predict tremors, for example , in China animals like chickens, fish, dogs and toads are used for this purpose. Possible abnormal behavior of such animals like chickens flying atop trees, fish leaping out of water or toads moving in a group could indicate imminent earthquakes.

viii) **Food security** – Biodiversity provides vast majority of foodstuffs, e.g., fish, and food products from animals, such as eggs, meat, etc Wild biodiversity provides us fruits, nuts, mushroom, honey, spices, etc. Nearly one-third of world's land area is used for food production. United Nations has recently found a new weapon to fight hunger. The Food and Agricultural Organization hailed the likes of grasshoppers, ants and other insects as an underutilized food for people, livestock and pets. In a world grappling with food shortages, insect dishes like silkworm soup, cricket protein bars and red ant chutneys are most welcome assets. Red chutney or Chaprah is a celebrated insect based dish of Dhruva tribes in Chhattisgarh in India. The nutty-flavored polu leta fry, is a delicacy made from silkworm pupae in tribal Assam. The Assamese are also fond of the ant and ant egg dish, amlol parua, eaten especially during the spring festival. Caterpillars, termites, grasshoppers, crickets and beetles, all are dried and consumed by Assamese on daily basis. Majuli farmers in Guwahati, Assam, India have started eating beetles (Lepiota mansueta) which otherwise destroy vast swathes of crops every year between April and May. This resulted not only in creating a new human food but also controlling this pest through such social engineering activity. Dried karol puk, a stink bug found under stones in river beds, is a popular source of protein for the Adi people of Arunachal Pradesh. Two Icelanders have created 'Crowbar', an energy bar made from ground-up insects which are then mixed with other ingredients like almonds, coconut and cacao. This insect super food breaks down into amino acids and is also rich in calcium and vitamins – a great energy source for humans. Several western counties are heralding with idea of "Insects as Food for the Future". The European Union is investigating the potential of insects as healthy, protein stand-ins to bolster the continent's food supply. Belgium is already using insects in food. The first insect-laced food product across the country is a bread spread branded Green Bugs made from tomatoes, carrots and almost 6% mealworms. In United Kingdom, wine merchant Laithwaite's Wine has created the world's first insect and wine-matching guide. The Asian forest scorpion and the full-bodied Transylvanian Pinot Noir is a match made in oenological heaven. United Nation's recent pre-thank giving nudge in the direction of consuming insects, FMCG brands like Chirps (made with cricket flour), Bitty (cricket flour baked goods) and Chapul (cricket energy bars) seem to be jostling for space with more pedestrian potato chips and trail nut mixes. The practice of eating insects and other creepy crow flies were strong dietary habits of Sub-Saharan Africa and sub-East Asian countries. Whippey, one of a handful of UK entrepreneurs, is building businesses around entomophagy– human consumption of creepy – crawlies. He hopes of having business of around $73 million by 2020. Energy rich cockroach "milk" may someday be transformed into a food supplement worthy of human consumption. It is rich in protein, fat and sugar as per Director and educator programmers at Fernbank Museum of Natural in Atlanta, USA.

xi) **Medicines** – Nearly 80% of people in the developing world rely on medicines obtained from plants to treat malaria, stomach ulcers, syphilis, etc. New medicines are continuously being derived from various plant materials all over the world. Many animal species and their products provide cure for many human diseases. Recently, a study in snails has identified a painkiller which is thousand times more effective and without any addictive properties. Scientists of Lund University, Sweden have found lactic acid bacteria in honeybees showing promising results as an alternative to antibiotic. In studies at viral research institute, Pondicherry (Puducherry), the bacteria in bird droppings have been found to kill mosquitoes. It is now proposed to develop an anti-mosquito formulation using these bacteria.

Such bacterial anti-mosquito formulation have been launched and used in Sri Lanka, Indonesia, Nigeria and China. Since bacteria and viruses are becoming resistant to antibiotics, scientists are turning to jungle to find new sources of the next generation drugs. They perhaps have found them in the hair of one of the worlds slowest mammals – the sloth in Panama. They have found, among the flora and fauna of sloth hair, some fungi growing. These fungi are resistant to the parasites that cause malaria, human breast cancer cell lines, etc. The American scientists have recently found that spider toxin from reaper spiders or brown spiders holds key to anti-venom vaccines. These Laosceles spiders are most prevalent in Brazil. The engineered protein made from three pieces of a venom toxin from these spiders is effective protection against the effect of spider venom in animal models. It has also been found that tiny spikes on wing of insect cicada shred bacteria to pieces. Some celebrities (women) in Hollywood have done bizarre facials using different products obtained from various animals like Bird Droppings-Facial, Semen Facial, Vampire Facial, Human Placenta Facial, and The Leech Facial (Times of India, Trend). A team of researchers from the Technical University of Munich found Jellyfish to help fight heart disease in human beings. The extracted fluorescent protein from jellyfish is used to build biological sensors, which when inserted into lab-grown heart cells at a specific wavelength of light hits the heart cells with added sensors, they produce light. This will tell a lot about the electrical activities of the heart cells.

x) **Economic base** – The biodiversity is very important for the functioning of global economy. Bioprospecting (search for previously unknown biotic products of specific utility, e.g. natural anti-fungal toxins and oil-eating enzymes) is one example. Most of our building materials, fibres, fuels, wax, aromatic dyes and gums are also obtained. Besides, biodiversity also does great service through 'ecotourism'. People visit, on holidays, various ecosystems around the world contributing billions of dollars per year. Corals are marine invertebrates. Coral reefs begin to form when free-swimming coral larvae attach to submerged rocks or other hard surfaces. They are some of the most diverse ecosystems in the world, housing tens of thousands of marine species. Corals protect more than 4000 species of fishes in the world and act as a natural barrier protecting islands from rough ocean waves.

Great Barrier Reef of Australia is the biggest example of importance of biodiversity in tourism. People from various countries visit Australia to enjoy such a great beauty of Nature. The spring spawning of these corals is a grand affair with vast expanses of the Pacific Ocean turning red as millions of sperms and eggs are released in a spectacle that is visible from space. These Reefs are 5000,000 years old and still relatively young. The current reef structure is much younger at less than around 8,000 years old. Every year millions of tourists who visit Australia indulge in scuba diving and snorkeling at this reef. This much sought after activity takes you on a tour to a place where you see like ship wrecks, coral gardens, the thousands of species of marine animals and under water canyons. Seven per cent of the Great Barrier Reef, which attracts around $3.9 Billion in tourism every year, has been untouched by mass bleaching that is likely to destroy half the coral. Bleaching occurs when the water is too warm, forcing coral to expel living algae and causing it to calcify and turn white. The mildly bleached coral can recover if the temperature drops, otherwise it may die. The United Nations most recently voiced its concern about the stake of these Reefs urging the Australia to initiate efforts to conserve the 2300 km reef off the Queensland coast (UNESCO World Heritage Panel). Experts believed that the reef, the world's largest living ecosystem, is threatened by climate change, waster water, fishing and coastal developments. It is reported that ocean heat wave could kill 5% of world's corals. Even sunscreen is also killing coral reefs around the world. Coral bleaching in Australia, especially in the World Heritage listed Great Barrier Reef, is a matter of concern in northern region. It is believed that El Nino may be responsible in en masse killing of coral reefs.

Some animals are used for various recreation activities like bull fight in Spain; regular crab race in Tobago Island, etc.

Sheep in Australia can produce up to 40 kg wool in one shearing. This quantity of wool is sufficient to produce 30 quality sweaters (Plate D).

As Marijuana has become legal in most parts of USA, the effectiveness of drug sniffing dogs is declining; honeybees that have been trained to detect illegal substances are being raised. Learning from the great intelligence, cooperation and communication of ants in formation of a colony, an approach like "Robotic Ants" are being developed by the developers, German technology firm FESTO, to carry collectively heavy loads that they would not be able to achieve individually in the factories.

Recently, a worm that eats up plastic has been discovered by scientists in China. It belongs to *Tenebrio molitor* (yellow mealworms), the larvae of which can be fed with polystyrene, one of the most stubborn in the plastic family. The plastic can be fully digested by the worm and degraded into carbon dioxide or nutrition for the worm.

xi) **Ethical importance** – Human-beings like to live in varied natural environments with open space to walk and play in, trees for shade, colourful flowers, birds and animals. So there is an ethical side to maintain biodiversity. It is also our duty to pass on the same natural world to our children, Even the species themselves have their own value and right to exist whether humans need them or not.

Presently, the stuffed animal biodiversity industry has grown beyond expectation. These are mostly polyester-fibre filled replicas of endangered species available in large numbers in the open market. The number of species available in this form went up to 800 per cent since 1990 (as per WWF, world wildlife fund). The rise in stuffed-animal biodiversity shows growing human interest in environmental issues. This is a very good way of educating general public about the utility of biodiversity in our life.

Recently, scientists in United Kingdom are genetically altering dairy cows to make them without horns in a bid to make them safer and to cut the risk of injury to farmers and animals.

CONSERVATION OF BIODIVERSITY

Biodiversity actually boasts ecosystem productivity where each species, no matter how small, all have an important role to play and in that combination the ecosystem is able to prevent and recover from a variety of disasters. The conservation of biodiversity thus becomes primordial in relation to existence of mankind. Every species in nature is not only supportive to each other but also cooperative directly or indirectly. Thus, biodiversity is the resource on which families, communities, nations and future generations depend. It serves as link between all organisms on earth, binding each into an independent ecosystem, in which all biological species interact with each other. In other words—**IT IS THE WEB OF LIFE.**

International Union for Conservation of Nature (IUCN) has access to many different kinds of information on species. The "Red List of Threatened Species" provides global assessments of the conservation status of species. The **"IUCN Species Survival Commission"**, together with the **"Species Programme"** and their partners (see pages 146–147 also) has developed a number of approaches to build up a comprehensive picture of the status and trends in species and biodiversity at global, regional and national levels. It is also necessary to note that the importance of biodiversity and natural processes in providing benefits to people is very much ignored by the financial markets. If the full economic value of these services was taken into account in decision-making, the degradation of ecosystem services could be significantly slowed down or even reversed. The economic value of the biodiversity and ecosystems is still poorly recognized and there is urgent need to build support for and guide efforts to reduce ecosystem degradation and halt biodiversity loss. In view of this the conservation of biodiversity becomes all the more important.

THREAT TO BIODIVERSITY

The established biodiversity is presently under great threat due to various factors. The unchecked rising human population requiring more food and shelter is one such factor. This led to monoculture of crops and trees, extensive use of agrochemicals, intensive tillage etc; industrialisation and urbanization at the cost of most valuable wild habitats. Our impact on biodiversity has been greater than any other species The loss of biodiversity is irreversible. One in six species on Earth could face extinction due to climate change. Recent dry spell in Ran of Kutch in Gujarat forced flamingoes to abandon thousands of eggs due to dry spell. After the flamingoes laid eggs, the water in the particular breeding site located between Wasraj Bait and Dhrangadhra dried up than anticipated resulting in scarcity of food and water and forcing the flamingoes to abandon their eggs, the number of which goes up to 18000. About 5000 sheeps in Iceland have died recently (TOI, Trend, July 30, 2015) this spring. It is believed that the deaths were caused by sulphur released by the Holuhraun volcanic eruption. The sulphur could have contaminated vegetation on the island nation on which the sheep feed. Scientists have already documented five mass extinctions in Earth's history. Today, many believe that we may face the sixth, the greatest mass extinction since the appearance of dinosaurs. Within the beginning of this century we will loose over a quarter of the world's existing plant and animal species. Among well studied groups, the figures are quite alarming—it is estimated that a third of all reef-building corals, a third of shark and rays, a quarter of all mammals, a fifth of all reptiles and a sixth of all birds are vulnerable to extinction. The Western Ghats of India is one of the worlds' 34 "Biodiversity Hot Spots", harbouring threatened species. India as whole is one the 17 identified mega diverse countries of the world. It has 172 (2.3%) of IUCN designated threatened species, like the Asiatic Lion, the Bengal Tiger and the White-Ringed Vulture. The majestic Great Indian Bustards are vanishing from sight and their dwindling numbers have put them in IUCN critically endangered category because less than 200 are left in the country now. Britain Wildlife Vets International (WVI) is concerned of big cats (tigers) in India. They are facing a new threat in the form of the canine distemper virus (CDV), an emerging pathogen threatening tigers worldwide, including Sunder bans in India. The virus makes big cats less afraid of humans thus increases the chance of human-tiger conflict, leaving them vulnerable to poaching. At least three endangered Amur (Siberian) tigers, the world's largest big cat, have died in recent years after containing Canine Distemper. In the latest report, i.e. Third Edition of the **UN's Global Biodiversity Outlook based** on data obtained from 120 countries across the World, it is mentioned that the World's biodiversity is threatened due to the economic growth of countries like China, India and Brazil, putting one third species of plants and animals at risk of extinction. Recent studies (Dipannita Das, 2011) show that several freshwater species of fish, invertebrates (incl. insects like dragonflies, damselflies) and aquatic plants in the Western Ghats, in India, may soon become extinct because of water pollution from agricultural and urban sources, over fishing and invasive species (as per survey done by Zoo Outreach Organization, Coimbatore, India and International Union for Conservation of Nature (IUCN). The study has warned that at least 16% of the 1146 freshwater species face extinction. The developing world's appetite for raw materials is destroying vulnerable ecosystems. Population growth, pollution and the spread of western-style consumption are also blamed for hitting plant and animal species at risk. The good old house sparrow which is becoming rarer by the day has been declared "STATE OF THE BIRD" of Delhi by Govt. of Delhi with the slogan "nurtures and protects it". The mad rush for global development and growth is thus creating serious environmental problems which could endanger the very existence of the **Planet Earth** itself. There is presently greatest need for thinking on **Green Domestic Product** rather than **Gross Domestic Product (GDP).** This will not only be useful to sustain economic growth for the present but also for our future generations. Due to not striking balance between development and environment sustainability, the latter is now at the edge of collapse.

OZONE AND BIODIVERSITY

With regard to climate change, a lot is talked about the depletion or OZONE, almost 90% of which is concentrated in the stratosphere, the layer of atmosphere between about 10 and 50 Km altitude. The ozone acts like a sunscreen lotion and absorbs a large proportion of the sun's ultraviolet radiation which would otherwise sterilize earth and eliminate life from the planet, in the atmosphere, oxygen and ozone molecules are continuously converting into each other and a balance is maintained. But because of increasing man-made pollution and the release of gases like chlorofluorocarbons (CFCs) the natural balance is broken and the amount destroyed is far higher than the amount naturally formed. To overcome such disastrous depletion it was decided in Montreal protocol in 1987 to phase out hydrochlorofluorocarbons (HCFCs) by 2030 in the developed world and by 2040 in the developing world. Such acts are very necessary not only to facilitate the increase in our biodiversity but also in halting the decrease in the number of those organisms which are facing extinction.

Main ozone depleting substances (ODS) like Chlorofluoro-carbons (CFCs) and Hydro-Chlorofluorocarbons (HCFCs) and Hylon are to be phased out in a time frame mentioned above. CFCs and HCFCs are mainly used as refrigerants (refrigerators and air conditioners). The developed countries have moved to high cost new (non-HFC) technology for fridges and ACs and they want developing countries, including India, China and others, should also phase out HFC. But India and other developing countries want HFCs should rather be dealt with under UNECC which put the onus of phasing out greenhouse gases only on developed countries. Huge and fast growing market of India for fridges and ACs are main attraction for developed countries. Moreover, if developed nations share their non-HFCs technology with developing countries without the use of Intellectual Property Rights and Patents clause to replace HFCs, it can be possible everywhere. Scientists in Canada have recently discovered that the chemical "perfluorotributylamin (PERTBA)" is most efficient and extremely long-lived in the atmosphere and has a very high radioactive efficiency over CO_2. This has been in use since mid 20th century for various applications in electrical equipment and is currently used in thermally and chemically stable liquids marketed for use in electronic testing and as heat transfer. It does not occur naturally and so its impact on globally can be more than carbon dioxide. India reiterates its commitment to the protection of the ozone layer on the thirtieth anniversary of the Vienna Convention and 21st International Day for the Preservation of the Ozone Layer held on on 16th September, 2015, in New Delhi, India.

Climate change is one the most significant challenges facing humanity. Our reliance on heavy use of fossil fuels like coal and oil, releases greenhouse gas emissions in the atmosphere, where they trap heat and warm the planet, kicking of change of devastating impacts. Conservation Law Foundation (CLF) was founded in 1966 in New England and CLF venture in 1997. It is involving ways to reduce our dependence on fossil fuels by promoting renewable energy (like wind, solar and certain types of biomass as critical inputs which are naturally occurring and so sustainable) advocating and smart growth, working to responsibly shut down use of coal.

The Paris Agreement of 2015 has year marked setting up health insurance for the planet. This agreement is a triumph for people, the environment and for multilateralism. It is for the first time that every country of the world has pledged to curb their emissions, strength resilience and act internationally and domestically to address climate change. Thus, what was once unthinkable is now unstoppable.

Clean Energy Ministerial group or CEM was set up in 2009 to accelerate a global clean energy revolution for a group of countries encompassing around 90% of global clean energy investment. CEM will work with the World Bank's Energy Sector Management Assistance Programmed (ESMAP) to deter targeted technical assistance that will help developing and middle-income countries integrate large share of wind and solar energy into their electricity grids. Since 2011, CEMs Clean Energy Solution Centre has provided no-cost, high-quality and real-time expert advice to more than 80 countries of the world. Here, the energy is produced from non-fossil resources. The ultimate aim is to have Clean Energy from non-fossil resources so that the carbon-dioxide emission in the atmosphere are brought to the minimum most level and thus, eliminating air and water pollution (Goyal et al., 2015, Times of India-The Times Ideas). United States of America has made a bold move towards renewable energy. The carbon emissions from power plants are to be reduced by 32%. The renewable energy from solar and wind are to be used in India as an opportunity from this and because of 80% reduction in prices of equipments used in these areas to purchase cheap and non-polluter energy. The government of India is very close to give go ahead for hundred percent use of biofuel. It is setting the stage of vehicle engines that can function on 100% bio-diesel soon. Such vehicles are already plying in Brazil and cities like Berkeley in California, USA. These vehicles will be known as "B100" in India.

Recently, India has drawn up a plan for CLIMATE CHANGE for 2030 (Times of India-Times Nation-Mission Emission Cut; October 3, 2015, page 18). It promises to propagate healthy and sustainable way of living based on traditions and values of conservation and moderation. It proposes to reduce emissions intensity of country's GDP by 33-35% by 2030 from 2005 level. This will involve 40% power generation from renewable (solar, wind, biomass, hydro and nuclear). It will create additional carbon sink of 2.5 to 3 billion tons of carbon dioxide by adding to forest and tree cover. This will further be added by adapting to climate change by investments in development programmes in sectors vulnerable to climate change, particularly agriculture, water resources, Himalayan region, coastal regions, health and disaster management. A preliminary estimate suggests that at least USD 2.5 trillion (at 2014-15 prices) will be required for meeting India's climate change actions between October, 2015 and 2030. Stepping up its climatic action voluntarily, India on 13th October, 2016, made it mandatory for manufacturing companies to control emission of climate-damaging refrigerants (HCFC-22) through incineration. HFC-23 is a byproduct in the production of a chemical (HCFC-22) used primarily in ACs and refrigeration. This will be applicable for those companies which manufacture HCFC-22. HFC-23 is 14,800 times more damaging to the climate than carbon dioxide. HFC-23 is among the most abundant HFCs (Hydro fluorocarbons) in the global atmosphere (Times of India, p. 14, Oct. 16, 2016).

The climate report in 2014 (Times of India Trend, July 20, 2015) shows record heat, record sea levels, more hot days and lesser cool nights, surging cyclones, unprecedented pollution and rapidly diminishing glaciers. The US National Oceanic and Atmospheric Administration have issued the report compiling the latest data gathered by 413 scientists from 58 countries around the world. The report highlights – glaciers retreat (ice loss) for the 31st consecutive year (most alarming is the rate of loss which is accelerating overtime); temperature set a new record (e.g. last year was the hottest in 135 years, more than 20 countries in Europe set new heat records, with Africa, Asia and Australia also experiencing near – record heat; the almost mean temperature for India was two degree centigrade above the 1963-90 overage. The east coast of North America was the only region to have cooler than average conditions). The sea level surges to a record. The global mean sea level continued to rise, keeping pace with a trend of 3.2 mm per year over the last two decades. This is posing great danger to many small islands being vanished altogether. A recent report about Washington DC sinking is a cause of worry (Times of India-Trend, July, 30, 2015). Scientists have found that land under the Chesapeake Bay, the largest estuary in the US, are sinking rapidly and the country's capital, Washington DC, could drop by six or more inches by 2100, adding to the problems of sea-level rise. The falling land will exacerbate the flooding that Washington DC faces from rising ocean waters due to a warming climate and melting ice sheets. The record greenhouse gases fill the atmosphere due to burning of fossil fuels by more than 40% since the industrial revolution. The carbon dioxide reached a concentration of 400 parts per million for the first time in May, 2013. The oceans are absorbing even more global warming than the surface of the planet, concentrating to rising seas, melting glaciers and dying coral reefs and fish population. In 2015, the world has moved into "El Nino" warming pattern in the Pacific Ocean. El Nino phases release some of the ocean's stored heat into the atmosphere, causing weather shifts around the world. The El Nino has not peaked yet this year (2015), but by some measures it is already the most extreme ever recorded for this time of the year. This is how such a climate change is adversely affecting our biodiversity—both on land and in sea. It is now believed that burning all the world's deposits of coal, oil and natural gas would raise the temperature enough to melt the entire ice sheet covering Antarctica, driving the level of sea up by more than 160 feet and a subsequent sea level rise of 200 feet will see London, Paris, New York, Hong Kong and Tokyo being submerged.

Most recent report (TOI, Feb. 14, 2016) on increasing water on land slowing down rising seas is welcome and positive one. New measurements from a NASA satellite, Pasadena, California and University-of California- Irvine reported for the first time that climate-driven increases of liquid water storage on land have affected the rate of sea level rise. What we did not realize until now is that over the past decade, changes in the global water cycle more than offset the losses that occurred from ground water pumping, causing the land to act like a sponge—at least temporarily.

These data are vital in understanding decadal variations in sea level change.

DEGRADATION OF ECOSYSTEMS AND BIODIVERSITY LOSS

Before the beginning of humanity, each species and its descendant lived for somewhere between 1 and 10 million years. Any animal species from amoeba to human-being contains from 1 to 10 billion nucleotide pairs (genetic letters). These occur in unique traits in anatomy and behaviour and biochemistry. All these help the species to adapt to the ecosystem in which it lives. When a species becomes extinct, all that heritage is also lost. Human activity also blocks new species from evolving and due to this, sometimes speciation runs in reverse. The research scientists of British Colombia, University of Bern, Switzerland and McGill University, Montreal recently found three-spine stickleback fishes following the same evolutionary path in different lakes in Western Canada-with all of them sharing similar features (Hindustan Times, 27th June, 2006). New species usually evolve when old species split apart and some mutate to get new physical characteristics not shared by the rest. Similar transformations found in Darwin's finches on the Galapagos Islands also showed that the bird populations were diverging on similar lines. These findings, also prove that the presence of human settlements at the sites discouraged speciation, making easier for hybrids to survive. The introduction of signal crayfish into the lakes, for instance, apparently disturbed the habitats of sticklebacks, driving them into more contact, leading to interbreeding. These findings are disturbing since much of the world's biodiversity is made up of fragile young species that are threatened by successfully exploding human population; or human-beings are too successful at the expense of other species.

Man has survived and prevailed by exploiting all ecosystems and resources like oil, natural gas, and minerals. With six billion, crowding the Earth and driving out biodiversity that human activity's destructive impact on the world's food chains has already been duly acknowledged. Such activities also inhabit growth in the biodiversity of a species. Species extinction is a natural process. Since life began on Earth, five major mass extinction became responsible to fast and sudden drops in biodiversity. Massachusetts Institute of Technology's (MIT) scientists have found enough evidence that methane producing microbes that suddenly bloomed explosively in the ocean are to blame for the mass extinction around 252 million years ago. The Phanerozoic eon (last 540 million years) marked a rapid growth in biodiversity via the Cambrian explosion (when nearly every phylum of multicellular organisms first appeared); the next 400 years showed massive biodiversity losses and so called mass extinction events. In the Carboniferous period, rainforest collapse led to a great loss of plant and animal species while The Permian or Triassic period (251 million years ago) was the worst of all the previous periods. The most recent, the Cretaceous-Tertiary period (65 million years ago) is best known because it was the period when the famous, dinosaurs became extinct. In the Holocene period (when humans appeared) there started marked reduction of biodiversity primarily due to human impacts, particularly habitat reduction. However, over the past 100 years human activities have increased the extinction rate by at least 100 times compared to the natural rate, resulting in a net loss of biodiversity. At this rate, it is being rightly feared that human impact could wipe out a fourth of the world's remaining species in the next 40 to 50 years. Already an estimated two of every three birds decline worldwide, one in every eight plant species is endangered or threatened, and one-quarter of amphibians and one-fifth of reptiles are endangered or vulnerable to decline. Presently, it should be our prime duty to discover more and more new and missing species as a continuous process.

GAIN IN BIODIVERSITY

Recently, conservationists have discovered two species of African frogs (Omaniundu reed and Mount Nimba reed) last seen in 1941 and cave dwelling Mexican salamander (Splayfoot) which was feared to have become extinct (Plate A). Similarly, a team of systematic zoologists from Delhi University, Bombay Natural History Society, Zoological Survey or India and Brussels' Vrije University have most recently discovered twelve new frog species in the Western Ghats, India. These findings included three frog species (coorg night frog, forest night frog and spinular night frog) which were not seen for more than 75 years since their original description by C R Narayana Rao in the 1920s and 930s. These discoveries show lot of hope on green front. Besides, due to tremendous efforts of scientists all over the world a very large number of new species are being discovered. Some rare and colourful such species are shown in Plate B.

Although fish are cold-blooded animals yet recently a warm blooded fish, Opah was discovered by scientists of the National Oceanic and Atmospheric Administration (NOAA) from the waters of USA. It is fully warm-blooded, circulating heated blood throughout its body much like mammals and birds. Residents of Australia's southern Tablelands were shocked to see millions of tiny spiders falling from the sky; along with mounds of silky thread (phenomenon also called 'spider rain' or 'Angel Hair'). This is a manifestation of a form of spider transportation called "ballooning". The spiders climb some high areas and stick their butts up in the air and release silk. Then they just take off. Recently, a frog that gives birth to tadpoles instead of eggs was discovered in the Rain forest of Sulavesi Island of Indonesia. Another interesting species, looks like a cross between a house cat and a teddy bear was discovered by the scientists of Smithsonian Institute of Washington. It is a first carnivoure, *Olinguito*, discovered in western hemisphere in 35 years. Another interesting new species of shark that walks along sea beds using its fins as tiny legs has been discovered from Eastern Indonesia. Surprisingly, vultures, which had practically become extinct in and around Delhi over the past couple of decades are slowly making a comeback. During 2015, there have been three sightings of the primary or nominate species of endangered Egyptian vulture, a migratory bird. In Britain, a rarest fresh water fish and relic of last ice-age, long thought to have been extinct has emerged in Lake District. A team of scientists in India came across an adult *Enhydrus setboldi* in Yamuna Biodiversity Park. This venomous water snake was last seen in 1940s. The Indian Stripped Hyena, once thought to be doomed, is making a grand come back in Western Ghats in India. It is a great scavenger and thus a great bonus for conservationists.

MARINE BIODEVRSITY

Our marine life is also on the brink of mass extinction due to global warming and pollution. The potential extinction of species, from large fish to tiny corals, is comparable to the five great mass extinctions in history; during each of which much of the world's life died out (Times of India, Michael McCarthy, June 22, 2011). The World Wildlife Fund has also cautioned that spread of so-called marine dead zones where nothing can survive due to lack of oxygen, could cause collapse of Baltic Sea Ecosystem. The rising temperature of ocean water is causing a proliferation of the Vibrio genus of bacteria, which can cause food poisoning, serious gastroenteritis, septicemia and cholera in European countries (Times of India, Sept. 15, 2011, p. 21). With regard to the release of carbon dioxide in the atmosphere, the wealthy western nations like United States of America and European Union have "over-polluted" the atmosphere These countries have emitted enormous amount of carbon dioxide over the last century and the developing countries like India and China will never be able to emit a fraction of their share they are entitled to in order to industrialize.

Due to the negative impact of carbon dioxide in the environment, there are various thoughts to minimize the impact of CO_2 around us. One such proposal is of storing the carbon dioxide in the depths of the ocean to clear the air. The environmentalists are opposed to it because this practice would be killing the organisms at the bottom of the sea which forms a vital link in the chain of life on the only home we know, Planet Earth. Deep-sea ocean beds are probably the last frontier on earth, relatively free of toxic human footprints. We are poisoning the air we breathe, the water we drink and the soil we live off. With global warming the ocean waters are turning acidic, bleaching coral reefs and compromising marine life. The world's highest mountain peak, Mt. Everest is also the world's highest garbage dump. So leave ocean beds alone, if we believe the current reports, the part of the planet from Alaska to Antarctica with a rising tide of trash of plastics (up to 95% of total debris) is suffocating the seas.

Due to already such a high level of pollution in the sea, one-third of reef-building corals would be facing extinction by 2050.The death of a coral reef would jeopardize the survival of marine organisms that depend on reef for sustenance. As rain forests of the ocean, coral reefs are pivotal to the health of underwater ecosystems; their degradation would adversely impact life under water. They are rich source of seafood besides being tourist's attraction. Coral reefs, like mangroves, act as buffers against erosion in coastal areas, protecting human habitations and livelihoods (Place C). The most visible and glaring example of continuous depletion of ecosystem can be seen with the threat to the destruction of mangroves in Mumbai, India, due to proliferation of slums along its CREEKS, As mentioned above mangroves protect coastal areas from erosion, storm surge (especially during hurricanes), and tsunamis. Their massive root system dissipates wave energy and slows down tidal water to the extent that its sediment is deposited as the tide comes in, leaving back all except fine particles during ebb.

The Canadian scientists have found that "oceans" flow deep beneath Earth's crust. Ocean's deep secret surprisinginly has been revealed by a team of scientists of the university of southern Denmark (Times of India, trend, March 13, 2013). A specially designed underwater robot, about 4 meters tall and weighing 600 kgs was put in "The Marine Trench" in the Pacific Ocean having maximum depth of 36000 feet (11000 meters), big enough to swallow up Mount Everest. It was once thought to be too hostile an environment for life. The scientists through a video recording found very few large animals but a world dominated by microbes. The United States of America is carving out a wide swath of Pacific Ocean for expanded marine preserve, putting the waters off – limits to drilling and most fishing in a bid to protect fragile underwater life. The revamped Pacific Remote Islands Marine National Monument, will broaden the George Bush-era preserve, and will cover 490,000 square miles—an area roughly three times the size of California and will become world's largest marine preserve in the world. To honour President Barack Obama of United States of America due to his bold initiative for preserving marine biodiversity, a new species of gold fish was named after him. The fish was discovered 300 feet deep in the waters off Kure Atoll in the Pacific Ocean. Earlier, a dinosaur, spider and a parasite were also named after him. New Zealand has also planned a France-sized marine sanctuary to protect one of the world's most pristine ocean environments. This Kormaed Ocean Sanctuary would cover an area of 620,000 square kilometers. This sanctuary was home to thousands of important species, including whales, dolphins, sea-birds and endangered turtles. Half of the world's marine population (49%) has disappeared between 1970 and 2012 due to over-fishing and climate change.

This sanctuary will also prevent fishing and mineral exploration in an area where scientists regularly discover new marine species. A UN report said half of the world's wetlands – mostly in Asia have vanished in the last few years because of increased human pressure. There is thus grave risk to global food security because of unsustainable use of natural resources. Reduced flow of fresh water into the seas in India will also be responsible in enhancing coastal erosion affecting 25% of India's coast.

BIODIVERSITY AWARENESS

Thus, presently there is greater need not only to prevent the loss of biodiversity but also to preserve it. The general public need to be educated about the utility of our heritage in the form of biodiversity. Various countries have shown rising awareness with regard to protection and conservation of biodiversity around us. Some self driven ways go a long way in the protection of thousands of individual animals. The Estonia Nature Fund's efforts have estimated to have saved the lives of over 6,300 frogs over the last few years. Without human intervention, thousands of the amphibians would have been killed crossing busy roads. The frogs on their spring migration were to cross busy roads and thus subjected to being killed by motorists. Now this Nature Fund Organization is hoping to build tunnels to help these creatures to avoid traffic altogether. Recently, an American alligator, 12–13 feet long, was found walking into the seventh hole of golf club in Florida, USA. The golfers were very cautious to keep an eye on it for days together. Aiming to people's wrong perception towards foxes, there is a "fox café" in east London where the customers can dine and play with the animal and see its playful side. Sometimes slight mistake can become tornado mistake if we are not aware of the behavior of various species before finally getting rid of them. A pet owner in USA dumped a few goldfish into a Colorado Teller Lake few years back. Today, the fish has so much multiplied that it has taken over the entire ecosystem of the lake, thus putting threats of survival to the local fish species in the lake. Fishermen of Jaunpur village, about 85 kms from DehraDun, Uttarakhand in India take part in the Maun festival, an animal event where villagers kill thousands of fish in the Agiaad river, a tributary of river Yamuna. Animal right NGOs and forest officials have persuaded them to give up fish killing and mark this "Maun Mela" for fish conservation. A strange show was performed on Discovery channel on TV in USA where a man offered himself to be eaten by Anaconda. Animal activists were concerned about the harassment of animal than the safety of the man. On completion of the live show, unharmed man declared that the show was an experiment designed to raise money to save the animal's habitat. Recently, an ice-age mammoth's skeleton was auctioned in London for 150,000 pounds. The skeleton was found in Europe and is 30,000 and 50,000 years old; it is 18 ft high and 11.6 feet long weighing up to six tones. After making the local residents realized the importance of tourism in Periyar Tiger Reserve, in India, they changed themselves from traditional hunters to tiger conservationists. Similarly women in Gundlaba village in Orissa, India took steps to increase green cover and adopted ways to protect olive ridley turtles.The residents of Badopal village, about 10 kms from Fatehabad town in Uttar Pradesh, in India got the metal fencing of Nuclear Power Cooperation of India Limited (NPCIL) removed as this village is heaven to endangered blackbuck, deers, neelgai (blue bull) etc. This area has abundant food and other resources for the survival of these animals. The mountains of Western Ghats traversing from Kerala, Tamil Nadu, Karnataka, Goa, Maharashtra and Gujarat spread over 140,000 sq.kms and its forests are home to at least 325 globally threatened flora, fauna (like birds, reptiles, amphibians and fish species). The Ghats have already been listed as critically biodiversity hotspots and declared a UNESCO World Heritage Site.

It is surprising that in spite of so much awareness at public and government level to protect and conserve animal species, laws of the land are not clear enough to deter people from poaching and selling them. There is still a thriving illegal trade of endangered species on the web points using loopholes in Wildlife Act, 1972 in India. The Act covers only endemic species in India. Bird and other animals that are imported do not have any kind of protection. In such cases, the onus is on State Governments to keep tabs on sale of endangered animal species. But their role is restricted to ensuring that the traders have proper licenses and the animals are disease-free. Though once in a while action is taken against those who breed and sell exotic species but it does not seem to deter traders. Various countries are very active in dealing with such kinds of illegal trade. As many as 231 dinosaur eggs and a dinosaur skeleton were recently seized in China. The eggs date back to the Cretaceous Period. The skeleton was identified as belonging to a Psittacosaurus, a genus of extinct ceratopsian dinosaur. It is high time that the Law in India is amended to put a stop to illegal trade in all endangered species, irrespective of their origin.

There is need to create more protected areas like natural parks. sanctuaries and biosphere reserves, etc. National Park System claims some of the most complete ecosystem and biodiversity remaining in a country; e.g., over 60% of the endangered species in the United States are found in national parks. The habitat preserved within park boundaries provides many species an oasis of survival and some of the last bastions of intact ecosystems found today. In India the protected areas presently covers a little less than 5% of country's total land area which includes 94 national parks and 501 wildlife sanctuaries; of these, 100 cover both terrestrial and freshwater ecosystems and 31 are marine protected areas. Amongst them the top ten National Parks/Wild Life Sanctuaries are Corbett National Park in Uttarakhand; Kazirranga National Park in Assam; Bandipur

National Park in Karnataka; Bandhavgarh National Park and Kanha National Park in Madhya Pradesh; Gir National Park and Sasan Gir Sanctuary in Gujarat; Keoladeo Ghana National Park, Bharatpur, Rajasthan; Periyar National Park in Kerala; Pench National Park in Madhya Pradesh; and Ranthambore National Park in Rajasthan. In addition to these there are ten more which are equally Important like: Manas National Park in Assam; Nagarhole National Park in Karnataka; Sunderbans National Park in West Bengal; Sariska National Park in Rajasthan; Dudhwa National Park in Uttar Pradesh; Penna National Park in Madhya Pradesh; Tadoba Andhari Tiger Reserve in Meghalaya; Chinnar Wildlife Sanctuary in Kerala; The Great Himalaya National Park in Himachal Pradesh; and Dandeli Wildlife Sanctuary in Karnataka. In addition there are also 18 Biosphere Reserves and several Reserved Forests as part of the most strictly protected forests outside the above mentioned protected areas. These Biosphere Reserves are roughly as per IUCN category V Protected Areas mainly for the protection of natural habitat and often include one or more National Parks and preserves along buffer zones that are open to some economic uses. These places get protection to humanities living there along with the fauna and flora. These Biosphere Reserves are: Nilgiri in Karnataka; Gulf of Munnar in Tamilnadu; Sunderbans in West Bengal; Nanda Devi in Uttarakhand; Nokrek in Meghaya; Pachmarhi in Madhya Pradesh; Simlipal in Orissa; Achanakmar-Amarkantak in Chhattisgarh; and Great Nicobar in Andaman and Nicobar Islands. The first four mentioned above are recognized by UNESCO under World Network of Biosphere Reserves. The Guryal ravine site, in the vicinity of the Srinagar city (India: Kashmir), is one of the world's richest fossil sites, being rated by geologists as the world's premier site for the study of species from the Permian period (299–251 million year 5 ago). But even the biodiversity found in such protected areas is in jeopardy. From pollution to poaching, invasive species to habitat loss and fragmentation, these islands of life may not be enough to ensure the survival of endangered species. Recent reports from few countries have shown decline in vulture population, the great scavengers, due to the use of the veterinary drug, Diclofenac on livestock. Vultures feed on carcasses of animals exposed to this drug, leading to their poisoning and finally death due to renal failure within 2–3 days of consuming drugged carcasses. Due to the almost non-availability of vultures, the hygiene and sanitation conditions are becoming quite serious in our nearby surroundings. Moreover, vultures are slow breeding birds and lay just one egg every year. This breeding cycle further aggravates the situation of decreasing their number. There is great need to do a lot in the protection and conservation of these important scavengers. Recently, Australia has declared 580,000 hectares of remote land as a conservation zone to protect the endangered 'warm' or black footed rock wallaby, to be managed by its 'traditional owners' as only 100 of them are left at the moment. Punjab and Haryana have now planned a 'Vulture Safe Zone' over 100 km where birds are bred in captivity at the country's first Jatayu (vulture) centre in Bir Shikargah wildlife sanctuary in Pinjore, near Chandigarh.

NEW TECHNIQUES AND BIODIVERSITY PRESERVATION

It is also therefore necessary to explore the use of new techniques for better protection and preservation of our invaluable biodiversity. Biotechnology provides some long term methods for preserving biodiversity in the form of germplasm (genetic material). Seeds, pollen grains, vegetatively propagated parts are collected and stored in gene banks in specialized conditions. Similarly, endangered animal species are preserved in the form of their sperms. Equator has, recently, sent 201 tortoises to the Galapagos Islands. These tortoises represent a subspecies of *Chelonoidis* sp. that went extinct about 150 years ago. The replacements are from the *Chelonoidis hoodensis* subspecies and they resemble the extinct subspecies. The 201 tortoises are 4–10 years old. Of these, 30 have a radio transmitter that will allow park rangers to follow the animals in the wild. This is a good example of increasing the species closer to extinct ones and then increasing their number.

ROLE OF UNITED NATIONS IN BIODIVERSITY

Realising the seriousness of over exploitation of our environment, over 170 countries gathered together in 'Rio de Janeiro' in Brazil in 1992 for a meeting called "The Earth Summit" to discuss the future of the world. Together they agreed for a need to work together in "global partnership for sustainable development". Since then **"sustainable development** has been in use the world over. It is defined as the "development that meets the needs of the present without compromising the ability of future generations to meet their own needs Recognizing the fact of depletion of their ecological wealth, world leaders from 192 countries again gather together in 2002, to set their goal of stemming *biodiversity loss* under the Convention of Biodiversity. The main focus of this Convention was of getting all parties to agree on not only on access to genetic resources but also to fair and equitable sharing of their use. The assessment of the work done by all member countries was made in another meet of world leaders in Nagoya, Japan in 2010. On the basis of negative reports, United Nation rings alarm bell on saving species. It was realized that the world must stop the rapid loss of animal and plant species and the habitats they live in. It was recommended to stop the incalculable loss of biodiversity by avoiding deforestation and Sand degradation from a coordinated ecosystem approach. We must know that our prosperity and indeed our survival depend on healthy ecosystems. The Earth's forests, oceans and rivers are the very foundation of our society and economy, in other words it is our foremost duty to make sure that we must keep the world in good condition so that our children and our children's children inherit the same natural resources that we have. These natural resources are: fresh air, clean farmland, wildlife, forests, unpolluted seas and a stable climate. India being a party to Convention on Biodiversity (or Biodiversity Convention) (1992) recognizes the contributions of local and indigenous communities to the conservation and sustainable utilization of biological resources through traditional knowledge, practices and innovations with emphasis on equitable sharing of benefits with such people arising from the utilization of their knowledge, practices and innovations. There is an urgent to formulate the projects in such a way that their implementation should see minimum most risk to fauna, flora, their habitats and their ecosystems as a whole. In the year 2013, UNESCO designated Nicobar Islands as a World Biosphere Reserve under its Man of the Biosphere programme. This biosphere is home to about 1800 animal species and abode to world's most endangered species. United Nations declared

2010 as the year of Biodiversity to create awareness for the protection and conservation of our fast depleting biodiversity. On 22nd May, 2015, it declared international day for the protection of Biodiversity.

SUPPORT TO BIODIVERSITY

Due to tremendous interests created in the conservation of biodiversity the world over, various organizations have come out in its support. Few such organizations are discussed below:

i) **Biodiversity Conservation Information System (BCIS).** It is a consortium of ten international organizations and programmes of **International Union for the Conservation of Nature (IUCN) or THE WORLD CONSERVATION UNION.** The members of the organization collectively represent the single largest global source of biodiversity conservation information in the world. Here the members' networks work together for a common goal, i.e to support environmentally sound decision-making and act by facilitating easy access to data and information on biodiversity. Under its Conservation Commons, a key product of BCIS, a collaborative effort is developed to improve open access to data, information and knowledge related to conservation and sustainable use of biodiversity.

ii) **Global Biodiversity Information Facility (GBIF).** It was established in Germany with the aim to digitize and network the world's biodiversity data free and universally available. Scientists of more than 47 countries and 31 international organizations are participating and supporting the set up of data bases under GBIF. The international GBIF DATA PORTAL is now providing access to more than 93 million records from 173 data providers. GBIF thus encourages, coordinates and supports the development of worldwide capacity to access the vast amount of biodiversity data held in natural history museum collections, libraries and data banks.

iii) **Society for Conservation of Biodiversity (SCB).** It was established in 2005 and is an international organization dedicated to advancing the science and practice of conserving the Earth's diversity. It has more than 10,000 members worldwide.

Biodiversity is a broad subject involving varied activities and actions. India being a signatory of Un's Biodiversity Convention, formulated Biological Act in 2002 which includes; (i) To regulate access to biological resources of the country with the purpose of securing equitable share in benefits arising out of the use of biological resources; and associated knowledge related to biological resources; (ii) To conserve and sustainable use of biological diversity; (iii) to respect and protect knowledge of local communities related to biodiversity; (iv) to secure sharing of benefits with local people as conservers of biological resources and holders of knowledge and information relating to use of biological resources; (v) conservation and development of areas of importance from the standpoint of biological diversity by declaring them as biological diversity heritage sites; (vi) protection and rehabilitation of threatened species; and (vii) Involvement of institutions of state governments in the broad scheme of the implementation of the Biological Diversity Act through constitution of committees.

India's 41% of forest cover is at different stages of degradation. And the making limited use of its vast pool in agriculture and livestock. India has one of the world's richest biodiversities but stands to loose it all if it does not take proactive steps.

NATIONAL BIODIVERSITY AUTHORITY

In view of carrying forward the objectives of Biodiversity Act, 2002, The Government of India established **"National Biodiversity Authority"** in 2003 with fulltime Chairman, 10 official members and 5 non-official members with its headquarter in Chennai, India, under the Ministry of Environment and Forests, the nodal agency for implementing provisions of Convention of Biodiversity in India.

AGRICULTURAL BIODIVERSITY

Agricultural biodiversity or Agro-biodiversity has also come in wide use in recent years. It is a vital sub-set of biodiversity.

Although its examples have been included under biodiversity dealt above, it is discussed here In brief to give a clear idea about this term. Agro-biodiversity encompasses socio-cultural, economic and environmental elements. All domesticated crops and animals result from human management of biological diversity which is constantly responding to new challenges to maintain and increase productivity, in broad sense, agro-biodiversity includes—harvested crop varieties, livestock breeds, fish and wild resources of the field, forest, and aquatic ecosystem; non-harvested species within production ecosystems supportive to food provision, including soil-micro-biota, pollinators, etc.

Decline in Agro-biodiversity

Presently world's population is continuously growing at a fast rate, especially in the under-developed and developing countries. Agricultural production has to meet the demand of food for this ever growing vast population. This results into agricultural expansion at the cost of forest and marginal lands, combined with over grazing and urban and industrial growth resulting in substantial loss of biodiversity over significant areas. Today agro-biodiversity is under serious threat and its loss is extensive due to disappearance of harvested species, varieties and breeds. It is mainly due to loss of forest cover, coastal wetlands; and other 'wild' uncultivated areas; and destruction of aquatic environment. All this also account for substantial losses to wild relatives and 'wild' foods which otherwise become essential food provision in times of crisis. Sustainable agriculture should be given priority. It is in the best interest of our country to adopt forestry, tourism etc. to protect diverse species including humanity. India has rainforests from Assam to Western Ghats. Nearly, half of the world's rainforests have been cleared (i.e. nearly 32 million acres lost) and between 150 and 200 species of plants, insect, bird and mammal become extinct every twenty four hours. Besides, 15% of mammal species and 11% of bird species are endangered. Deforestation causes 12% to 15% of CO_2 emission globally and 70% of forests are lost by agricultural conversion. Depleting forest cover due to unsustainable agriculture threatens more than half of world's protected areas.

Genetic Erosion

Many countries have reported 'genetic erosion' of local crop varieties and animal breeds. This is mainly due to their replacement by improved or exotic varieties, breeds or even species. Since the beginning of agriculture about 12000 years ago, nearly 7000 plant species have been cultivated and collected by humans for food. To-day only 15 plant species supply major part of 90% of our food provision. Domestic animals contribute a lot as human food in the form of meat, milk products, eggs, etc. This major contribution is due to 4,500 breeds drawn from some 40 or more animal species. These breeds have been developed over the past 12,000 years, represent the remaining pool of genetic diversity for our future demands. Currently these breeds are dying at the rate of six breeds per month. It is estimated that 30% of the world's breeds are at risk of extinction. Since 1960, intensification of agricultural systems coupled with specialization by plant breeders and harmonizing effects of globalization have led to substantial reduction in the genetic diversity. Today, a third of the 6500 breeds of domestic species are threatened with extinction. Local breeds are genetically better adapted to their environment, more resistant to local parasites and are most adjustable to climate change while being productive. Greater use of local breeds will be most effective in achieving food and nutrition security. Moreover, very few know that biodiversity has the power to improve the capacity of a social-ecological system both to withstand perturbations (from climatic or economic shocks) and to rebuild and renew it afterwards. In view of this, the animal production geneticists, the world over, are searching for genes which can influence the production and quality of products together with the health and reproductive traits of animals. This depends largely on crosses between breeds with extreme characteristics for which high level of biodiversity is required. Local breeds play major role in this process.

CARBON CREDITS

It is a term used for any tradable certificate or permit representing the right to emit one tonne of carbon dioxide or the mass of another greenhouse gas equivalent to one tonne of carbon dioxide. One carbon credit is equivalent to one metric tonne of carbon dioxide. Greenhouse gas emissions are capped and then markets are used to allocate the emissions among group of regulated sources. It is meant to regulate the individual and commercial processes in the direction of low emissions or less carbon intensive approaches to protect the ozone layer which is a must for survival of all fauna, fauna and we. The carbon credits are generated by **UN-run** scheme called the **Clean Development Mechanism** (CMD). In developing countries the commercial firms get financial incentives to cut greenhouse gas emissions.

There are many different ways an individual or business houses can reduce their Carbon Footprints – "Reduce, Reuse, Recycle" is a great way to start cultivating mindset that takes your decisions into account, such as questioning the need to drive for a single trip compared to walking, or biking. A person can put a price on these carbon footprints (carbon offsets) and invest that equivalent amount of money in a renewable energy project out of the atmosphere—this is how one can balance the pull and keep carbon out of the climate system.

China, USA, European Union and India are the largest emitters of carbon dioxide accounting for 50% of greenhouse gas. In India, the South Delhi Municipal Corporation's plant in Okhla, Delhi has become the first to get carbon credits from United Nations Frame Work Convention on Climate Change. The civic agency has been given an advance of Rs. 25 lakhs against net Carbon Emission Reduction (CER) earning from the plant.

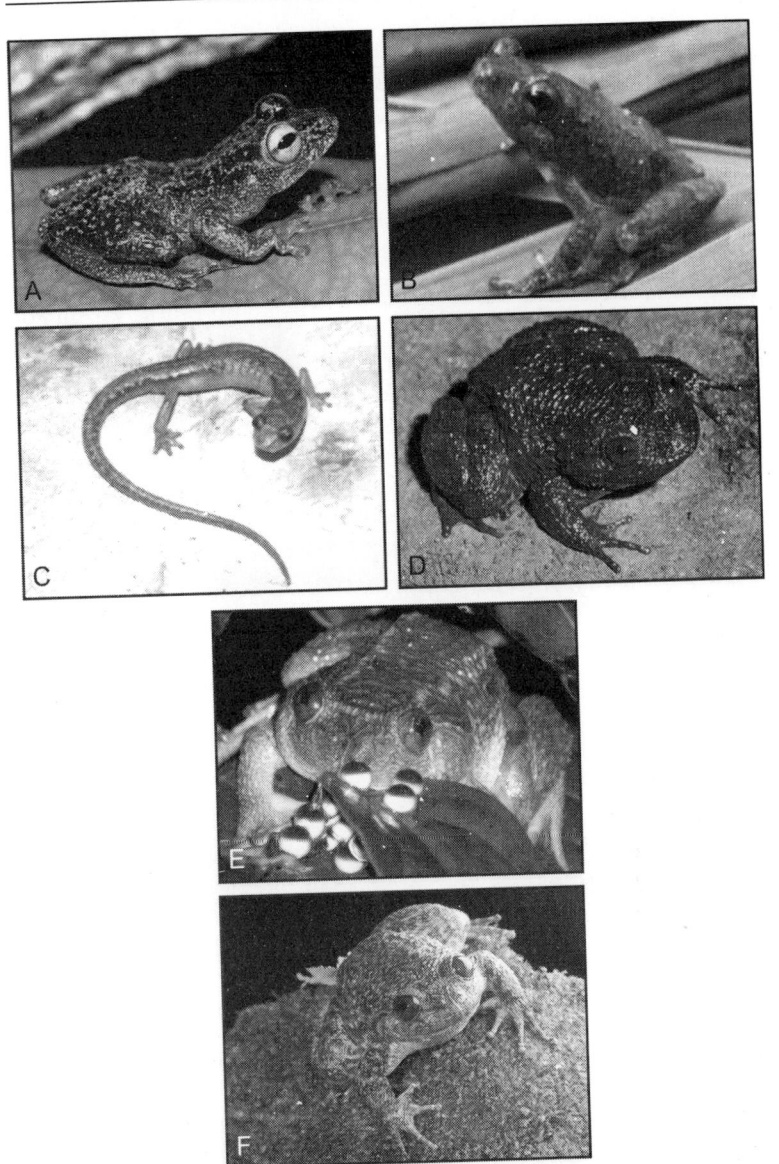

PLATE A: Some Missing Species Rediscovered Recently.

A) Omaniundu Reed Frog; B) Mount NImba Frog (Both from Mount NImba in Africa); C) Cave Dwelling Mexican Salamander (Splayfoot); D) Spinular Night Frog: E) Jog or Coorg Night Frog; F) Forest Night Frog; (D-F) From Western Ghats, India).

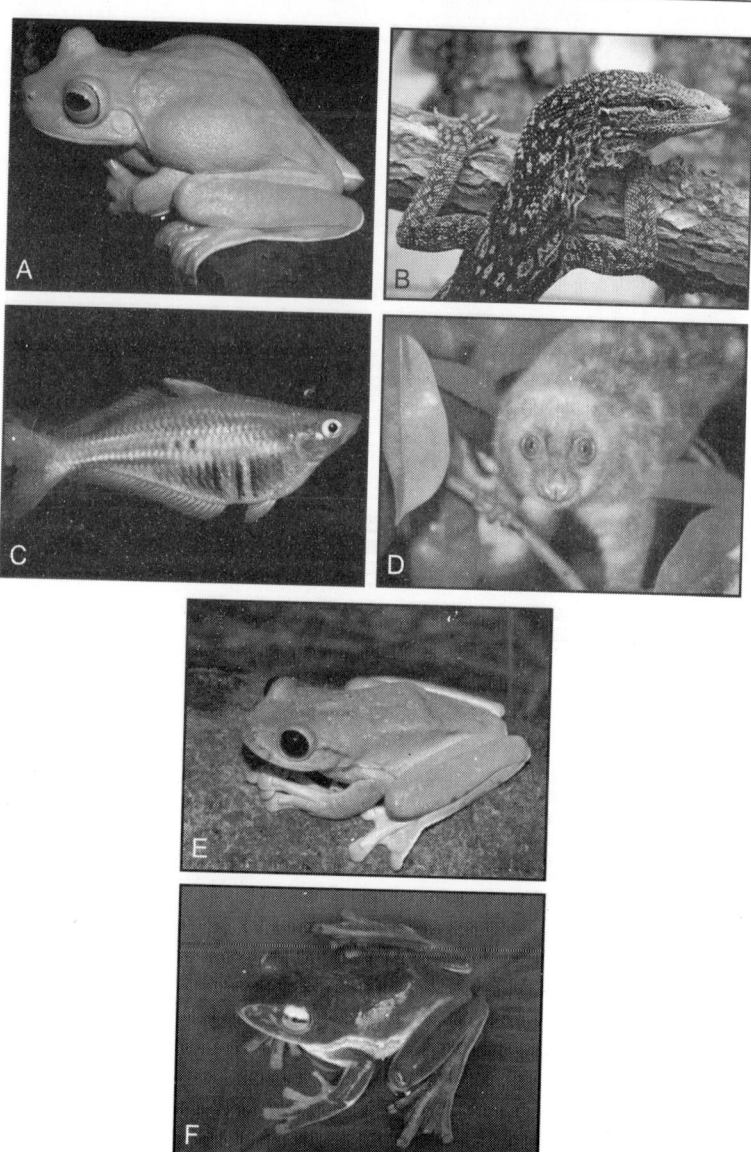

PLATE B

PLATE B: Same Colourful New Species Out of Hundreds Discovered During Last Ten Years from Around the Globe.

A) Large Green Tree-dwelling Frog; B) Monitor Lizard; C) Rainbow Fish; D) Blue-eyed Spotted Cuscus (From Melanesian Island in New Guinea); New Tree Frog (Papua New Guinea); F) New Flying Frog from Eastern Himalayas, G) A Lizard (Dubbed Varanus Bitatawa) Endowed with a Double Penis, from Philippines; H) Pink Eyed Caedicia from Papua New Guinea; I) Transparent Sea Cucumber (Engypanias from Ocean Depths in Gulf of Mexico).

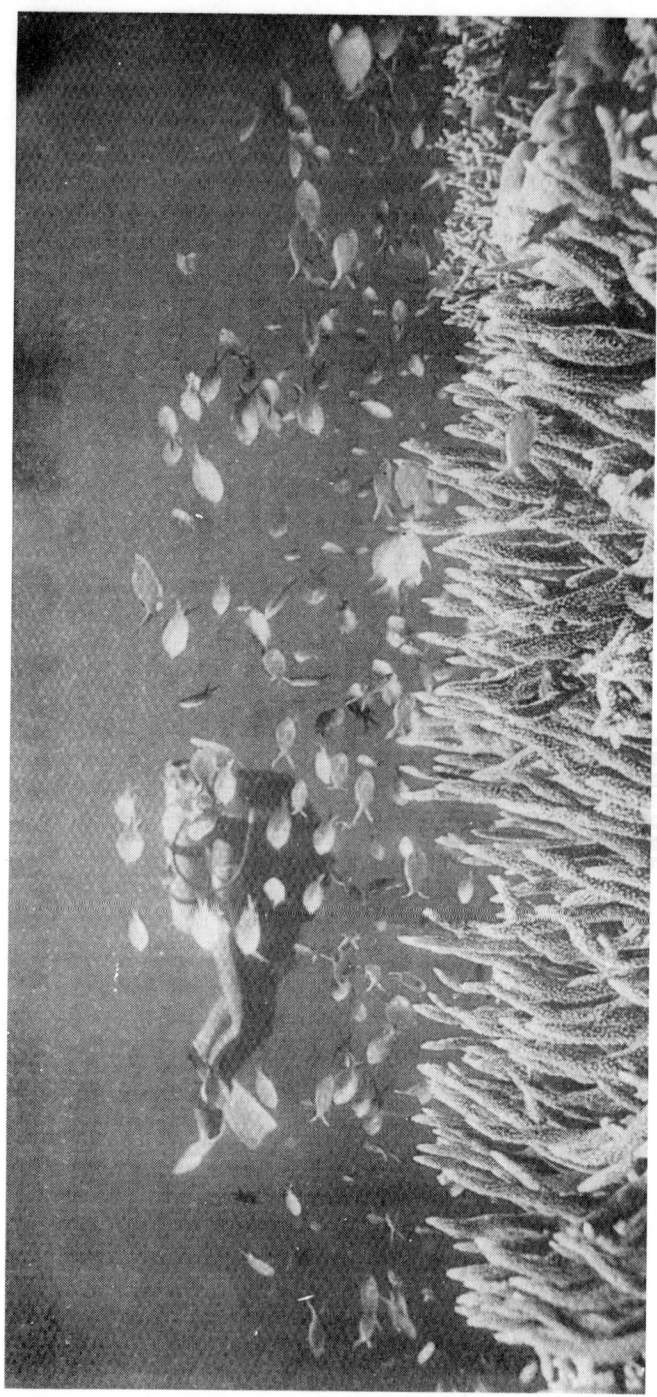

PLATE C: Coral Reef (Australia's Great Barrier Reef) Under Threat And So Are Coral Reef Fish (Picture Shows A Diver Also Swimming Over Them).

PLATE D: A Sheep is Readied to be Shorn in Canberra, Australia, Yielded About 40 kgs of Wool — The Equivalent of 30 Sweaters.

3

Rise of Taxonomy

The history of taxonomy dates back to the origin of human language. The western scientific taxonomy was in Greek. A good beginning of taxonomy was made by Aristotle (384–322 B.C.), a student of Plato and teacher or Alexander the Great. He made an excellent study of the comparative anatomy, embryology, habit and ecology. He was able to emphasise that animals can be classified according to their way of living, actions, habits, and body parts. He was the first to classify living things and some of his groups like vertebrates and invertebrates are still in use. The invertebrates were further divided by him into animals with blood and without blood. The former were then further divided into live-bearing and egg-bearing. He also divided animals into 3 groups according to how they moved—walking, flying or swimming (land, air or water). He was also able to distinguish mandibulate from haustellate types and winged from wingless forms among insects in addition to monitoring other animals like birds, fishes and whales. The insect orders like Coleoptera, Diptera, and Psychae (now Lepidoptera) were created by him. Thus Aristotelian taxonomy proceeds by logical subdivision, in which member of a pair of taxa is characterised respectively, by the presence and absence of a chosen feature. He described more than 500 species of animals in his book **"History of Animals"**; his innovation was **"binomial"** means 2 names. All present attempts to derive a taxonomy from an identification key reflect this process. Such great contributions to biology have earned him the title of "**FATHER OF BIOLOGICAL CLASSIFICATION**".

The first important work both on plants and animals was initiated by John Ray (1627-1705), a native of Notly. Essex, and a fellow of Trinity, Cambridge. His most interesting systematic work was *Synopsis Methodica Animalium Quadrupe-dum et Serpentini Generis* published in 1693. He followed Aristotle and divided animals into those with blood and those apparently without blood. The former was further divided into those with gills and those with lungs. He also used other characteristics like the production of eggs or living young ones, the possession of broad hooves or narrow claws, the existence of two or more incisor teeth, and so on. In this way he covered almost the whole animal kingdom. His most important contribution was **"the establishment of species as the ultimate unit of taxonomy"**. The biologists of his time found his system logical, practical and easy to follow. He produced the first rudiments of our hierarchical classification and the first good definition of the species as 'a reproducing unit'.

During the 18th century, the works of Linnaeus and his followers (Haartman, 1751, 1764; Kolreuter, 1761-66) helped systematics to blossom further. Carl Linnaeus later Carolous von Linne' (1707-1778), the son of a clergyman, and subsequently Professor of Botany at Uppsala, was born two years after the death of John Ray at Raschult in Sweden. He followed Aristotelian philosophy. He published his first small book of eleven pages in 1735 at Leyden. He was the first to introduce the hierarchic system both in animal and plant kingdoms. He followed four categories (class, order, genus and species) for the animal world. He published 12 editions of this book but his greatest contribution to systematics was the use of binomial (now binominal) nomenclature for all species of plants and animals in the tenth edition of *Systema Nature* in 1758. In this edition the name Mammalia was used in place of Quardrupeda for which binominal nomenclature was consistently used throughout for the first time. He conceived the species as a fixed and unchanging thing which was divinely created and persisted unchanged in the form in which they appeared. This concept is now greatly opposed in view of the biological species concept. He presented the first character-based classification which serves as basis for the arrangement of specimens in the collections and the binominal nomenclature for information storage and retrieval system from the great bulk of biological data. In his publications, Linnaeus provided a concise, usable survey of all the world's plants and animals as then known to him, about 7,700 species of plants and 4,400 species of animals. Due to such great contributions he is rightly called the **"FOUNDING FATHER OF MODERN TAXONOMY"**. Due to him the date of January 1, 1758 is of great importance in zoological taxonomy because It has been adopted as the starting point in nomenclature under Article 3 of International Code of Zoological Nomenclature. It was a tribute to his contributions that 'Linnaean Society of London' was founded two years after his death and it is still going strong as an International Society for the study of natural history. The society preserves the bulk of Linnaean's surviving collections, manuscripts, and the library.

Buffon (1707–1788) strongly criticized Linnaeus. He believed that one should think of describing the world rather than to classify it and his theories threw light on development of species, infraspecific variety and acquired inherited characters in species. This opened a pathway for an **"Evolutionary Theory"**. Michel Adanson (1727–1806), a French botanist and a young contemporary and rival of Linnaeus, strongly believed that classifications could not be based on single, or even few characters. He stressed that the most trustworthy method involved the observation and comparison of as many characters as possible. His concept, has now been extended to a new type of taxonomy called **"NUMERICAL TAXONOMY"**.

The criticisms and improvements of the Linnaean system continued but the first serious attempt to improve it was made by Lamarck. He (1744-1829) took over the charge of the newly established Museum National d' Histoire Naturelle in France and made a number of contributions to the field of systematics. He began his career as a botanist and later switched over to investigations on animal morphology and classification. He published seven volumes of *'Histoire Naturelle des Animaux sans Vertebres'* from 1815 to 1822. Lamarck launched an evolutionary theory including inheritance of acquired characters, named as **"Lamarckism"**. Here, the development of anatomy and physiology with improved optical instruments, opened a new era of taxonomy. This period marked the description of large number of new species of organisms. On the basis of evolution he arranged animals from simpler to complex ones showing that the latter evolved from the former. He divided the animal kingdom into three sections on the basis of their mental capacities. These were further divided into classes. His division of animals into four major types—vertebrates, molluscs, arthropods, and radiates, was the end of 'scala naturae' of Aristotle In this way even the bloodless animals were broken up and this led to the end of unnatural groupings. This was the greatest contribution to taxonomy in that period.

Lamarck's taxonomy was mainly static in nature and his classification does not show its true value to the development of modern taxonomy. He believed that the species are the creations of God (like Aristotle) and yet had reasons to believe that all the groups of animals originated through evolutionary process, He further stated that the process of evolution is s continuous one and its apparent interruptions were due to our ignorance. His most important support to systematics was his preciseness in the diagnosis of various taxa. Many of his generic and other names are still in use. He displayed the groups of animals in the form of a branching tree, which was the beginning of the use of phylogeny in systematics.

Cuvier (1769–1832) was critical of Lamarck's evolutionary concept which thereafter remained in oblivion for half a century. Cuvier is regarded as a **"FOUNDER OF THE COMPARATIVE ANATOMY"** which is an important part of systematics. He divided animals into four sections— Animaux vertebres, mollusques, articules, and rajonnes. But this did not throw any light on phylogeny and had even very little to recommend it. His greatest contribution to systematics was his insistence that extinct fossil forms should be included in the table of classification.

Cuvier's outright criticisms of Lamarck greatly affected the progress of animal taxonomy during that period. The following decades were known by the emergence of three great ideas which saw a lot of progress in animal taxonomy. The first of these was "von Baer's Law' which was put forward by von Baer, Professor at Konigsberg. According to this "the younger the embryo the more closely did it resemble other embryos of the same stage of development". The second idea was an explanation of this rule by Ernst Haeckel of Jena. He stated that during its development an animal recapitulated the stages of its ancestral evolution, or, more briefly, that "ontogeny repeats phylogeny". This theory is better known by the name of **"RECAPITULATION THEORY"**. The third theory of evolution through natural selection was put forward jointly by Darwin and Wallace in 1859. Darwin in 1859 published a summary of his ideas in his book *'Origin of Species Through Natural Selection,'* This theory supported the idea of both Lamarck and Cuvier. Thus, this theory gave the greatest support to systematic zoology. It is believed that a Scottish horticulturalist, **Patrick Matthew (1790–1874) proposed the idea of "evolution by Natural Selection" 27 years before Charles Darwin. Thus, Matthew deserves to be considered along with Darwin and Alfred Wallace as one of the three originators of the idea of large scale evolution by natural selection.**

The development of systematics was adversely affected for several decades due to extensive opposition to Darwinism. Later the supporters of Darwinism started gathering more and more information in support of the evolutionary theory and by the 19th century Darwin's ideas had been widely accepted. The empirical taxonomists were greatly influenced by his ideas. The theory of relationship by descent introduced by Darwin in 1859 was welcomed and followed as rationale for the tested system of taxonomic categories put forth by Linnaeus, a century ago. They started searching for missing links between seemingly unconnected taxa and finally, reconstructed the "primitive ancestors". The phylogenetic trees proposed by Haeckel (1834-1919) also stimulated the empirical workers. Haeckel is known for his contribution to establishing the term **"Phylogeny"**. Due to this a large number of species were discovered and described. This trend continued before the end of the 19th century. But the unending descriptions of new species with splitting of families and genera by few taxonomists created numerous synonymies. This brought a lot of discredit to taxonomy. Others who tried to uncover fine long-forgotten synonyms were also disliked by general biologists who rightly questioned the basic objective of nomenclature as an information retrieval system.

The greatest development of modern taxonomy started around the 1930s when the workers realised that the Linnaean species based on one or two specimens are not as perfect as those which are based on population studies. Due to this Mayr (1942) considered species as 'groups of interbreeding natural populations". This idea of population taxonomy was useful in establishing 'polytypic concept". This resulted in greater simplification of classification of many animal groups, especially birds. New terms like **'new systematics'** (Huxley, 1940) and **'biosystematics'** (Camp and Gilly, 1943) were added to extend the taxonomic theory in modern era. The new systematics led to the reevaluation of the species concept from the biological point of view. The publication of the book *New Systematics* became a landmark in the history of taxonomy. The taxonomists were forced to believe that all organisms occur in nature as members of populations and these cannot be studied and properly classified unless they are treated as samples of natural populations. This made taxonomists realise the importance of other characters in sound classification. The taxonomists started moving from the museum to the field to supplement the morphological characters with the various characteristics of the living animals, namely, behaviour, sound, ecology, genetics, zoogeography, physiology, and biochemistry. Thus, taxonomy got a new label—**'biological taxonomy'**—in its true sense. The importance of correct identification was also then felt by experimental biologists. At the same time the taxonomists, too, recognised their dependence on new characteristics in solving species complexes. This interdependence between orthodox taxonomists and experimental biologists brought more and more credit to it and which by about 1955 reached a state of such ebullience that it was called the **"taxonomic explosion".** The twentieth century mainly saw the development of cladistics (Hennig, 1966, etc,) and integration of information from many areas of biology into taxonomy (New systematics) with increased dependence on evolutionary relationships.

The present day taxonomic works are intended to include all available differences and similarities, not merely reflection of divergence such as the original basis, morphology. The phylogenetic adaptations, embryological patterns, biochemical variations, genetic similarity, behavioural characteristics, etc., are all applicable to taxonomic studies. The general concordance of the data from all such diverse sources mutually support the basic validity of the scheme of classification. Now a student of zoology, investigating any other aspect of animal life such as genetics, ecology, ethology, biology, bionomics, physiology, toxicology, economic zoology, zoogeography, etc., will be profited very much through a knowledge of the taxonomic position of the species under study. On the other hand, the data collected by each one of them can be readily made use of by a taxonomist for finalising his interpretations and conclusions. Thus, the present day taxonomist can no longer afford to remain in a watertight compartment as in the past, quite oblivious of the progress made in other aspects of biological sciences. The use of information technology has been a boon not only to the entire science as a whole but particularly to "Taxonomy".

4

Newer Trends in Taxonomy

The aim of modern taxonomy is not only to describe, identify and arrange organisms in convenient categories but also to understand their evolutionary histories and mechanisms. Earlier approaches were mainly based exclusively on observed characters without going into the question of infraspecific differences. Many of the species are, therefore, known by single or a few specimens. Presently great attention is paid to the subgroupings of the species, like subspecies and populations. Modem systematics uses data from many sources, like the fossil record, comparative homologies (similarity in structures due to shared ancestry), and comparative sequencing of nucleic acids (DNA and RNA), etc.

The old morphological species is now called a biological one which also includes ecological, genetical, biochemical and other characters All these new approaches have contributed a lot in explaining the true structure of the species and its evolutionary position. But since most of the new approaches require specific methods, these, too, involve some problems. A brief account of all these current approaches in taxonomy is given below.

MORPHOLOGICAL APPROACH

The important morphological characters like genitalia, antennae, wings, chaetotaxy, etc., of adults, mainly among arthropods, still dominate our taxonomic studies. In the era of new systematics various new techniques have been discovered to understand the fine structure of various morphological

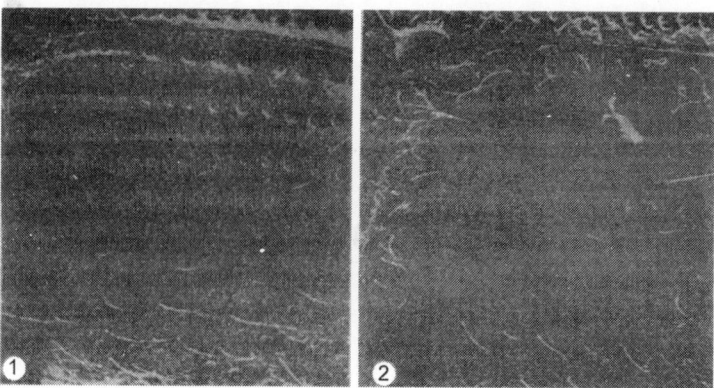

Figs. 1-2. Scanning electron micrographs of dorso-lateral view of elytra (after Lanier *et al.*, 1972); **Fig. 1.** *Ips calligraphus calligraphus;* **Fig. 2.** *I.c. ponderosae.*

characters. These new methods have resulted in the discovery of new and more reliable characters. The use of scanning electron microscope has been greatly intensified in the systematic studies of arthropods and other invertebrates. It gives an excellent information about the surface of a specimen by providing, a quasi-three dimensional image with a useful magnification of 50 to 10,000x. There are many minute characters in insects, mites, ticks and other small arthropods, which cannot be adequately studied under stereomicroscopes, that need to be studied only under such high magnifications. The highly magnified three dimensional pictures help a lot in the discovery of new characters in addition to revealing the unsuspected detail of various known characters. Lanier (1972) made use of such highly magnified pictures of some minute characters (Figs. 1, 2) in the biosystematic study of the genus *Ips* (Coleoptera: Scolytidae) and Krzysztof (1979) and Keirans *et at.* (1979) in the soft tick *Argas* (Figs. 3-21). The latter workers were able to find out new characters in addition to getting unsuspected detail of other such characters which led them to discover new species and also prepare dichotomous keys. Such works are still being continued in other groups of invertebrates and it is hoped that these will certainly lead in the breaking up of a number of species complexes and finally resulting in the simplification of classification to a great extent. Transmission electron microscopy (TEM) can also be of great value in groups like Protozoa where surface features appear

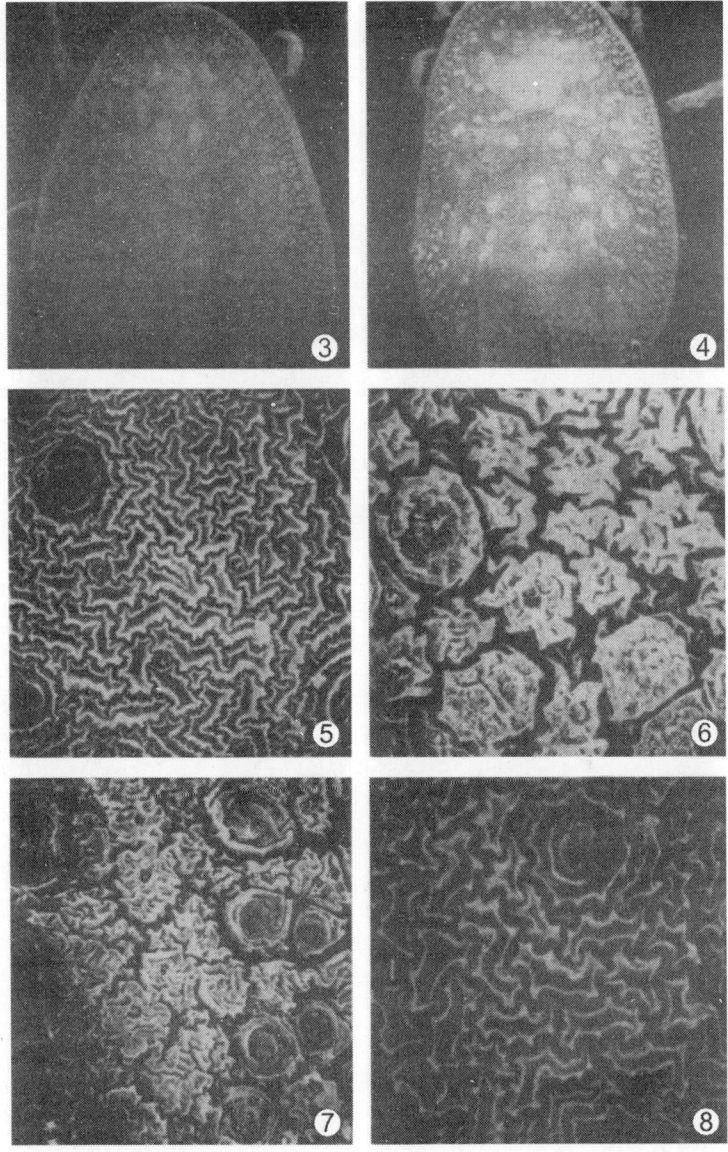

Figs 3-8. Scanning electron micrographs of some species of *Argas (Argas),* all. figs. 3-21 (after Krzysztof *et al.,* 1979, and Keirans *et al.,* 1979). Dorsal views—**3.** *polonicus* (x 3); **4,** *vulgaris* (x 26); postero-dorsal integument— **5.** *moreli* (x 161); **6.** *monachus* (x 158); **7.** *dul* (x 158); **8.** *neghmei* (x 182).

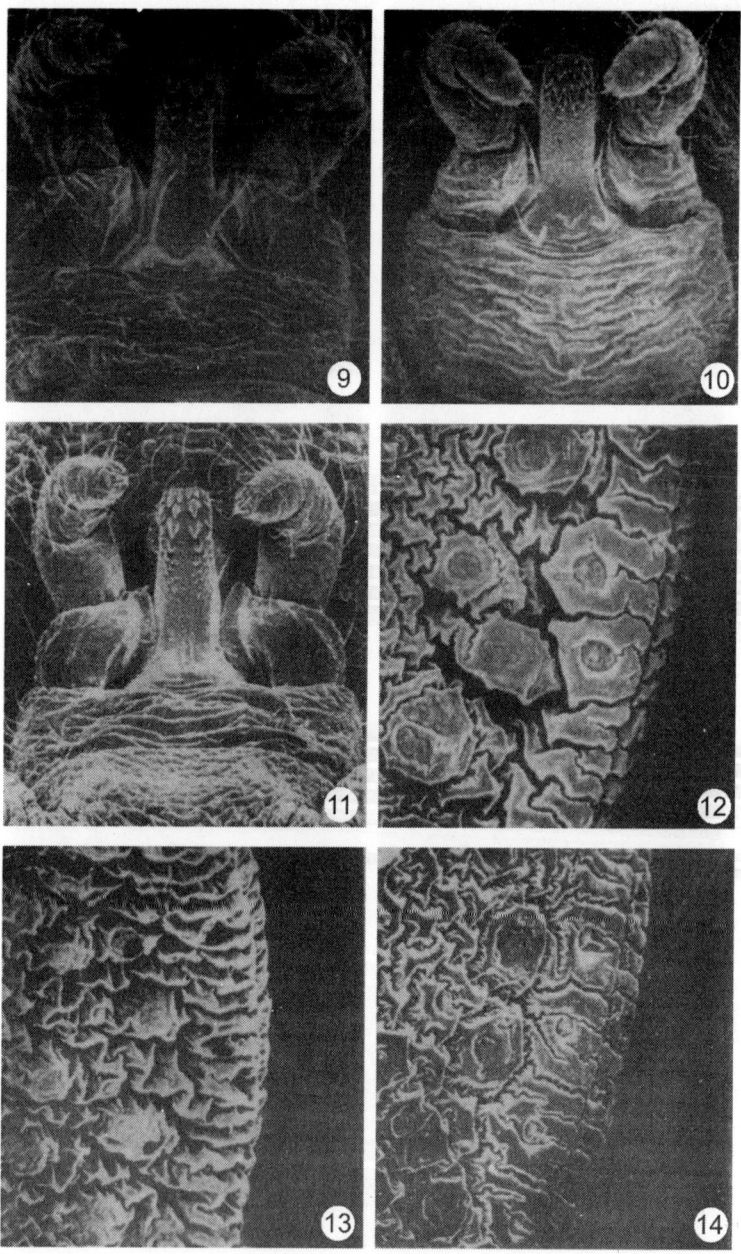

Figs. 9-14. capitulum (ventral view)—**9.** *polonicus* (x 221); **10.** *vulgaris* (x 182); **11.** *morli* (x 121; one Posthypostomal seta missing): peripheral striated area—**12.** *neghmei* (x 162); **13.** *dalei* (x 162); **14.** *moreli:* (x 161).

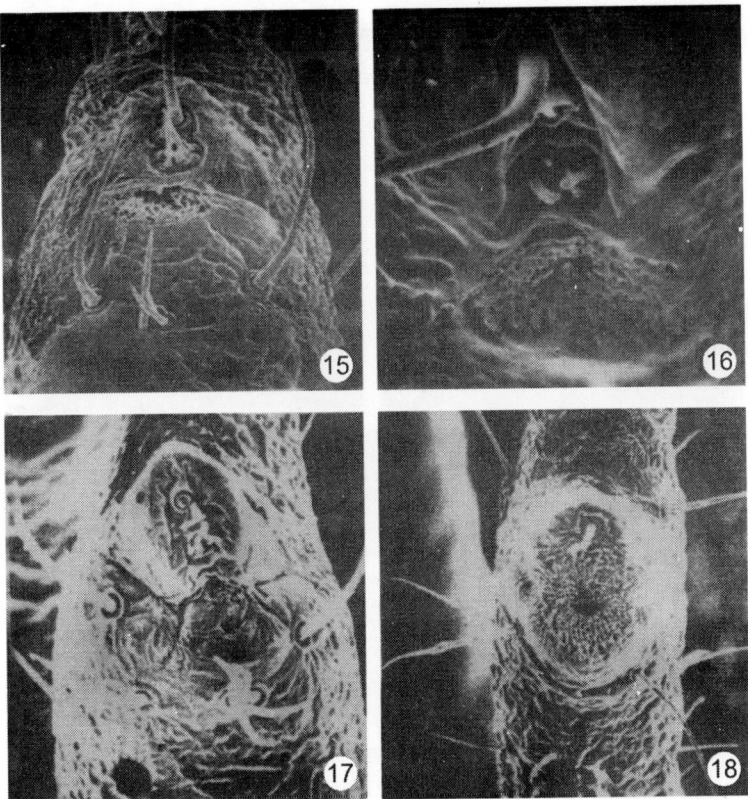

Figs. 15-18. Haller's organ area—**15.** *polonicus* (x 480); **16.** *vulgaris* (x 182); **17.** *dalei* (x 208); **18.** *cucumerinus* (x 208).

to be few. Gremigni (1979) supported the classification and phyllogenetic results of Ball (1977a, b) for planarian Turbellaria (free living flatworms) through his studies on the fine structure of cytoplasmic inclusions of their ovarian oocytes.

It is also now known that ultraviolet reflection patterns on the wings of butterflies sometimes bear little or no relation to those seen by humans. Such ultraviolet patterns have been used as taxonomic characters in some butterflies (Nekrutenko, 1964; Ghiradella et al., 1972) and can be of great value in the recognition of sibling and closely related species.

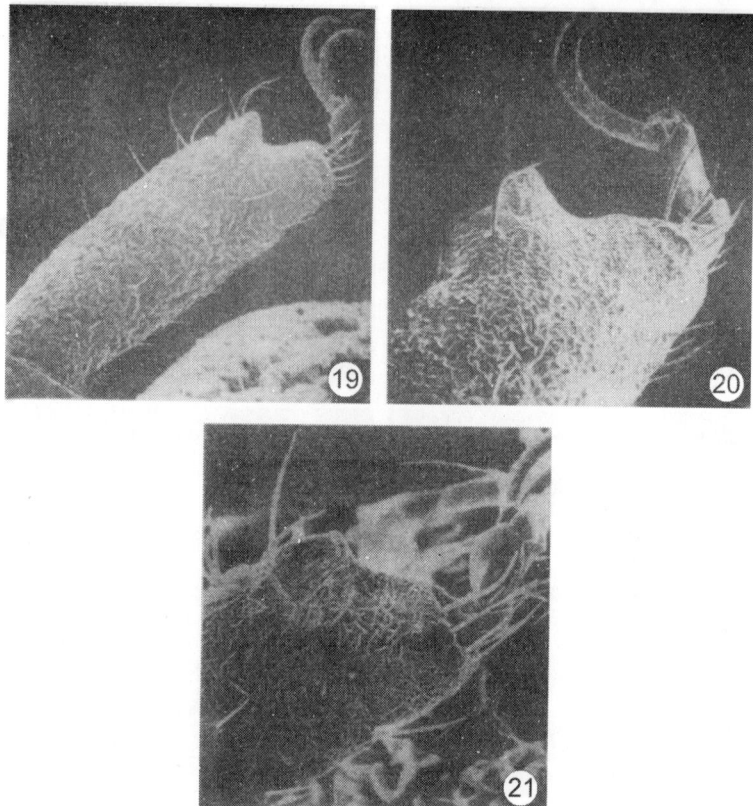

Figs. 19-21. Tarsus I, apical external view—**19.** *polonicus* (x 156), **20.** *magnus* (x 302); **21.** dalei (x 302).

IMMATURE STAGES AND EMBRYOLOGICAL APPROACH

In all individuals there occur changes in the patterns of characters during embryonic development. Such changes are quite obvious in groups where the individual passes through quite different morphological stages. Excellent examples are provided by organisms with useful polymorphic stages whose development includes several distinct juvenile stages, each separated by a moult. The taxonomic description is based not only on the morphological characters of the adult but also on the sum total of all characters of all stages.

There are various animal groups where the classification is usefully aided by the characters of their immature stages.

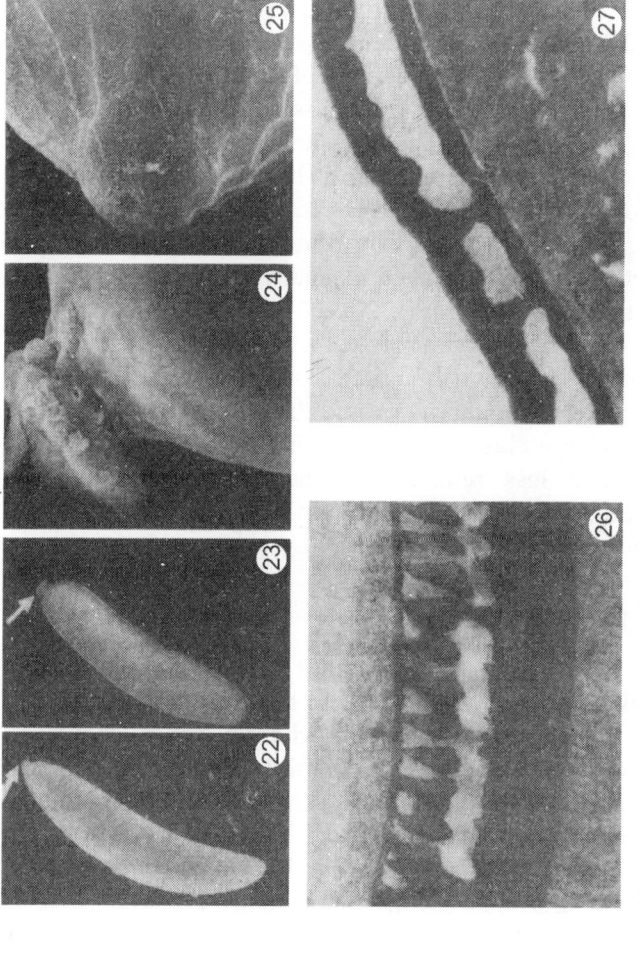

Fig 22-27. Whole eggs (Scanning electron micrograph) — **Fig. 22.** *Ceratitis capitata;* Fig. 23. *Bactrocera oleae.* Anterior pole of eggs, (shown with arrow) (SEM) Fig. 24. bactrocera oleae; **Fig. 25.** *Ceratitis capitata.* Thin sections through main body of egg (transmission electron micrographs) — **Fig. 26.** *Ceratitis cappitata;* **Fig. 27.** *Bactrocera oleae.* (Courtesy of L.H. Margaritis. Athens. Greece.) en = endochorion: vm = vitelline membrane.

The *Anopheles maculipennis* complex was broken into a number of sibling species on the basis of their egg structure. Although the egg-shells of two economically important fruit files, *Bactrocera oleae* and *Ceratitis capitata* (Diptera: Tephritidae) show superficial similarity (shape and size of the egg), their fine structural analysis reveals distinct structural differences in their egg-shell. The whole eggs of both these species are separated only on the basis of the character of anterior pole (Figs. 22-25) studied under scanning electron microscope. The thin sections through the main body of the eggs of these species (Figs. 26, 27) studied under transmission electron microscope show major differences. The egg-shell of *capitata* consists of very thick vitelline membrane and the endochorion complex composed of two trabecular layers, inverted in respect to each other (Fig. 26). The *oleae* has very thin vitelline membrane and a compact endochorionic layer; the main endochorion contains only one trabecular layer (Fig. 27).

The whole classification of whiteflies (Homoptera: Aleurodidae) is primarily based on the structure of their pupae. The cercariae larvae provide useful and most reliable characters in the classification of trematodes. In groups like sponges, with less reliable morphological characters, the study of embryology is of great help in the classification and separation of species.

ECOLOGICAL APPROACH

Perhaps the earliest recorded attempt to use ecological data in the classification was by Plato who used aquatic, terrestrial, or aerial habitation as the main characteristics. However, Aristotle was the first to seriously think of the importance of such characteristics in the animal classification. He even used them liberally in his classification.

It is now a well established fact that every species has its own niche in nature, differing from its nearest relatives in food preference, breeding season, tolerance to various physical factors, etc. When two closely related species coexist in the same general habitat, they avoid fatal competition by their species-specific niche characteristics. The closely related species are either found to live in different habitats or places, in which case their ecological characteristics may be extremely similar, or they share the same habitat and each having different type of food or otherwise avoid interspecific competition. The larvae of *Drosophila mulleri* and *aldrichi* though live simultaneously in the decaying pulp of the fruits of the cactus *Opuntia lindheimeri* yet both have specialities in their preference for certain yeasts and bacteria (Wagner, 1944). Similarly, *Anopheles maculipennis* complex was also broken into six independent species on the basis of ecological differences (Table 1; Mayr, 1969).

Table 1. Break up of six sibling species of *Anopheles maculipennis* Complex (After Mayr, 1969)

Species	Habitat	Water type	Hibernation
1. *melanoon*	Rice fields	fresh water	No
2. *messeae*	Cool standing water	—do—	Yes
3. *maculipennis*	Cool running water	—do—	Yes
4. *atroparrus*	Cool water	Brackish	No
5. *labranchiae*	Mostly warm water	—do—	No
6. *sacharovi*	Shallow standing water	Often brackish	No

BEHAVIOURAL APPROACH

Aldrovandi, a 16th century naturalist of Bologna classified birds on the basis of behavioural characteristics. Later, even Gilber White, an 18th century naturalist of Selborne, separated three very similar sympatric British bird species *(Phylloscopus collybita, P. trochilus* and *P. sibilatrix)* using song patterns. The use of behavioural characteristics in animal systematics is now seriously felt. It is one of the most important sources of taxonomic information. These characters are of great help in separating closely related species. The comparative ethology has proved very useful in improving the classification of birds, insects (especially crickets, bees and wasps, some beetles), frogs, fishes, etc. These characteristics are genetically determined and are transmitted from generation to generation like morphological and physiological characteristics. Thus, these characters play a great role as isolating mechanisms and initiating new adaptations.

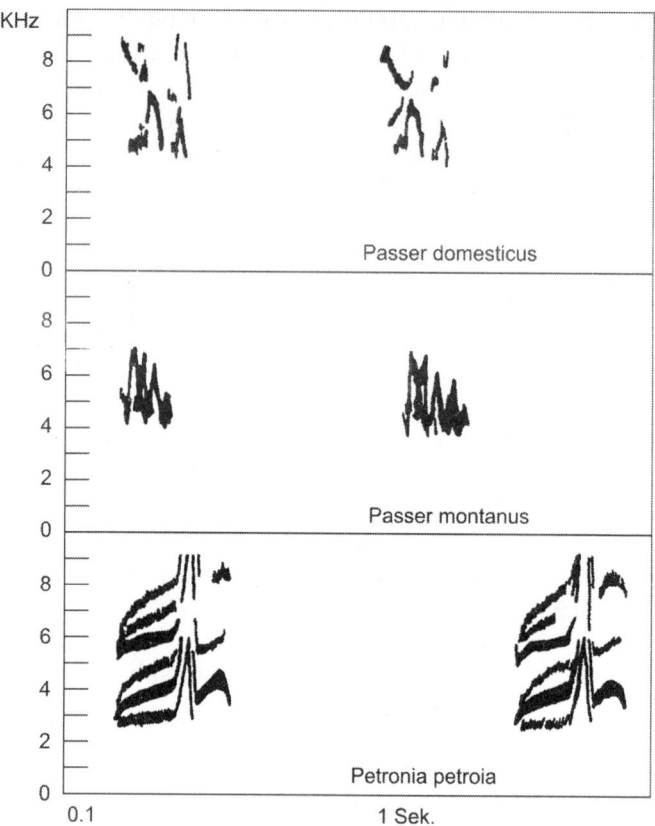

Fig. 28. Sound spectrograms of three species of birds (After Thielcke 1964).

The accurate sound recording devices and sonographs (Fig. 28) have been quite useful in segregating closely related species of birds (Thielcke, 1964; Smith; 1966; and other) and other animals. Alexander (1962) discovered about 40 species of crickets in North America on the basis of sound analysis. These species are still separated only by these characteristics. Walker (1964) estimated that about 50% of North American crickets (Gryllidae) and bush crickets (Tettigoniidae) (Orthoptora) were discovered on the basis of different song patterns. Ultrasonic sounds produced by a number of animals in various ways are used in both intra- and interspecific communication. These can also be exploited in the discovery of not only sibling species but also separation of closely related species and simplification of higher classification. Barber (1951) distinguished some species of fireflies of the genus *Photuris* (Coleoptera: Lampyridae) in North America on the basis of the height and length of the marks indicating intensity and pattern of flashes (Fig. 29). He discovered 18 sibling species in the genus *Photuris*. van der Kloot and Williams (1953a, b) used the characters of web construction in the classification of spiders, mites, and caterpillars. The bee genera *Anthidium* and *Dianthidium* are easily separated on the basis of the material used in the construction of nests by their species which in the former genus are of cottony plant fibers and in the latter of resinous plant exudations and sand or small pebbles. Similarly, various species of the termite genus *Apicotermes* are separated on the basis of structure of nests (Schmidt, 1955a, b). Even the extraneous materials used in the construction of nests or of larval or pupal cases are very useful characters in the classification of insects like Trichoptera and lepidopterous family Psychidae. Among molluscs the way in which the materials are attached to the shells provides useful taxonomic character in segregating the species, especially in the genus *Xenophora*. Various species of praying mantids (Mantodea) are easily identified on the basis of the forms of their egg cases. Similarly, eriophyiid mites (Acarina: Eriophyidae) are minute and very difficult to identify but are easily recognised from their peculiar type of galls produced on their specific hosts.

CYTOLOGICAL APPROACH

It includes the following studies:

DNA and DNA Hybridization

Deoxyribonucleic acid (DNA) is the essential material of heredity. If the DNA composition of all species is known, their evolutionary course would become quite apparent. It is believed that the amount of deoxyribonucleic acid per chromosome set is constant for each species but it is still not certain whether the ratios of the DNA content of the chromosomes are attributable to variation in the size of heterochromatic segments or they are associated with differences in the metaphase thickness. Even today it is not known that a given amount of DNA and proteins is stimulated at mitosis to become distributed into a particular number of chromosomes.

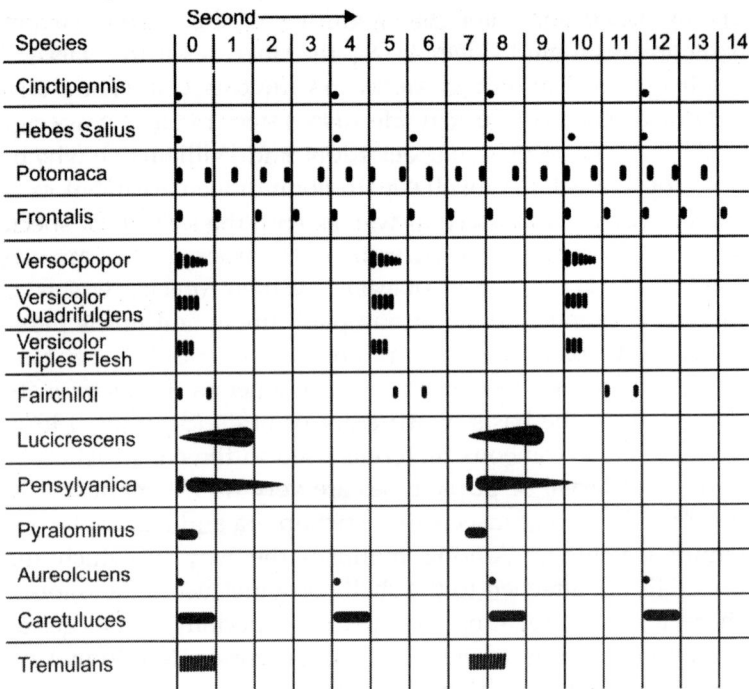

Fig. 29. Pattern of light flashes in some species of fireflies of the genus *Photuris* (Modified from Barber, 1951).

The discovery that "hybridisation" between single stranded DNA components from different origins can occur (Schildkrant *et al.*, 1961) provides a physico-chemical means for assessing genetic relatedness among species (Marmur *et al.*, 1963). In such studies the DNA is extracted from an organism and made to hybridise *in vitro* with the cell-lines of other organisms. These DNA matching techniques hold much promise in solving complex taxonomic problems. The taxonomic implications of these have been well reviewed by Hoyer *et al.* (1964). The technique of DNA-DNA Hybridization provides a way of comparing the total genome of two species. The procedure may be used to find out evolutionary relationship of two species (B to species A) as below:

First the total DNA is extracted from the cells of each species and purified. For each species, the DNA is then heated to make it denatured into single strands (ssDNA). The temperature is lowered allowing multiple short sequences of repetitive DNA to rehybridize back into double-stranded

DNA (dsDNA). Then the mixture of ssDNA (representing single genes) and dsDNA (representing repetitive DNA) is passed over a column packed with hydroxyapatite. The *dsDNA* sticks to hydroxyapatite while *ssDNA* does not and flows right through. The ssDNA of species A is made radioactive which is then allowed to rehybridize with nonradioactive ssDNA of the same species (species A) as well as with the ssDNA of species B. On completion of hybridization process, the mixtures (A/A) and (A/B) are separately heated in small (2°–3°C) increments. At each higher temperature, an aliquot is passed over hydroxyapatite. Any radioactive strands (A) on being separated from DNA duplexes pass through the column. The amount is then measured from their radioactivity. A graph showing the percentage of ssDNA at each temperature is then drawn. The temperature on which fifty percent of DNA duplexes (dsDNA) become denatured ($T_{50}H$) is noted.

Thus DNA-DNA Hybridization is quite useful in knowing the genetic comparisons integrated over the entire genome. It has so far cleared many puzzling taxonomic relationships. The technique can also be used to compare genomes of mixed populations of organisms, like populations in a soil sample.

Chromosome painting is another technique which allows comparison of entire genomes. It is done by attracting a fluorescent label to the DNA of individual chromosomes of one species and then exposing to it the chromosomes of another species. On taking up the fluorescent label the regions of gene homology are hybridized and the 'painted chromosomes' are then examined under the microscope. This method has shown that human **chromosome 6,** having hundreds of genes in the major histocompatability complex, have homologous genes in **chromosome 5** of chimpanzee: **chromosome B2** of domestic cat; **chromosome 7** of pig, **chromosome 23 of** cow.

The incomplete fossil record in many animal groups may pose problems in solving the evolutionary or phylogenetic problems through these studies.

DNA Barcoding or DNA Taxonomy

Since proteins are the expression of genes, it is good to compare the actual gene sequences. This process is useful in several ways. For example, DNA is much easier to sequence than proteins; and gene contains sites which are much free to change during evolution than protein sequences. Every animal has a mitochondrial cytochrome oxidase gene I and its sequence helps naturalists assign that animal to a given species to the great extent. DNA sequencing is a gift to taxbnomists and conservationists to change their way of working. DNA barcoding is a taxonomic method in which a short genetic marker is used in an organism's DNA to identify as belonging to a particular species, it is based on a relatively simple concept of mitochondrial gene of DNA. The mitochondrial DNA (mtDNA) has a relatively fast mutation rate, thus showing significant variation in mtDNA sequences between species and comparatively shows less intraspecific variation. A **658-bp region** of the **mitochondrial cytochrome C oxidase subunit (COI) gene** was proposed as a potential "barcode". This is now used as a new technology for cataloguing organisms. Here the specific **DNA tags** or **'barcodes'** are used to separate one species from another with the help of COI gene as unique **'finger prints'** for each species. Till 2009, database of COI sequences included at least 620,000 specimens from over 58,000 animal species. It is believed that this application has wide scientific utility. It can be of great use in conservation biology, i.e., including biodiversity surveys; also in cases of eggs, larvae and stomach contents or excreta to know food webs where traditional methods are not of much help. Thus, it can provide treasury of biological information which is currently underutilized.

Herbert and co-workers (2004) sequenced DNA barcodes of 260 of the 667 bird species of North America and found that every single species had a different COI. Surprisingly, in four cases there were deep intraspecific divergences, indicating possible new species, three of these four have already split into two species by some taxonomists. This DNA barcoding is also helpful in breaking cryptic species as done in case of cryptic Neotropical butterfly, *Astraptes fulgerator* in north-west Costa Rica. It was broken into ten different species.

The taxonomic crisis in recent years is supposed to be due lack of popularity or lack of interest and more due to major financial interest in Biodiversity. This makes the situation more difficult in knowing the undiscovered species. Two and half centuries after Linnaeus, there are 1.5 to 1.8 million of described animal species, with an estimate of 5 to 100 million species still await discovery (Wilson, 2003). Herbert et.al (2003) mentioned the following limitations in finding more and more undiscovered species:

1. Phenotype plasticity in the characters employed for species recognition leading to incorrect identification;

2. Morphologically cryptic species are often overlooked;

3. There is lack of taxonomic keys to identify immature specimens of many species; and

4. Traditional Taxonomy requires high levels of expertise in any given group and is therefore restricted to specialists.

The DNA barcoding specialists believe that identifications can be performed quickly and at low cost without the need of a taxonomist in the group. Additional advantages of this method are also attributed by them would be the possibility of identifying individuals at any stage of the development and the prospect of discriminating between morphological identifiable species. It is, therefore, advocated as a tool to revitalize taxonomy through DNA barcoding. At the same time traditional taxonomists believe that DNA-based methods would be unavailable for preserved biodiversity. Comparatively, the cost is also more besides other issues.

Although DNA barcoding specialists have even proposed **"Integrative Taxonomy"** as united move for both traditional taxonomy and DNA-barcoding, it cannot replace the former but can act as a support in solving complicated issues.

This current approach can be a useful tool when used along with the traditional taxonomic methods and alternative forms of molecular systematic so that the problems can be addressed positively and errors eliminated. Moreover, when most of the global biodiversity remains unknown, molecular barcoding can only hint of a possible occurrence of new taxa but can neither delimit nor describe (Desalle, 2006; Rubinoff, 2006). COI gene varies so much from one individual to another in amphibians that it cannot be used reliably to mark species. It should also be understood that science is not a supermarket where all items are marked with a barcode. Thus, DNA barcoding can not be a substitute for taxonomy and may not be allowed to go as **"Anti-Taxonomy"**.

Karyological Studies

Chromosomal cytology has been manipulated more extensively by plant taxonomists rather than animal taxonomists. The foundation of karyotaxonomy was actually laid down when the principles of chromosomal individuality were accepted and the chromosome theory of heredity became established. The karyotype, characterised by chromosome number, size and morphology, is a definite and constant character of each species. The number, shape, and banding of chromosomes can be determined by using various dissecting and staining techniques. Chromosomal taxonomy can be quite useful both in determining the phylogenetic relationships of the taxa as well as in the segregation of sibling or cryptic species.

The improved techniques evolved during the past 30 years have made chromosome work much less laborious. Now it is also possible to work with the difficult groups like mammals, birds and insects like Lepidoptera. There are now more reliable karyotypes for about 1000 species of mammals, several hundred species of fishes, amphibians, reptiles and birds. A number of species complexes have been broken up, especially in mammals and urodeles. The dipterous flies, particularly with giant or polytene chromosomes and orthopterans, are the most suitable groups for chromosomal studies. Patterson and Stone (1952) differentiated 16 species of the genus *Drosophila* on the basis of number and shape of chromosomes. Kiauta (1968) was able to demonstrate the phylogenetic relationship among the various families of the order Trichoptera on the basis of number of chromosomes (Fig. 30). Mittal *et al.* (1974) were able to separate two synonymised species of the earwig genus *Labidura* on the basis of number and morphology of their chromosomes (Fig. 31). Grewal (1982) separated some important fruit fly species (Diptera, Tephritidae) on the basis of the shape and number of chromosomes (Figs. 32, 33). He also discovered another population of *Bactrocera zonata* (Figs. 32, 33) on the basis of these characteristics.

Polyploidy is not common in animals except among wholly parthenogenetic forms. But the variation in the number of chromosomes occurs in many ways. One of the ways for reduction in the number of chromosomes is through chromosome fusion. In some groups the chromosome pattern has remained substantially constant through long

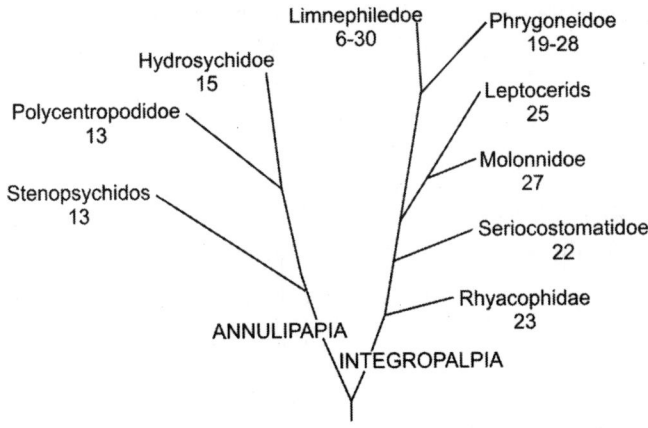

Fig. 30. Chromosome numbers plotted against the phylogeny of Trichopteran families (From Kiauta. 1968).

Fig. 31. Male karyotypes—A. *Labidura riparia;* B. *Labidura bengalensis.*

Fig. 32. Differences in the number and shape of chromosomes in some species of fruit flies (Diptera: Tephritidae), courtesy Dr. J.S. Grewal—A Main *Bactrocera zonata,* B. another population of *zonata.* C. B *dorsalis.* D. B. *cucurbitae* E B. *tau.* F. B. *scutellaris.* G. *Diarrhegma modesta* H. *Anoplomus flexuosus* I. *Sphaeniscus atilus.* and J. *Carpomyia vesuviana.*

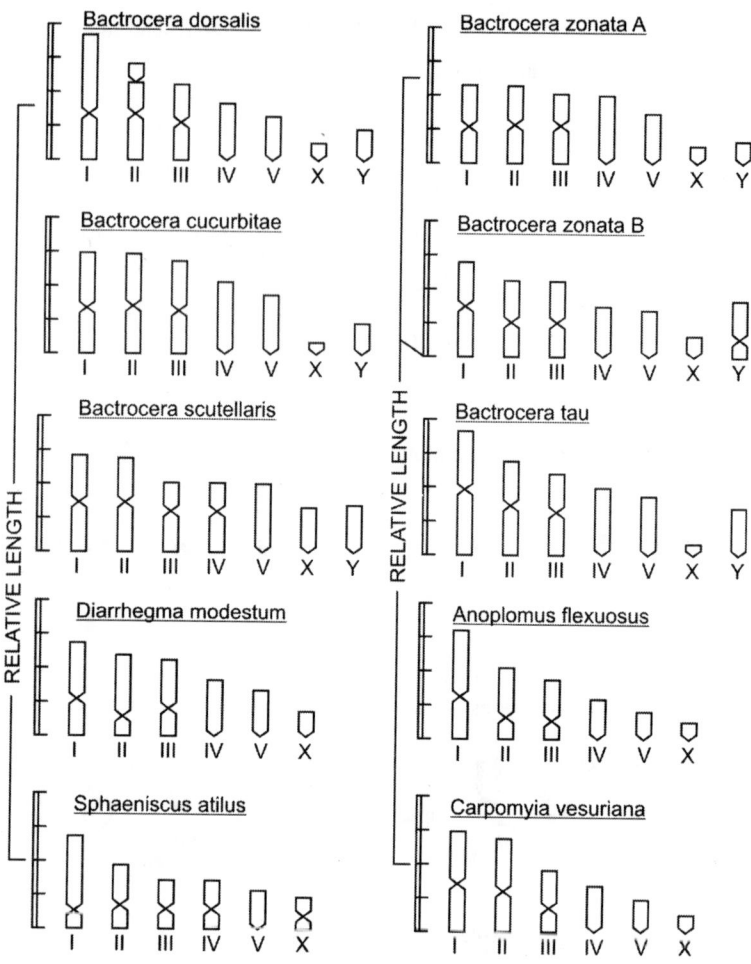

Fig. 33. Diagrammatic presentation of the idiograms of some fruit fly species of Fig. 32.

evolutionary stages (for example among grasshoppers, there is great uniformity in the Acrididae (2n z = 23 acrocentrics); Pamphagidae (2n z = 19 acrocentrics) and Pyrgomorphidae (2n z = 19 acrocentrics), while in others it has undergone distinct changes even in closely related species. For example among insects like Odonata, Diptera and Coleoptera, the chromosome number is fairly constant while in Lepidoptera, Trichoptera (insects), scorpions and fishes, it has shown marked variations. Moreover, it is also not true that the amount of chromosomal change reflects the amount of genetic change even though hereditary patrimony is carried by the chromosomes. The closely related species may show considerable rearrangement and many species are polymorphic for the very chromosomal differences that in other cases differentiate closely related species. Conversely well-defined and reproductively isolated species may completely agree structurally in their chromosomes and differ only in their gene contents. Geographical races among many insect species differ in the banding patterns of their polytene salivary gland chromosomes. One should be very careful in using chromosomal characteristics in taxonomy not only for geographical races, individual abnormalities, clines and polymorphic morphs but also for taxa exhibiting seasonal differences. Thus, karyomorphology cannot be taken as the only answer to solve all systematic problems. It can be used in selective cases. However, the value of the karyological data can be better utilised if combined with the highest possible taxonomic elements for the diagnosis of species.

BIOCHEMICAL APPROACH

This, too, has been more extensively studied in plants than in animals. The use of such characters in taxonomy was first initiated by de Candolle in 1813 to differentiate closely related species of plants. Even Lankester (1871) speculated that 'The chemical differences of different species and genera of animals and plants are certainly as significant for the history of their origin as the differences of form. If we would clearly grasp the difference of the molecular constitution and activities of different kinds of organisms, we would be able to form a clearer and better grounded judgement on the question how they have been developed, one from the other, than we now can from morphological consideration". It is now a well-known fact that the metabolism of an organism is a complex of chemical change and all morphology, behaviour and ecology of an organism must depend on its metabolism. The animal contains a large number of complex compounds like hormones, enzymes and other proteins with peptides, nucleic acids, amino acids and sugars. The biochemical taxonomic techniques are probably less subject to direct environmental influences and thus are more likely to reflect genetic divergence than many of the classic morphological analyses. The principal work of a biochemical taxonomist concerns the comparison and contrasting of compounds of the same class and performing the same function in different animal species, with regard to their properties as well as to their distribution in different organs of the body. Thus, the species can be differentiated on the basis of the amino acid sequences in the proteins of an organism and on differences between these as found in different species. Crick (1958) called it **'Protein Taxonomy'**. It is also believed that the changes in the enzyme structure can also help in the discovery of new species. Lahni (1964) calls it **'Molecular Taxonomy'**. Molecular taxonomy was primarily the nucleotide sequences of polynucleotides. Turner (1966) preferred to divide it into two—Micromolecular- and Macromolecular-Taxonomy. The former lays stress upon the distribution and biosynthetic interrelationships of small molecular weight compounds such as free amino acids, terpenes, flavonoides, etc., commonly referred to as secondary compounds. This approach is especially useful in resolving systematic problems where hybridisation has been a factor. The latter is concerned with the polymeric molecules like DNA, RNA, polysaccharides and proteins. This approach is useful in resolving some of the more intractable systematic problems especially those involving relationships among higher categories.

There are many good examples where biochemical characters have been found extremely useful in solving taxonomic problems. Basu Chaudhury and Chatterjee (1969) demonstrated phylogenetic relationship among various orders of birds on the basis of the quantitative analysis of ascorbic acid (Fig. 34). In some birds it is produced in the kidney; in some, in the liver; in some, in both liver and kidney; and in others, in neither. Accordingly they clarified that the ancestral enzyme systems involved occurred first in kidney, were later somehow transferred to the liver, and finally, in some of the more evolved passerine birds, completely lost. Similarly, Brand *et al.* (1972) established the phylogeny of a group of fireants using biochemical characters of the highly unique fireant venoms. Walbank and Waterhouse (1970) corrected the phylogenetic affinities (based earlier on morphological data) of certain genera of Australian cockroaches after analysing their defence secretions.

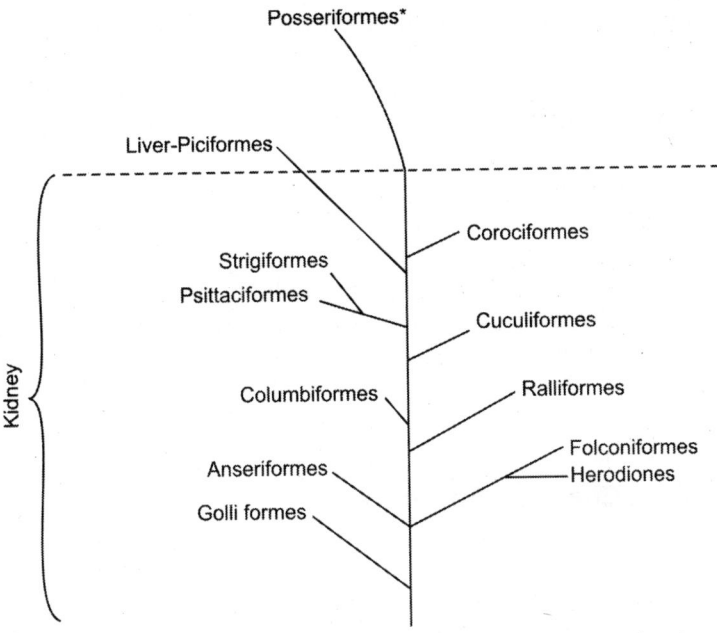

Fig. 34. Probable phylogenetic presentation in birds based on the sites of synthesis of **L-ascorbic** acid; in few both kidney and liver, in some only liver, and majority incapable.

Thus, the present biochemical approach is definitely helpful in solving many taxonomical problems. But this, too, is not useful in many cases. Moreover, such studies are possible only in the existing organisms and therefore it is difficult to trace the course of evolutionary history. It cannot lead to definite judgements with regard to the phylogeny of any organism whose fossil records are inadequate or lacking. Most of the biochemical taxonomic works are based on qualitative and quantitative differences in single chemical constituent of whole organisms or one of their tissues. Like morphological characters, chemical characters are also variable. A proper understanding of the taxonomic relationships of organisms requires comparison of a number of biochemical characters in combination with one another to reveal the diversity on biochemical patterns rather than on a single biochemical character. Since proteins and nucleic acids provide a reliable, though indirect, estimate of the degree of genetic homology among animals (Wilson and Kaplan. 1964), comparison of various characteristics of these chemical constituents as such are more suitable than other constituents for understanding their taxonomic relationships. The distribution of free amino acids in different organs of insects is of greater taxonomic value than their mere presence or absence or concentration in whole animals or in one of their tissues (Seshachar *et al.*, 1966). Brown and Heffron (1928) did not find the distribution of metabolic amino acid 3-hydroxy-L-Kyneurenine, a wing pigment, as a useful taxonomic character in nymphalid butterflies. However, in mammals the classification of species, based on amino acid sequences of the peptides, agrees in general with the accepted one based on the morphological data (Blomback and Blomback, 1968).

Kinds of Chemical Approaches

These studies are taken up in five ways—immunological, chromatographic, electrophoresis, infrared spectrophotometry and histochemical. All these are concerned with elucidating the chemical composition of the tissues and the serum of the blood which carries the necessary chemicals to feed the cells both in the development and reproduction.

Immunological

This approach is based on the precipitin reaction preferred for the study of soluble antigens, such as those contained in animal sera or tissue extracts from plants or animals. It was first discovered by Rudolph Krauss (1897) in respect of micro-organisms. Nuttal (1901) was the first to extend its use in animal systematics. Boyden (1943, 1959) further elaborated its use in animal systematics with refined techniques. Its application is based on the fact that "the proteins of one organism will react more strongly with antibodies to the proteins of a closely related organism than to those of one more distantly related". An antigen (usually a protein), when injected into an animal, will stimulate that animal to generate compounds, and the *antibodies* will react with a high degree of specificity to the material that was injected. The animals with the antibodies are considered to be *immunised* and the process of immunisation is detected by the formation of a precipitate, the *precipitin*, when the soluble antigen is mixed with its immune serum in optimal proportions. It is also possible to determine whether an antigen is unique to a certain genus, species within the genus, or even to a particular strain within a species or it shows cross reactions.

Although this practice has been in use for over half a century, it has not yet benefited us as much as was expected. Irwin's work (1947) on the blood group genes for the specific classification of pigeons has been remarkable and has been extensively applied to the study of primates. Some of the important achievements made possible through these studies in taxonomy have been discussed in the proceedings of the 'Kansas Symposium' (ed. Hawkes, 1968) and in Biochemical and Immunological Taxonomy of Animals' (ed. Wright, 1974).

Chromatography

It is a technique by which the constituents of a complex mixture can be separated and subsequently identified. It depends on the "different rates at which the compounds in a double mixture move along a porous medium, i.e., a piece of paper (paper chromatography) or a column of powdered chalk (column chromatography)". Paper chromatography has been widely used for comparing the chemical composition of closely related species, especially with regard to amino acids and peptides through ninhydrin treatment (spray), and purines, pyrimidines, or other compounds which either fluoresce or absorb ultraviolet light. The material to be analysed is prepared through two general approaches. Either pieces of tissues or small whole animals are squashed directly on to the filter paper or extracts are prepared from which soluble proteins precipitate so that the resultant solution contains only amino acids and small peptides. A minimum of 21 amino acids have been detected in homogenised adult mosquitoes. But one should be very cautious in using total homogenates, often hydrolysed as these show differences due to age, sex and physiological state.

Buzzati-Traverso and Rechnitzer (1953) were the first to apply this practice to animal systematics. They studied the amino acids of the muscle protein in different species of fish and found these characters extremely useful in segregating them. Kirk *et al.,* (1954) distinguished seven species of land snails by their fluorescent patterns. Florkin and Jeuniaux (1964) discovered that the primitive hemimetabolous insects have low concentration of free amino acids in their haemolymph as compared to high concentration in holometabolous insects. Fox (1956), Duchateau and Florkin (1958), Wyatt (1961), Chen (1962, 1966), Ball and Clark (1953), Micks (1954,1956), Throckmorton (1962), Saxena *et al.* (1965), Micks *et al.* (1966), Seshachar *et al.* (1967), Harlow *et al.* (1969) and Stephen (1974) are other workers who have shown, the importance of these studies in animal systematics.

Electrophoresis

This is another technique involving a similar movement of dissolved substances through a fixed medium, but here the movement is brought about by electrical potential differences. It is based on the fact that the "components of mixtures carry electric charges of varying amounts and so will move at different rates in salt solution through which a current is passed".

Such techniques were first used by Tiselius (1937) to distinguish multiple fractions of serum proteins migrating through solution under the influence of an electric current. Since then these techniques have been greatly refined to permit even large numbers of different proteins in the same cyanins (Smithies, 1955; Hubby, 1963; Hubby and Throckmorton, 1964; Hubby and Lewontin, 1966; Lewontin and Hubby, 1966; Williams and Chase, 1967, 1963). Now there are various types of electrophoretic methods to study the molecular composition of complex proteins.

In paper electrophoresis the mixture to be analysed is poured on a strip of paper after it has been moistened with the salt solution. Each end of the strip is put into a container filled with solution. An electrode is submerged in each container and direct current is passed through the solution. The different components migrate at different rates according to their electric charges. When they are separated, their identification is made by various means. The paper, due to its high molecular absorptive qualities, variable pore size, and high electro-endosomatic buffer flow, was first replaced by agar gel, then starch and more recently by acrylamide gel. Each of these provided an increase in macromolecular differentiation. The media starch and acrylamide introduced a second dimension to protein separation. The use of polyacrylamide gels for the separation of proteins as a substitute for starch gel was first reported by Davis and Ornstein (1959) and Raymond and Weintraub (1959). Since then the technique has been extended and modified in various ways.

The electrophoretic investigations of insulin from oxen, horses, and sheep have shown that they are different; the ACTH of pigs is different from that of oxen; the vasopressin of oxen contains arginine but that of the pigs lysine. Such biochemical studies are of great help in solving the phylogenetic problems which otherwise do not receive enough support from taxonomy. Sibley (1960) analysed the egg-white protein of 359 species of non-passerine birds by paper electrophoresis. He was not only able to corroborate the standard classification of Mayr and Amadon (1951) and Wetmore (1960), but also to raise doubts concerning previous agreements and make suggestions for the relationship of taxa previously considered highly isolated. Wright (1974) made a breakthrough in molluscan taxonomy when he separated some species of the genus *Bulinus* through electrophoretic analysis of their egg protein (Fig. 35).

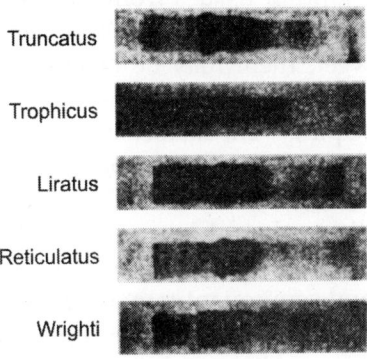

Truncatus

Trophicus

Liratus

Reticulatus

Wrighti

Fig. 35. Electrophoretic analysis of egg proteins of some species of the molluscan genus Bulinus (Modified form Wright, 1974).

In another biochemical approach to taxonomy a single complex molecule, for example haemoglobin, of one species is selected and its amino acid composition is compared with that of closely related or more distantly related species (Sande and Karcher, 1960). Manwell and Baker (1963) reviewed much of the systematic literature in this field and themselves discovered a sibling species of sea cucumber using these techniques. Manwell et al. (1967) also reviewed the utility of haemoglobin bands for the deduction of hybrids. Crenshaw (1965) also claims to have detected introgression in turtles through this approach. Other important works in this field are those of Handler (1964) and Bryson and Vogel (1965).

Infrared Spectrophotometry

It is based on the principle of absorption of infrared light by biological materials. The patterns thus formed depend upon their chemical composition and bring to light many features of taxonomic importance. So far this approach is mainly applied to micro-organisms (Randall et al., 1951; Stevenson and Bolduan, 1952; Norris, 1959). Micks and Benedict (1953) for the first time applied this technique in the identification of mosquitoes. It is hoped that this technique, if applied extensively to other animal groups, can yield useful taxonomic information.

Histochemical Studies

When the same kind of tissues from different animal species may exhibit apparently the same functions, histochemical differences between them may be observed which could be of taxonomic value. This can also help in the recognition of infraspecific groupings. The histochemical approaches involve distinctive microtechniques and specific staining reactions. The mode of fixation of material of such studies is of great importance as there should be no chemical alteration from what exists in life. These techniques have been employed in the qualitative and semi-quantitative analysis of proteins, free amino acids, enzymes, carbohydrates, lipids and nucleic acids including metal ions. Various dyes for staining are used for better perception of these characteristics. The use of cryostat microtomes, etc., and above all electron microscopes have made such studies more meaningful. This approach, when combined with other characteristics, can also be of great help in inferring taxonomic relationships amongst various animal groups.

NUMERICAL TAXONOMY

It is the numerical evaluation of the affinity or similarity between taxonomic units and the ordering of these units into taxa on the basis of their affinities. Some workers prefer the term 'Taximetrics' (Rogers, 1963). Two other terms, 'Taxonometrics' and 'Taxometry' are also in use but the competition exists between Taximetrics and 'Taximetry'. Some prefer only the former term due to its derivation from the Greek word but at the same time doubt its existence over the latter term. If the term, 'Taximetry' is finally accepted, the numerical taxonomists will be called Taximeters'.

Numerically it is actually based on the principles of Adanson (1727-1806). This concept is based on the use of maximum number of characters and that, too, all are given equal weight. In general the larger the number of taxonomic characters, the better will be the result. The characters may not be necessarily derived only from external or internal morphology but may now include any attributes of the OTU (biochemical, behavioural, cytological, ecological, developmental, etc.). Distinct taxa can be constructed due to the diverse character correlation in the various groups under study (Sneath and Sokal, 1962). But still there are differences of opinion regarding the number of characters used in this approach. Sokal and Sneath (1963) prefer the use of at least 60 characters; Moss (1967) 135 to 146 characters, while Steyskal (1968) at least 1000 characters, especially in insects.

Majority of the work in numerical taxonomy is concerned with the classification in the group-recognition sense. These workers, have greatly supported their studies and claim too much in solving the problems of biological classification. Moss (1967) reviewed the work of Funk who applied numerical methods to a large group of trigynaspid mites, Calanopsoidea, and designated a number of new families and genera. Moss himself made similar studies on the mite genus, *Dermanyssus*, and used 135 characters belonging to 15 OTUs (Operation Taxonomic Units, conventional species or taxa) from all areas of the body (Fig. 36). He suggested the creation of some new subgenera and subgroups for the various species of this genus. Other important works on numerical taxonomy are those of Michener and Sokal (1957), Smirnov, Crowson (1970), Lerman

(1970). Green and Carmone, Jardine and Sibson (1971) and Blackith and Reyment (1971). Bossert (1969), Pankhurst (1970, 1975) and Morse (1971) made use of computerised methods in devising dichotomous keys for the identification of various taxa. Dr. J.A. Peters of the United States National Museum, Washington had already devised a computer programme for the identification of the genera of Central and South American snakes (Ross, 1974).

Table 2. Break up of number of characters (After Moss, 1967)

Body parts	Qualitative	Quantitative		Total	%
		Mer-istic	Conti-nuous		
Gnathosoma	1	0	14	15	11.1
Idiosoma dorsal	8	3	34	45	33.3
Idiosoma ventral	5	4	31	40	29.6
i. Sternal shield	0	4	14	18	13.33
ii. Epigynial shield	1	0	7	8	5.92
iii. Anal shield	1	0	7	8	5.92
iv. General	2	0	1	3	2.22
v. Peritreme	1	0	2	3	2.22
Legs	1	25	9	35	25.9
Total	15 5	32 4	88 31	135 40	99.9
29.6					

The orthodox taxonomists do not accept the claims of numerical taxonomists regarding their ability to establish real blood relationships between organisms. Their approach is strictly typological and has no relation to the historical reality and to the casuality of the differentiation of organisms. Simpson (1961) feels that the approach of comparing characters in common including minuteness of resemblance and multiplicity of similarities is very useful in all taxonomy because the conclusions based on the affinities become stronger and stronger when more and more characters are used. But the use of large number of characters probably does tend to reduce the effect of homoplasy on the result. Blackwelder (1967) and many others also doubt the usefulness of numerical methods. The approach is exposed to the great risk of reaching unsound classification, because in giving equal weight to all characters it does not allow for mosaic evolution,

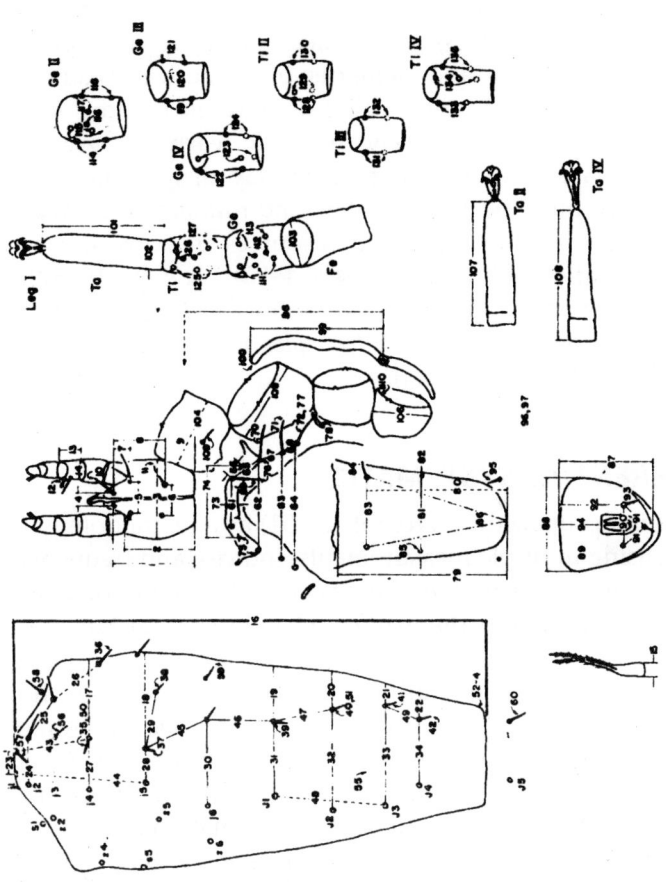

Fig. 36. The 125 characters utilised from dorsum, venter and legs (detail in table 2) of a dermanyssid mite (Courtesy: W.W. Moss, Academy Natural Sciences, Philadelphia, U.S.A.).

special adaptations, convergence, parallelism, development and genetic homeostasis in addition to evolutionary, genetic and developmental phenomena that disturb the expected close relation between phenetic similarity and phytogeny (Mayr, 1969). Moreover, the use of compiex mathematical and statistical methods by numerical taxonomists has further gone against its due recognition because of great difficulties faced by the biological taxonomists to follow them. Moss and Hendrickson (1973) and Schlee (1975a) have discussed in detail the merits and demerits of numerical taxonomy. The development of numerical taxonomy, in broader sense, has not followed very closely the basic (early) programme, except perhaps in Microbiology. Yet today all systematics is to some extent numerical. Computers and numerical taxonomic programmes are now standard resources in every museum and systematics laboratory (Jensen, 1993). Sneath (1995) has given the history of numerical taxonomy during the last thirty years (since the publication of *Principles of Numerical Taxonomy* in 1963 by Sokal and Sneath).

DIFFERENTIAL SYSTEMATICS

It was first proposed by Womble (1951) as a methodology for synthesising multiple measurements, indices, and frequencies into a composite variable, the systematic function of which, for all loci, evaluates the average change with distance of a total reality, it will allow one to sum up the rates of change with distance (differential) of several characters to show zones of differentiation within a taxon. It is one of the most objective methods for investigating the EVOLUTIONARY WORKS, in order to apply Womble's technique (called after him as WOMBLING), a good deal of information first must be accumulated on the distribution of points on the geno-, phene-, or ecoline. Differential systematic can be a powerful tool for the population worker in revealing areas of rapid differentiation using not one, but several characters simultaneously. Differential systematics has been applied to gene frequency in a population of organisms and to evolution of language.

Apparently only a few biologists (Hagmeir, 1958; Adams, 1970; Horovitz and Meyer, 1995: Williams, *et al.*, 1999;

Rosenberg, 2001) have made use of this approach, presumably due to the laborious procedures, outlined by Womble to obtain the differential of a set of characters. With the advent of modern computing methods this drawback can now be eliminated.

Differential systematics can also be used in ecological studies since the variables need not be the characters of an organism but might be species densities within different associations, percentages of fauna and flora with certain attributes.

CONCLUSION

Taxonomy, especially traditional taxonomy, is presently facing controversies from new technological advances. The new technologies like DNA taxonomy, Phyla code, new statistical methods, etc. are being promoted at the cost of traditional taxonomy when it has already stood the test of time since Linnaeus. New approaches are emerging, sometimes without understanding. The replacement of traditional taxonomy with new approaches would be suicidal not only to the traditional taxonomy but also to the development of biodiversity and conservation. Our main concern presently is to know and understand the species around us when about 90% of which still await discovery. Moreover, a large number of which is becoming extinct before they are discovered. Thus, it should be our endeavour to see the importance of taxonomy as a basic science for understanding life around us. The new approaches can be taken as supplementary to the traditional taxonomy rather than to kill it. Taxonomy is taxonomy and let it live as such.

5

Zoological Classification

Presently over one and a half million species of animals are known in addition to more than 12,000 species of fossils and the list grows every year, especially for insects in the tropical rainforests. The process of describing species is continuing and it is estimated that, three to ten million of living species and a fantastic array of fossil species still await discovery. This implies that most of the diversity of life has not even been documented in terms of a simple description and naming of species let alone taken further in understanding of the ecological role, geographic distribution, evolutionary relationships of that species. Most of these unknown species are believed to occur in tropical forest and coastal habitats. The average rate of discovering species is accelerating in palaeontology; in birds it is approaching zero and in insects it is quite high, about 8000 per annum. Recently, 10-20 new species of tiny creatures were found in the depths of the Atlantic during a survey of tropical waters between the Eastern United States and the Mid-Atlantic Ridge using special nets while catching fragile zooplankton-animals like shrimp, jellyfish and swimming worms-at lightless depths of 1-5 km (Reuters, Oslo report in Hindustan Times, dt. 5th May, 2006). Similarly, another discovery of the fossil *Tikataalik rosae* (a link between water and land species) is very exciting in being a proof of what evolutionary scientists could only "predict" till date. The fossil perfectly fits in the "tree of life". This 380-million-year-old fish was caught between two worlds, land and water, well defined as compared to poorly preserved ones of shat period not enabling authentically to establish the transition of animals from land to wafer. This fossil has fins that can be flexed, wrists, and a mobile neck; with ribs like a land animal. It was a sharp-toothed predator (Hindustan Times, April 8, 2006). It would be practically impossible to work with them, unless they were organised around a standard. The differences and similarities among the organisms point to the necessity of having such an organisation, Classification is certainly as old as language for categories are basic to vocabulary, in a sense it originated with reacting organisms. The primary function of the classification is to construct classes about which one can make inductive generalisation. A classification is constructed by grouping the animals on the basis of the degree of similarities among them. This classification is not static but changes continually in tire light of new information and ideas and the addition of new species. The validity of the classification is confirmed when new species fit into the accepted classification. Generally most of the changes occur only at specific levels of grouping species and genera. There are two distinct processes in the classification which are frequently carried out at the same time, i.e. the establishment of equivalent features of the objects to be classified and their grouping on the basis of their features.

The main purpose of the classification is to keep track of the diversity and numbers of more than one million species and provision for the discovery of new ones. It is a communication system which provides greatest information content with greatest ease.

KINDS OF CLASSIFICATIONS

The arrangement of animals in biological classification has been described in various ways. But all of them belong to any of the following types:

1. Phenetic classification
2. Natural classification
3. Phylogenetic classification
4. Evolutionary classification
5. Omnispective classification

Phenetic Classification

This system is based exclusively upon face value of observed characters without direct reference to phylogeny. The taxa are either classified on the basis of few characters or overall characteristics. When based on few characters the groupings are subjected to change on the discovery of natural affinities of the taxa. The idea of overall characters, as discussed earlier, was orignated by Adarson as early as 1757. Since the discovery of electronic computers, the idea of numerical classification was further extended by workers like Sneath, Sakal, Moss and others and named it **'Numerical Taxonomy'** or **'Phenetic Cladistics'**. These workers believed that organisms should be grouped on the basis of similarity, independently of whether these groupings represent phylogeny or not. They have devised various methods for obtaining similarity-dissimilarity data using large number of characters, without any weighting in establishing a classification.

The greatest weakness in phenetic approach is that it demonstrates false claim in establishing natural groups as products of the human mind (or of the computer) rather than of evolution. It is now a well recognised fact for all natural taxa, especially species with their genetically programmed isolating mechanisms which safeguard their reproductive isolation, that they are not an arbitrary, subjective, man-made phenomena. This approach is quite useful for groups with immature classifications and to those with numerous non-redundant characters. It is of no use when applied to higher organisms. The computer methods of pheneticists can be of great help in taxonomy if combined with philosophy of evolutionary taxonomy together with the proper weighing of characters.

Natural Classification

All will agree that a classification should be natural, i.e., the classification based on the natural characters of the taxa. Some consider natural classification a phylogenetic one reflecting the evolutionary relationships among the groups that comprise it (Smith, 1965; and others). Blackwelder (1967) opposes this concept on the ground that phylogeny is not known but merely hypothesised. It is not an attribute but only a guess at an attribute and it is based on previous classifications. In the natural system of classification, the animals are placed into as many groups and subgroups as are the similarities and dissimilarities. He defines **"natural classification"** as one in which the groups are recognised by having a maximum number of attributes in common, with their limits set by discontinuities in the diversity, and capable of yielding the maximum number of correct deductions about correlations of other features.

Phylogenetic or Cladistic Classification

Phylogeny plays a great role in classification. It is the appropriate theoretical background for taxonomy and is quite essential in explaining all the associations involved in classification. Phylogeny provides new ways to measure biodiversity, to asses conservation priorities, and to quantify the evolutionary history in any set of species (Mace et al., 2003; Guerra-Garcia et al., 2008). Some workers consider phylogenetic and evolutionary classifications similar to each other since they both are based on the features derived from a common ancestor. Others consider them as different approaches. Cladistics is a type of systematics developed by Willi Hennig (1950), who attempted to develop a more objective method of classifying organisms. Cladistic group of organisms are based on shared derived characters, not the overall similarity of potential group members. Cladistic classification is exclusively based on phylogenetic branching. **Cladistic phylogeny,** in opposition to numerical phenetics, includes an attempt to map the sequence of phyletic branching through a determination of characters that are shared-primitive (symplesiomorphic) and that are shared-derived (synapomorphic). But there is to date no true phylogenetic classification for any group of animals, except (to some extent) that of horses. This is due to incomplete fossil record and also because the comparative data collected through other approaches fail to possibly give a clear picture by itself. Such studies are quite useful when there is substantial fossil record. Presently, lot of interest is being created all over the world to discover fossils which can throw more light on the past animal groups and their impact on the present ones. Recently, United Kingdom paleontologists have unearthed 48 prehistoric species including Dinosaurs from cliffs at the Isle of

Wight, dubbed Britain's Jurassic Park. This discovery includes 8 dinosaurs, 6 mammals and 15 different types of lizards dating back to 130 million years. Similarly, scientists in United States of America discovered fossils in the Utah desert of two new dinosaur species (*Utahceratops* and *Kosmoceratops)* closely related to Triceratops, including one with 15 horns (Plate E). A well preserved fossil of a bird-like dinosaur with four wings was unearthed in China suggesting as missing link in dinosaur's evolution into birds. A 150 million-years-old fossil of the *Archaeopteryx,* long thought to contain just bone and rock, has been hiding remnants of the animal's original chemistry. X-ray scans show that trace metal contents in the fossil bone are similar to those in birds. Scientists have also discovered the remains of the largest and bloodiest known cheetah which stalked the planet thousands of years ago. Similarly, fossilized remains of a large predatory fish in the Canadian Arctic which once believed to be prowled North American Waterways some 375 million years ago has been discovered. This lobed-finned fish, now called *Laccognathus embryi* had a wide head with small eyes and robust jaws lined with large piercing teeth.

Recently, four new species of plant-eating dinosaurs, spanning a period of about 10 million years, have been unearthed in Alberta, Canada. Paleontologists have recently unearthed the fossil of new crocodilian ancestor, *Carnufex carolensis* in North Carolina, USA. It was 9 ft long and roamed the earth on its hind legs around 231 million years ago and was the top land predator in late Triassic period. Similarly, a previously unknown species of a crocodile-like "super salamander" which roamed the earth more than 200 million years ago was found in Portugal. A five year old boy, in USA, has found a rare dinosaur from Texas, believed to be of land dwelling *Nodasaur,* a pony sized creature. Another fossil, 'Platypus' dinosaur because of its bizarre combination of characters, named as *Chilisaurus diegosauranzi,* was discovered from Chile. Another fossil of lizard which has ability to walk on water has been discovered in USA. It is a extinct member of Jesus lizard group, Corytophanidae. Recent research on pandas has created confusion with regard to their evolution. It has been reported in an online journal **mBio** that pandas appeared to have a digestive system "entirely differentiated" from other herbivoures. They still retain the gut bacteria of the omnivorous bears they evolved from. The giant panda still retains a gastrointestinal tract typical of carnivoures. The Oxford scientists believed that Dinosaurs are alive even today. Although most dinosaurs were extinct 65 million years ago, one dinosaur lineage survived and lives today as a major evolutionary success story- the birds. The shrinking of the bodies of dinosaurs may have helped the group that became lizards to continue exploiting new ecological niches throughout their evolution and become highly successful today. The world's first super massive dinosaur, even largest than six *T. rex* put together, has been discovered- *Dredonoughtus sehrani*, its weight equal to a dozen African elephants put together. Scientists from the Smithsonian Museum of Natural History, the Carnegie Museum of Natural History and the University of Utah have discovered a new species of unusual bird-like dinosaur, *Anzu wyliei*. It resembles a cross between a emu and a reptile.

On November 24, 1974, a team of archaeologists working in Ethiopia discovered a 3.2 million-years-old skeleton. It was nicknamed "LUCY" – the skeleton of a hominid from the *Australopithecus afracnsis* species – was the oldest known example of a bipedal primate and a crucial stepping stone between apes and *Homo sapiens*.

Christoffersen (1995) has dealt cladistic and phylogenetic classifications as distinct. In cladistic classification one uses a cladogram as the graphical model for constructing biological system. A cladogram is a predominantly bifurcating, asymmetrical, *nontruncate* dendrogram, with no defined vertical and horizontal axes. In phylogenetic classification one uses a phylogeny or a temporalised cladogram as a graphical model for constructing biological system. A phylogeny is a predominantly bifurcating, asymmetrical and *truncate* dendrogram, with time as its vertical axis. Here ancestor-descendant relationships are recognised in addition to the sister-group relationships of cladistics because ancestors belong ontologically to our evolutionary models. The time extended lineages of ancestor-descendant populations are conveniently named as species. **Phylogenetic taxonomy** (also used by some), like modern linnean taxonomy, was modeled on a phylogentic tree rather than a cladogram and like its predecessor, perpetuates the use of morphology as a means of recognizing clades (Serene, 2005).

Phylogenetic taxonomy or systematics is strictly founded on the logic of scientific augmentation as per Karl Popper; hypotheses on homologies and monophyly can be substantiated or falsified (Wagele, 2005).

Evolutionary Classification

It is still not dearly distinct from phylogenetic classification. Simpson (1961) and others prefer evolutionary classification because it commonly needs information which is still largely phylogenetic but practically impossible to include in a tree diagram. It does not express phylogeny as based on it but as consistent with it. Thus, a consistent evolutionary classification does not contradict the classifier's views as to the phytogeny of the group. It snows objectivity, reality, arbitrariness, and the likes; monophyly, polyphyly; clades and grades; different kinds and degrees of affinity involved in phytogeny; and relative antiquity of taxa. It is a combination of phylogenetic branching with the amount of evolutionary divergence existing between different taxa. It is based on the evolutionary relationship of organisms not just their phylogeny. This classification provides foundations of all comparative studies in biology through the degree of genetic similarity existing between organisms and the phylogenetic sequence of events in their history.

The whole concept of this Classification is based on Darwinism. Darwin's ideas influenced the workers to a great extent when they started believing that the groups are created through evolution. They then started classifying organisms rather than characters which were regarded merely of independent existence in nature. A biologist, thus, understands that he is classifying populations, not individuals or phena. This classification, thus, is useful for the groups of organisms which are the result of divergent evolution. All living animal species are related to one another by way of evolutionary descent and this type of relationship helps us to establish correct systematic groupings. However, its claims are doubtful in examples of reticulate evolution resulting from the fusion of previously separated lines or when total convergence leads to groupings with unexplained polyphyletic origin.

Some workers even call it **Orthodox Taxonomy or Systematics.** The traditional classification is mainly based on evolutionary taxonomy, which accepts paraphyietic groups.

Thus, evolutionary systematics is essentially traditional systematics or taxonomy with evolution taken into account. Phylogenetic trees are often misinterpreted and caution is always required in inferring phytogenies. So there is still a long way to arrive on a final **'Tree of Life'** (Crisp and Cook, 2005).

Omnispective Classification

This is the extension of the concept of natural classification put forward by Blackwelder (1967). This approach seems quite realistic and pragmatic. Here an experienced taxonomist includes all the readily available features of the organisms but uses only those for classificatory purpose which are helpful in establishing groupings and distinctions. This practice is currently used by most of the animal taxonomists.

Future of Classification

Classification has to be viewed as an organisational system which is continually undergoing a sort of evolutionary development in the light of new groups of animals still being discovered and also their relationship to one another. Since about three to ten million species still await discovery, the animal taxonomists have to continue their struggle of discovering and classifying species, especially the insect taxonomists whose group dominates the whole animal kingdom with about three-fourth the number of species known so far out of the total of one and a half million animal species. While species across the world are rapdily going extinct due to human activities, there is also rapid evolution and the emergence of new species as per report. For example, a mosquito adapted to the environment of the underground rail in London now no longer interbreed with it's above the ground counterpart and is thought to be a new species now.

PHYLETIC LINEAGES

There are various types of phyletic lineages in use by different taxonomists, including both phylogeneticists (cladistics) and evolutionists. *Monophyletic lineage* is defined as that evolutionary origin which derives from one common ancestor or one parent; or pertaining to a single phylum. Hennig (1950) defined it as that monophyletic group which included all the descendants of the common ancestor. Ross (1956, 1974) explained that all members of a monophyletic group arise from a common ancestor, but not all lineages arising from that ancestor must be members of the designated group. Throckmorton (1962) has shown this in case of the genus *Drosophila* (Fig. 37) which gave rise to about a dozen genera, some of which are now excluded from this genus.

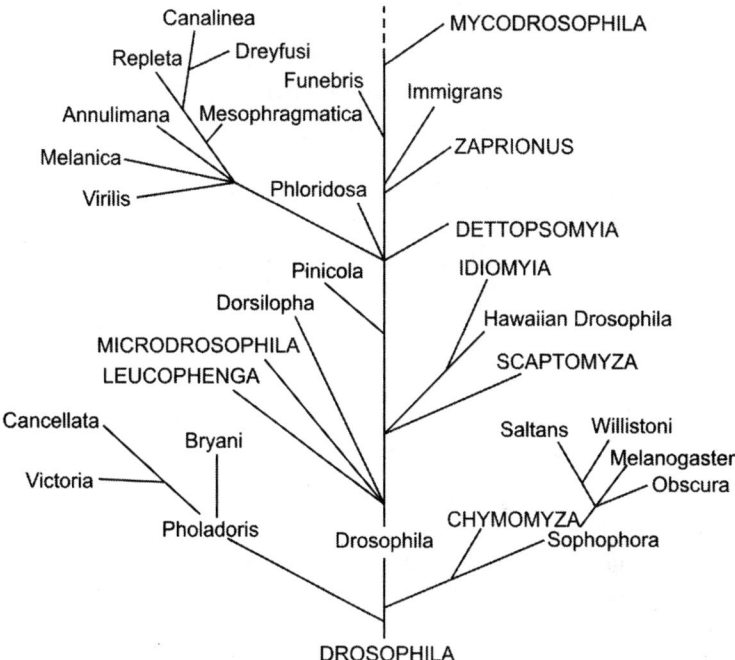

Fig. 37. Phylogenetic break-up of the genus *Drosophila:* genera in capital now excluded from this genus (slightly modified from Throckmorton, 1962).

Simpson (1961) and Mayr (1969) agree that "monophyly means derivation of a taxon through one or more lineages from an immediately ancestral taxon of the same or lower rank". But this definition does not explain the sequential branching of a phytogeny in a classification (Hull, 1965) and if this definition is followed, Aves and Mammalia may be regarded as one monophyletic group. A monophyletic group includes (only) a species and all its descendants (Bonde, 1975). But under this definition the euentomate insects like Monura and Thysanura would not be considered monophyletic, because they did not descend from a single species but because they led to the evolution of pterygote insects. Taking this fact into consideration Bergstrom (1979) slightly modified this definition as "monophyletic group restricted to a species and its descendants". According to Ashlock (1971, 1972) "monophyletic group is an ancestral species and its unexcluded progeny". He split it into two— holophyletic and paraphyletic. The former group includes all the progeny of the original ancestor while the latter includes *not all* of the progeny of the original ancestor.

The *polyphyly* is explained as that group whose species arise from two or more immediate ancestors, but which does not include the ultimate ancestor of these multiple immediate ancestors (Ross, 1974). Here the holophyly becomes identical with monophyly of Hennig. It also combines with Simpson's views that holophyletic should not necessarily reflect but remain consistent with phylogeny.

Nelson (1971) differentiates monophyly and polyphyly in the following way:

Monophyly is a quality of a group including all species, or group of species, assumed to be descendants of a hypothetical ancestral species. The members of such a group are all interrelated, from a sister-group system, and include all species of that system. Such groups may be considered complete sister-group systems.

Non-monophyly is a quality of a group not including all species, or group of species, assumed to be descendants of a hypothetical ancestral species. The members of such a group are all interrelated, but they form only part of a sister-group system, and do not include all species of that system. Such non-monophyletic groups may be considered complete sister-group systems of two parts—*Paraphyletic* which is an incomplete sister-group system lacking one species (or monophyletic species group); and *Polyphyletic* which is an incomplete sister-group system lacking two or more species (or monophyletic species group) that together do not form a monophyletic group. Osterbroek (1987) defined paraphyly and polyphyiy as below:

"A group of taxa is paraphyletic if their most recent common ancestor has given rise to one or more excluded taxa or monophyletic groups of excluded taxa of which sister group(s) is (are) completely included in the group".

"A group of taxa is polyphyletic if their most common ancestor has given rise to excluded taxa of which at least one of the sister groups is only partly included in the group".

There are still not clear and unanimously acceptable definitions of these terms. Moreover, the various explanations put forward by many workers have made the problem more acute. Thus, there is presently great need for the establishment of an International Committee to take up the task of defining these terms so that they are followed by one and all in every part of the world.

COMPONENTS OF CLASSIFICATION

It is now well-known that the similarity of organisms is not a freak of nature, but is due to the fact that similar species and genera are descended from common ancestors. Thus the first and foremost step in the classification is simply the grouping together of individual organisms on the basis of relationships or associations among them. These groups then have to be delimited and arranged in an orderly manner. For this it is necessary to understand the names like *hierarchy, taxon, category,* and *rank* which are the components of biological classification.

Animals occur in nature as distinct species. The individuals of a species show similar features, and recognisably distinct from other species. The species which share most of the features are placed into large and more inclusive taxa called the genera. Genera similarly are included into families, and so on. This sort of arrangement of taxa into an ascending series of ever-increasing inclusiveness is called 'hierarchic system of classification'. A hierarchy is thus a systematic framework for zoological classification with a sequence of classes (or sets) at different levels in which each class except the lowest includes one or more subordinate classes (Simpson, 1961). This disjoint classes at each level are called taxa and the ordinal levels are called ranks. The taxa of a given rank constitute a category. Taxon is a taxonomic group which is sufficiently distinct to be worthy of being distinguished by name and to be ranked in a definite category (Mayr, 1969), or a group of real organisms recognised as a formal unit at any level of hierarchic classification (Simpson, 1961). It was first proposed by Meyer (1926) but formally adopted by the 14th International Congress of Zoology held in Copenhagen in 1953. A taxonomic category is a class, the members of which are all taxa assigned to a given rank (Mayr, 1969). The species, genus, family, order, class, etc., are the names of the categories. The species and subspecies are lower categories while the rest above the species are called higher categories. Thus, systematic categories have a valuable and necessary function in taxonomy. These permit a semblance of order for the bewildering array of natural entities composing the animal kingdom. The main role of these categories has been to permit groupings of these natural entities in a form indicating their natural relationships in the best possible manner. The classical taxonomist prefers to delimit them on the basis of distinctive morphological characters while the experimental taxonomist does this thinking of natural interbreeding populations primarily on the basis of presence or absence of barriers to gene-exchange and ecological adaptations.

The arrangement of taxa into a taxonomic hierarchy not only serves as an aid to memory, but also has a true biological basis as the different levels in the hierarchy can be regarded to throw light on different degrees of evolutionary divergence. The taxonomic categories and ranks are purely arbitrary in nature but the taxa as a group of individual living organisms alone have any true basis. The words robins, blackbird, chiff-chaff and blue-tit signify taxa to be placed in the species category. Taxa belonging to a category are often described by the category name. Thus, in the name Rana tigrina, tigrina is a taxon of specific rank, or one belonging to the category 'species'. Likewise Rana is a genus, a taxon of generic rank and assigned to the category 'genus'. Ranidae is a family, a taxon of family rank, assigned to the category 'family'.

LINNAEAN HIERARCHY

Linnaeus recognised only five hierarchic levels within the animal kingdom. These were classis (class), ordo (order), genus, species and variety. Later, two additional categories—family (by Butschli in 1790) and phylum (by Haekel in 1886) were added. The six categories—phylum, class, order, family, genus, and species form the basic taxonomic hierarchy of animals and any given animal must belong to these six categories. The term variety was eventually discarded. More recently two other categories have become widely used. These are tribe between genus and family; and cohort between order and class. The increase in the number of known species together with the degrees of relationship between them influenced taxonomists to assign precise taxonomic position to species. This resulted in splitting of original basic categories and also additions of some among them. With the result there are as many as 33 categories presently in use in the hierarchic classification. Of these 33, only 18 (marked with asterisk below) are generally followed. The standardised endings are shown in parantheses on the facing page.

Kingdom*
Subkingdom
Infrakingdom
Superphylum
Phylum*
Subphylum*
Infraphylum
Superclass*
Class*
Subclass*
Infraclass
Supercohort
Cohort*
Subcohort
Infracohort
Superorder*
Order*
Suborder*
Infraorder
Superfamily* (-oidea)
Infrafamily
Supertribe
Tribe* (-ini)
Subtribe (-ina)
Supergenus
Genus*
Superspecies
Species*
Subspecies*

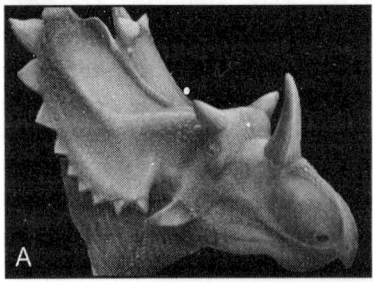

PLATE E: Recent Discovery of Some of Most Interesting New Fossil Species From Various Parts of the Globe.
New Dinosaur Species from Utah Desert of USA—A) Utahceratops With Massive Seven Feet Skull; B) Kosmoceratops with Fifteen Horns on Head, C) Feathered Dinosaur (Anchiornis Huxleyi) from North-east China; D) 150 Million Year old Fossil of Archeopteryx Dug Up Recently in China.

6

Concept of Species

The foremost task of a taxonomist is to know the different 'kinds' of animals occurring in nature. These 'kinds' are actually the species. History tells us that the existence of species has been recognised by man since the dawn of civilisation. To define a species has been one of the major problems of a taxonomist. The main purpose of defining species is to make it a historical, temporal and spatial entity. Various definitions have been put forward by various workers and yet the uncertainty persists. Even a major symposium held in 1957 on "the species problem" failed to arrive at a definite conclusion. This confusion is also related to the fact that some workers believed them as man's creations while others believed them to be the products of evolution.

Cuvier (1829) defined it as "the assemblage descended from one another or from common parents, and of those who resemble each other". Darwin (1859) also could not clearly define the species, and so he was not able to solve the problem of speciation. Since then there has been a tendency among the biologists to doubt the fundamental importance of the species and to deny its central position. Thompson (1937) defined species as "the group of individuals distinguished by an irreducible set of constant properties and connected by descent and genetic relationship". According to Dobszhansky (1937a), "there is a single systematic category which, in contrast to others, withstood all the changes in nomenclature with an amazing tenacity and that is the category of species". Again in the same year (1937b) he defined it "as that stage of evolutionary process at which the once actually or potentially interbreeding array of forms becomes segregated into two or more separate arrays which are physiologically incapable of interbreeding". In 1951, he (p. 262) again defined species as "groups of populations the gene exchange between which is limited or prevented in nature by one or by a combination of several reproductive isolating mechanisms". According to Huxley (1942), the species can be regarded as "a geographical definable group, whose members actually interbreed or are potentially capable of interbreeding in nature, which normally in nature does not interbreed freely or with full fertility with related groups, and is distinguished from them by constant it morphological differences".

Zimmermann (1948) presented a very confusing definition when he defined a species as "groups of organisms which are so similar that on superficial observation they are considered to be of the same kind". Borgmeier (1963) gave a good definition when he explains "species as natural phenomenon and biological unit based on objective facts". Wilmoth (1967) explains species as "a well defined autonomous and persistent organic unit, living in a free state of nature, not grading freely into any other unit, and generally of less perfect fecundity outside than inside its limits". It moreover differs from all other species not only morphologically, but also in its physiological manifestations, and its "psychic behaviour", in another statement it is explained as "a group of individuals of common descent, with certain constant specific characters in common which are represented in the nucleus of each cell by constant characteristic sets of chromosomes carrying homozygous genes, causing as a rule intrafertility and intersterility".

Modem workers speak of a biological species and considered it an invention of the 'New Systematise'. They have made it clear that the species is a unit as important and meaningful in biology as the cell or as is the atom in physics. In fact the species now occupies a central position in the hierarchy of organisms It is not regarded as a particular kind of organism but a kind of population.

Mayr (1957) reviewed the works of others on species problem and came to the conclusion that all these definitions given so far by earlier workers can be broadly grouped into three main concepts—*Typological or essentialist species concept; Nominalistic species concept;* and the *Biological species concept.*

Besides, the *Evolutionary species concept of Grant* (1971) and *Recognition species concept of Paterson* (1985) and Lambert *et al.* (1987) and others have also been ached here. Hausdorf (2011) made an excellent attempt to discuss all available species concepts to arrive at one common "**General Species Concept**".

TYPOLOGICAL SPECIES CONCEPT

According to this concept, the observed diversity of the universe reflects the existence of a limited number of underlying "Universals" or types, individuals do not stand in any special relation to each other being merely expressions of the same type, if two individuals or groups of individuals appear sufficiently different, they are deferent species and variation is considered a trivial and irrelevant phenomenon. This concept, going back to the philosophies of Plato and Aristotle, was the species concept of Linnaeus and his followers. This school of philosophy is now usually referred to as *essentialism* following Karl Popper and the above-mentioned species concept as the *essentialist species concept.* According to it the species can be recognised by their essential natures or essential characters, and these are expressed in their morphology or species in a group of organisms whose physical characteristics, colour, size, habitat etc, segregate them from all other organisms; it is, therefore, also called the *morphological species concept.*

Criticisms: Even though the morphological evidence is still used for inferences on the delimitations of biological species, a morphological species concept is no longer accepted by the modem biologists. However, in addition to various conceptual reasons for its rejection, two most important ones are, one, individuals are frequently found in nature that are clearly conspecific with other individuals in spite of striking morphologicol differences due to sexual dimorphism, age differences, polymorphism and other forms of individual variation are found in nature. Two, it is equally helpless in the case of so-called sibling species, that is perfectly good genetic species but lacking conspicuous morphological differences.

NOMINALISTIC SPECIES CONCEPT

This concept is of Occam and his followers who believed that only individuals exist, while species are man's own creations. This concept was popular in France in the 18th century but surprisingly it is still used (though very rarely) by some, especially botanists. Nature produces individuals and nothing more; species have no actual existence in nature. They are mental concepts and nothing more and these nave been invented in order that we may refer to great numbers of individuals collectively (Bessey, 1908).

Criticisms: No biologist can agree with the idea that the species are man-made when it is now an established fact that they are the products of evolution. The basic drawback with the nominalists was their misinterpretation of the casual relation between similarity and relationship. Members of a species taxon are similar to each other because of common heritage. It is not true that they belong to this taxon because they are similar, as claimed by these workers, The situation is the same as with the identical twins. Two brothers are identical twins not because they are extraordinarily similar, but because they are both derived from a single zygote. Thus, anyone who understands evolution will outright reject this concept.

BIOLOGICAL SPECIES CONCEPT

When it was realised in the 18th century that none of the above-mentioned concepts was applicable to biological species, an entirely new species concept emerged around 1750. But it took a number of years in finding its base in biology. The bioiogical species concept was introduced by Mayr. This concept is completely different from the earlier ones. According to this concept, "species are groups of interbreeding natural populations that are reproductively isolated from other such groups" (Mayr, 1940). Thus, this species has three separate functions: (i) it forms a *reproductive community* i.e., the individuals of an animal species recognise each other as potential mates and seek each other for the purpose of reproduction. The species-specific genetic programme of every individual ensures intraspecific reproduction; this is not absolute because sometimes different species can interbreed (the horse and donkey, for example, producing a mule when bred together), (ii) it is an *ecological unit* regardless of the individuals composing it, interacts as a unit with other species with which it shares the environment; and (iii) it is also a *genetical unit* consisting of a large intercommunicating gene pool, whereas the individual is merely a temporary vessel holding a small portion of the contents of gene pool for a short period.

The ability of one or more populations to interbreed or to hybridise denotes that its combined members belong to a single species, whereas a failure of hybridisation denotes populations belong to different species. The individual members of a population possess some cytological, genetical, or ecological isolating mechanism which maintains the gene pool of the species. Each species is thus limited by its genetic endowment (Grant, 1960). Due to their reproductive habits and potentials, the contemporary sexual species are natural self-defining classes.

Although this concept has wide following, it poses problems when applied to certain groups. The species are mainly determined from their visible characteristics and in some cases such characteristics are either imperceptible or poorly developed. Under such circumstances one has to depend on invisible or less convenient characters because the species do exist irrespective of being visibly expressed or not (Mayr, 1940). The major hurdles are created in the following cases when this concept is applied.

i. *Apomictic or asexual groups.* The asexually reproducing forms do not fulfil the criteria of interbreeding which is foremost characteristic of biological species concept. Parthenogenesis is regarded as a degeneration process of sexual mode of reproduction. Some characters of a sexually reproducing species are always present in the apomictic group. Such cases pose great problems to a taxonomist unless he is thoroughly aversed with the whole complex, i.e., apomictic entity and also the sexual ones before attempting to classify them. Mayr (1940) gave a good reasoning about such groups in being descendants of a single sexual species as a "collective species" including all strains arising due to mutations except polyploids; and these are likely to terminate sooner or later, either by extinction or fusion with another line through a sexual process.

ii. *Sibling and cryptic species.* Such species pose the greatest problems to a taxonomist (unless he is well aversed with his group) as these are feebly or not at all separated morphologically.

iii. *Gradual speciation.* It is very important for a taxonomist to have a thorough knowledge of all stages of differentiation between the individual variant and the well characterised distinct species as many species do pass through intermediate stages like biotypes, races, subspecies, ecotypes or semispecies. Moreover, we still do not have any sure method of designating an entity as a subspecies or species. Therefore, this criterion at the moment solely depends upon the knowledge of an experienced taxonomist.

iv. *Rings of races.* These, too, pose problems especially when extreme races are intersterile. These are thus theoretically good species like two allopatric ones not related by intergrading populations. If intermediate populations are taken into account, each such population irrespective of differences sometimes interbreeds with the members of populations on each side of their own population. Under such cases it becomes difficult for a biologist to regard such end-populations as different species or of the same species on the basis of reproductive isolation.

The problem also arises if one takes into account the reproductive criterion following the theory of evolution explaining the fact that all organisms have descended from relatively few initial individuals (Ruse, 1969). In majority of such cases, the parent and offspring generations can potentially interbreed producing temporal chain of races except asexually reproducing forms. If the principle of interbreeding is applied in these cases, one will be forced to believe all or nearly all animals, both of the past and the present, belong to the same species. Thus theoretically, if one insists on depending on the criterion of interbreeding and reproductive isolation of biological species concept, he will be forced to arrive at some decision which may be arbitrary with regard to reproductive criterion (Ruse. 1963).

v. *Hybrid complexes*. The biological species concept also fails to give satisfactory answer when applied to hybrid complex. Lotsy (19251) called such forms as "syngameon". Mayr (1940) redefined it as the one which reflects the sum total of species or semispecies linked by frequent or occasional hybridisation insisting that hybrids produced by the members of the syngameon can be either fertile or sterile; in a syngameon the original biological species may either persist as discrete entity or comes to occupy a definitely subordinate position to the total complex when it is found in varying degree of dissolution. Sibley (1954) called these species-like entities as "semispecies" which are discernible on the basis of partial reproductive isolation externally contrasting with free gene exchange internally.

In spite of these shortcomings the biological species concept has become the working definition of species among most population and evolutionary biologists.

EVOLUTIONARY SPECIES CONCEPT

Grant (1971) supports the biological species concept but only for sexually reproducing forms. It fails when applied to uniparental organisms where absolute self-fertilisation renders interbreeding impossible thereby suppressing the physical link which is a must for the individuals of the same species. The same viewpoint was also expressed by Meglitsch (1954) and Simpson (1961). Simpson then defines an evolutionary species "as a lineage (an ancestral-descendant sequence of populations) evolving separately from others and with its own unitary evolutionary role and tendencies". The evolutionary species concept of Grant is based on this fact and is applicable to not only *apomictic* populations but also fossil lineages. Wiley (1978) reconsidered the evolutionary concept and concluded that "a species is a single lineage of ancestral descendant population of organisms which maintains its own evolutionary tendencies and historical fate". Willis (1981) on the contrary believes that each species is an internally similar part of a phylogenetic tree. A species may be branched or not; it originated and perhaps ended at some intermediate (in characters) plane or cross section across a branch in the case of allopatric speciation or at a plane across the base of a branch in the case of sympatric speciation.

PHYLOGENETIC SPECIES CONCEPT

It can be applied to uniparental organisms. Here, species is "**a diagnosable cluster of individuals within which there is a parental pattern of ancestry and descent, beyond which there is not, and which exhibits a pattern of phylogenetic ancestry and descent among units of like kind**" (Eldredge and Cracraft 1980; Hausdorf, 2011). "This concept has been preferred for prokaryotes by Staleg (2006) but has also been applied to DNA taxonomy/barcoding by various workers (Kelly, *et al.*, 2007; Sarkar *et al.*, 2008; Monaghan *et al.*, 2009).

This concept creates problems when applied to both uniparental as well as biparental organisms. In the former, there is little or no gene exchange and thus each clone with a mutation can be classified as a separate species (Coyne and Orr, 2004; Hausdorf, 2011). Hybridization between closely related species is a natural process and the species can be differentiated despite the ongoing interbreeding. This means that many currently recognized species do not show separate parental patterns of "ancestry and descent" but that some descendants belonging to another simultaneously existing species and vice versa. Besides, polyphyletic species originating by parallel speciation will also now show " parental patterns of ancestry and descent", separate from its ancestral species and will ultimately lumped under this concept, too (Hausdorf, 2011).

RECOGNITION SPECIES CONCEPT

This concept has been put forward by Paterson (1985) and Lambert *et al.* (1987) as a replacement to biological species concept. They have even advocated its superiority over the latter. According to this concept (Paterson, 1985), a species is "that most inclusive population of individual biparental organisms which share a common fertilisation system". The fertilisation system includes all aspects of an organism's biology that "contribute to the ultimate function of bringing about fertilisation while organism occupies its normal habitat". The organisms recognise each other as mates (this recognition may be active as in courtship rituals and responses) or passive (as in mechanisms of gamete fusion); in other words, a species is set of organisms that recognize one another as potential mates: they have a shared mate recognition system; for example, within 30 or 40 species of crickets living in the same habitat, each species recognizes the song of its male counterpart. This concept does not include all barriers to gene flow that act after fertilisation, including zygotic or hybrid inviability and hybrid sterility.

This concept, too, has not received support. Butlin (1987) stated that Paterson and his colleagues misrepresented Dobzhansky and Mayr's original ideas. Coyne *et al.* (1988) did not support this concept and found it having more problems than anticipated. They had compared its disadvantages over biological species concept in detail.

ECOLOGICAL SPECIES CONCEPT

In this concept, a species is called "**a set of organisms adapted to particular set of resources, called a niche, in an environment**". Here, the populations form the discrete phenetic clusters that are called species because the ecological and evolutionary processes controlling how resources are divided and tend to produce these clusters. Here, the differences between species in form and behaviour are often related to differences in the ecological resources the species exploit.

This concept is controversial as the ecological processes can explain the existence of phenetic clusters and is not sound. Moreover, it is common for the different life stages of an organism to have utterly different ecologies; e.g. an adult toad can occupy a different environment as compared to its tadpole and as per this concept these both stages of the same species will be classified as distinct species.

GENOTYPIC CLUSTER SPECIES CONCEPT

In this concept a species is a "**Genotypic cluster that can overlap without fusing with its sibling**" (Mallet, 1995). This concept was on the idea of gene flow between independently evolving units and added genetics to the phenetic species concept (Hausdorf, 2011). It can be applied only to uniparental organisms like phylogenetic species concept. If this concept is applied, it will create undesirable consequence that each genetically different clone will be identified as a separate species (Coyne and Orr, 2004). Moreover, this concept defines species in Darwinian way as distinguishable groups of individuals which have few or no intermediates when in contact, to extend the definition to cover polytypic species and to incorporate new knowledge from genetics as well as morphology (Mallet, 1995).

Avise and Ball (1990) arrived at **"genealogical concordance"** method, whereby species are recognized if there are correlated molecular character sets that do not form intermediates when in contact. This concept supports that species can also be lost by hybridization which is not acceptable to biologists.

COHESION SPECIES CONCEPT

This concept defines a species as **"the most inclusive group of organisms having the potential for genetic and/or demographic exchangeability"**. This concept fails to apply with regard to species status of two groups when critical of genetic and demographic exchangeability conflict. In other words two groups of biparental individuals are genetically non-exchangeable but demographically exchangeable. Thus, as per this concept, all individuals connected by any **"cohesion mechanism"** are classified as one species and so this concept is also not acceptable (Hausdorf, 2011).

GENIC SPECIES CONCEPT

It was recognized by Wu (2001) who defined species as "groups that are differentially adapted and, upon contact, are not able to share genes controlling these adaptive characters, by direct exchanges or through intermediate hybrid populations".

This concept, too, fails as it exclusively focuses on **"Differential adaptation caused by mutation in genes"** (Britton-Davidian, 2001; Orr, 2001; Rundel et at., 2001; Noor, 2002; Hausdorf, 2011). They believed that in addition to differential adaptation, other processes (like genetic drift) might also cause formation of species.

DIFFERENTIAL FITNESS SPECIES CONCEPT

Here the species can be defined as "**Groups of individuals that are reciprocally characterized by features that would have negative fitness effects in other groups and that cannot be regularly exchanged between groups upon conflict** (Hausdorf, 2011).

The **benefits** of this concept are that **it classifies groups that keep differentiated and keep on differentiating despite interbreeding as species (i.e. not restricted to species mutations or mechanisms causing speciation and can be applied to both uni-as well bi-parental organisms.** Similar to **Genic Species Concept**, it allows for the exchange of genes as far as they are not important for the features that have negative fitness effects in other species (Hausdorf, 2011) .

This concept comes close to Darwin's (1859) definition of species than to **biological species concept.** According to Darwin (1859, p 485), "the only distinction between species well marked varieties is, that the latter are known or believed, to be connected at the present day by intermediate gradations, where as species were formally thus connected" (Hausdorf, 2011).

CONCLUSIONS

These species concepts considerably overlap each other; for some organisms one definition is more suitable than another and for some the definitions will coincide. It becomes even more difficult to estimate the populations of species In the living world or in special groups of organisms to which any two, three, or all species definitions apply, in any case the biological distinctness is primary and the morphological difference secondary (Mayr, 1957). The morphological distinctness is not an essential attribute of a species as this status can be acquired with or without the simultaneous or delayed acquisition of differential morphological characters.

The evolutionary species concept, recognised for uniparental organisms, is also not very promising. It does not possibly give positive clue to the rest of either interfertility of nonspecific populations or intersterility of heterospecific individuals Moreover, our knowledge of the number of uniparental organisms is too incomplete to permit an estimation of the proportion of evolutionary species that are not also biological species. There may even be some biological species, whose evolutionary role appears identical to the observer, but the number of such very close sibling species is impossible to estimate and by their very nature such cases would be extremely difficult to detect.

In the recent years, the development of cladistic analysis has given the species problem a particular new vitality (Wiley, 1978; Mishler and Donoghue, 1982; Cracraft, 1983; de Queiroz and Donoghue, 1988, 1990a, b; Nelson, 1989; Wheeler and Nixon, 1990; Ereshefsky, 1991; Baum, 1992; O'Hara, 1993). In addition to the new species concepts rooted in the ideas of cladistics, other new concepts based on behavioural (Paterson, 1985 as mentioned above) and ecological criteria (Ehrlich and Raven. 1989; Van Valen, 1976) were proposed. Mishler and Donoghue (1982) even suggested that no one species concept is applicable across all taxa and that to search for one is futile and systematists should adopt a pluralistic approach. Queiroz and Donoghue (1988) analysed the interrelations of various species concepts without advocating any one of them over the others.

Thus, it is still not settled as to which species concept should be accepted and it cannot be decided until a way is found to correlate the species composed of populations supposedly phylogenetically related, if we accept one, we cannot deny the other. Presently the zoologists can deal with them separately, but yet cannot arrive at a final single concept by combining them. The confusion still persists and if it is not settled early, it would not be surprising when zoological taxonomy may again be forced to surrender and abandon a well established term, **THE SPECIES,** as has already happened with the well established term, the **GENOTYPE.** Emerson (1941) had attempted to combine the biological species concept and evolutionary species concept and defined a true species as that "which has evolved or evolving, reproductively isolated and genetically distinct groups of natural populations".

Supporting the concept of interbreeding as primary interest of phylogenetic systematics of de Queiroz and Donoghue (1988) and then considering it the basis of species concepts, Christoffersen (1995) defined his *theoretical (ontological) species concept,* as a "Single lineage of ancestor-descendant sexual populations, genetically integrated by historically contingent events of interbreeding". This concept is closest to evolutionary species concept of Wiley (1978). In addition to this Christoffersen also provided *operational (epistemological) concept of species* based on that of Cracraft (1983, 1987), Wheeler and Nixon (1990) and Davis and Nixon (1992) "as an irreducible cluster of sexual organisms within which there is a parental pattern of ancestry and descent and that is diagnosably distinct from other such clusters by a unique combination of fixed characters". Thus according to Christoffersen—theoretical species concept is transformational (i.e., dynamic) where species are viewed as stages in the evolutionary process and ranged sequentially at the nodes of phylogenetic tree while epistomological concept is taxic (i.e., static), where species are viewed as the smallest discrete entities that can be recognised as an outcome of the evolutionary process. Similarly, the other concepts mentioned above, too, cannot be applied to all the organisms completely. In some characteristics, one is favourable while in other characteristics, the other is favourable. Sometimes we see that the combinations of some of the concepts did make some sense but in general none is as close in acceptance as the well known and most accepted one—the **BIOLOGICAL SPACIES CONCEPT.**

The modern fields of biology are also giving us much more valuable information and it would not be too late to presume when such fields, particularly biochemistry, would reveal many interesting and hidden characteristics of a species. Florkin (1964), gave a biochemical definition of a species as *"groups of individuals with more or less similar combinations of sequences of purine and pyrimidine bases in their macromolecules of DNA, and with a system of operators and repressors leading to the biosynthesis of similar amino acid sequences".*

HOW MANY SPECIES?

There is another very pertinent question which is often asked. How many species shall be described? The logical answer is: just as many as are present in nature, no more and no less. Few nontaxonomists have some conceptions of the magnitude of biological diversity. More than one and a half million species of animals have already been described and still our knowledge is highly uneven. Ornithologists frequently speak of "simplification of the system" by reducing the number of species. The species which are not species are thereby eliminated, and this tendency can only be welcomed. Only about three new species of birds are described every year, a very small addition to 8600 species previously recorded. In other groups the situation is quite good. In *Drosophila*, among dipterous insects, more than 1000 species are now recognised, almost as many species were described in the last 17 years as in the 170 years preceding 1950. Among the *chiggers*, a group of mites, only three species were known in 1900, 33 in 1912, 517 in 1952 and about 2250 in 1966. It is estimated that several hundred thousand species of mites still await discovery. What the total number of species is no one knows. It may be three million, five million, even up to ten million or even more. It is estimated that over 13,000 new species and 1000 new subspecies are named every year and most of them are of marine invertebrates. Thus, there cannot be imposed any limitations on taxonomists in describing species other than those which are imposed by nature itself; and scientific systematics is more than a nomenclatural affair.

OTHER KINDS OF SPECIES

In addition to true taxonomic species there are many other kinds of species which pertain to evolutionary or ecological concepts. It is, therefore, necessary for a student of taxonomy to know all such names for a clear understanding of true species. All such types are discussed below:

1. *Sibling species.* Pairs or groups of similar or closely related species which are reproductively isolated but morphologically identical or nearly so (good species).

   Steyskal (1972) objected to this Mayr's term and preferred to replace it by the term *Aphanic species* to describe cases where species differences are inapparent. McCafferty and Chandler (1974) agreed with Steyskal but instead suggested the terms, *Symmorphic* for morphologically indistinguishable species and *Allomorphic* for obviously distinguishable species from each other on a morphological basis, regardless of phylogenetic relationship. But majority of the taxonomists prefer the name sibling species.

Some confuse the term sibling with cryptic. Sibling and cryptic species are extremely different from one another. The term cryptic is defined as "serving to conceal, as the form or colouring of certain animals". The cryptic species camouflage with the other non-palatable ones (belonging to altogether different family or order) to protect themselves from their potential enemies.

2. *Sympatric species.* Those occupying the same geographical area (good species).

3. *Allopatric species.* Those normally inhabiting completely different geographical areas.

Rivas (1964) reinterpreted the concept of these two terms by adding 'Syntopic' and 'Allotopic' for clear understanding. All these four terms are explained by him in the following way.

Sympatric. To be used when two or more related species have the same or overlapping geographical distributions, regardless of whether or not they occupy the same macrohabitat (whether in the same locality) (Substantive form: Sympatry).

Allopatric. To be used in reference to two or more related species which have separate geographic distribution (Substantive form: Allopatry).

Syntopic. To be used in reference to two or more related species which occupy the same macrohabitat. The species occur together in the same locality, are observably in close proximity, and could possibly interbreed (Substantive form: Syntopy).

Allotopic. To be used in reference to two or more related species which do not occupy the same macrohabitat. These species are presumably not in close proximity, cannot interbreed and do not occur together in the same locality although they may have the same geographic distribution (Substantive form: Allotopy).

4. *Continental species.* Those living on the large land masses, as distinct from insular species.

5. *Insular species.* Those living on isolated islands which owe their fauna to dispersal methods other than overland migration.

6. *Cosmopolitan species.* Widely distributed species over the earth, in all majority zoogeographical regions.

7. *Tropicopolitan species* or *Pantropical species.* An ambiguous term, used *for* species found throughout the tropics.

8. *Montane species.* Those which occur at high elevations on mountain ranges.

9. *Morpho-geographical species.* Species known from Linnaean times to the present and based on morphological and geographical data (basic species of taxonomy).

10. *Agamospecies.* Those species which consist of uniparental organisms, i.e., all those animal species which reproduce parthenogenetically (obligatory parthenogenesis). Such species pose major problems for the accepted single definition of true species.

11. *Panmictic species.* Species in which each sex is produced by a different individual (dioecious, having two homes) or species in which the two sexes are produced by the same individual (monoecious, having one home, hermaphrodite) are panmictic if some of the progeny are the results of cross-fertilisation between different individuals, e.g., lumbricid worms.

12. *Apomictic species.* Those in which there is no mixing of gametes between different individuals; mostly unisexual, i.e., producing only ova; other reproduce completely asexually by budding or fission and have no functional sexual stage in any part of the life history

13. *Parapatric species.* Species whose ranges abut with at most a narrow area of overlap. Most of such parapatric ranges are due to competition, combined with critical ecological boundaries. White (1975) recorded a different kind of parapatric distribution in South Australian grasshopper, *Moraba viatica*—having two forms, an eastern 17-chromosome form and a western

19-chromosome form, Members of the forms interbreed readily, but the progeny of hybrid crosses have reduced fecundity.

14. *Contemporaneous species.* Those which occur at the same time level, whatever it is These species indicate the number of phylogenetic lines or lineages occurring at any particular time..

15. *Polytypic species.* Consisting of two or more subspecies; first defined by Huxley (1940).

16. *Monotypic species.* Species with no subspecies

17. *Transient species.* Those existing contemporaneously, as a cross section of the lineages of evolutionary species.

18. *Palaeospecies* or *Successional species.* Temporally succession species in a single evolutionary line or lineage and intergrade smoothly with each other.

19. *Palaeontoligical species.* A fossile species.

20. *Philapatric species.* Those which do not show tendency to extend their range.

21. *Incipient species.* Geographical subspecies or other segregates, which if presumably become isolated, will be distinct species.

22. *Morphospecies* (also includes form species and paraspecies). Those which establish by morphological similarity regardless of other considerations.

23. *Form species.* Group of fossil object not identifiable like any particular biological species, such as fragments or isolated parts.

24. *Paraspecies.* In palaeontology these are parataxa at the species level. Fragments (isolated parts of animals) are named equally as species and genera although such fragment species or paraspecies may include objects belonging to several species. Some include the study of parataxa under a new term "parataxonomy" But in zoological nomenclature parataxa have no status.

25. *Non-dimensional species.* The taxonomic species, lacking dimensions of space and time, are called non-dimensional species; applicable to only non-evolving

animals and so not accepted biologically. It is used only by speciationists.

POLYTYPIC SPECIES

It is now a well accepted fact that some species are widely distributed and form many local populations. If these populations are sufficiently distinct from each other, they are called subspecies. Species which contain two or more subspecies are called *polytypic* species and those which are not divided into subspecies are *monotypic* species. Beckner (1959) was the first to put forward the concept of polytypic species. Sneath (1962), the proponent of numerical taxonomy, used *Polythetic* and *Monothetic* terms in place of polytypic and monotypic, respectively.

The concept of polytypic species was most important development in the classification of animals. Certain local species that had been described from various parts of the world can be combined into species groups (allopatric species) because they were obviously more close to one another than to any other species and live in mutually exclusive geographical areas. When gaps between the ranges of allopatric species are explored (i.e., allopatric species integrated with one another), these allopatric species are included into a single polytypic species (Fig. 38).

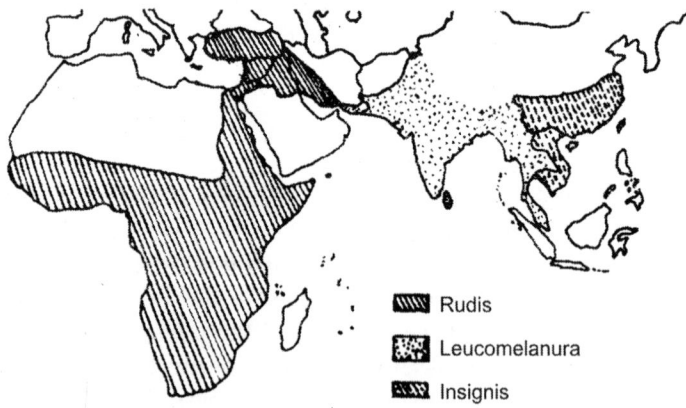

                                      ■ Rudis

                                      ■ Leucomelanura

                                      ■ Insignis

Fig. 38. Range of sub-species of small pied kingfisher, Ceryle rudis Linn. (modified form Dementev and Gladkov, 1996.)

This concept has greatly simplified the classification of groups, like birds, mammals, snails, butterflies among insects, etc. The reclassification of all allopatric forms, originally described separately as monotypic species, has resulted in tremendous simplification of the classificatory system. Such works are almost complete in birds and in full swing for mammals; snails; butterflies and beetles among insects. The monotypic species of birds which numbered about 11,000 in 1870 and 19,000 in 1970, have now been reduced to about 8600 species. This concept gave rise to a new thinking in taxonomy called the "Population Taxonomy". But it is not easy to apply polytypic species concept because still there is no foolproof method of recognising subspecies although good works have been done by Wilson and Brown (1953), Hedberg (1958), Inger (1961), Mayr (1963, 1969), and others on this subject.

SUBSPECIES

It is a basic and universal fact that all species vary and it has been known since long that certain species split into subspecies and races. Linnaeus and Fabricius preferred to call them "varieties". Kant (1775) was the first to differentiate species, subspecies and a variety. The ornithologist H. Schlegel in 1844 (Borgmeier, 1963) was the first to introduce the term "subspecies" into zoological nomenclature giving rise to trinominal nomenclature, later approved by International Congress of Zoology for inclusion in International Rules of Zoological Nomenclature.

But with the publication of Darwin's 'Origin of Species' (1859) there began a movement leading to the devaluation of species. Numerous infraspecific names were added indiscriminately leading to a sort of inflation of infraspscific nomenclature. It was then argued that systematics has become the science giving great many names to the same thing Wilson and Brown (1953) even proposed the abolition of the trinominals as they considered the subspecies concept as the most critical and disorderly area of modern systematic theory. Blackwelder (1967) believed them as 'Pseudotaxa' of subordinate status but in reality cannot be segregated and so cannot be classified.

Presently subspecies is the lowest taxonomically nameable category and is defined as "a geographically separate aggregate of local population of the species". In other words, subspecies are geographical races within the same species which are sufficiently different in some regard for them to be classified in this manner. Or the subspecios are groups of interbreeding populations with strong morphological differences combined with geographic, ecological, edaphic or physiological distinctions which give to such groups a species-like distinctiveness (Grant, 1960). Different subspecies of a species should be able to interbreed if given an opportunity to do so. Only a professional taxonomist can make a decision that the two races of a species are "taxonomically different". Nature does not follow a rigid scheme and *so* only characters which are relatively constant and sharply discontinuous can, be used to separate races or subspecies.

If one wants to find out that actual 'subspecies' do exist in nature, he must proceed from the species Experience has shown that not all species have tendencies to form subspecies. Complete discontinuity of the populations are necessary in order to be called subspecies (Rothanaler, 1954). Since the subspecies are primarily isolated gene pools, they would be in many cases incipient species. The use of the subspecific category indicates important elements and the possible phylogenetic trend, within a polytypic species (Grant, 1960). Some do not rely on basic differences between subspecies and species, either morphologically or in mode of inheritance. They believe that if two populations are distinguishable, they are different species (Rensch, 1929; Kinsey, 1930; Burma, 1949; and others). But the majority of the workers do not agree with them (Goldschmidts, 1934, 1940; Simpson, 1961; Mayr, 1969; and others). They strongly advocate that the races or subspecies and species are not equivalent categories. Species are essentially different. Races are essentially alike, because they agree in all basic structures and are linked together genetically. They differ only gradually by accidental characters or slight hereditary modifications of basic characters. The race depends upon splitting up of a species, it represents an "epiphenomenon" of the species. The species is primary and the race secondary. Every race shows the "facies" (complex specific characters common to all races of the same species) of the species from which it has been split off. A subspecies can arise abruptly by mutation in any place within the range of the species.

Thus, there is indeed a primary difference between species characters and race characters, and the fact that the specific characters are retained in the formation of every race is clear evidence for the formation of races from species but not species from races. The subspecies are therefore neither incipient species nor models for the origin of species but more or less diversified blind alleys within the species. The true subspecies exists in nature. It is, thus, a partial subcategory of the species and is a biological unit existing as an objective reality independent of man's contemplation which forms the basis for the concept of subspecies (or race). Although subspecies and race are two names of the same taxon, the latter has no nomenclatural status in taxonomy.

Some workers name subspecies differently on the basis of morphological, geographical, and ecological characteristics. Edwards (1955) explained all such types in the following way:

Geographical Subspecies

These are synchronic infraspecific populations (or aggregates of populations) which are isolated amacrogeographically during their mating times but whose respective members would cross-breed freely and normally if the populations were sympatric under natural conditions.

Temporal Subspecies

These are sympatric infraspecific populations or which are temporally isolated during mating season but whose members would crossbreed freely and normally if the populations were to become synchronic under natural conditions.

Seasonal Subspecies

If two distinct sympatric populations or aggregates of populations within a given species are composed of individuals which mature at different respective times during the same calendar year (e.g., one in spring and the other in the fall) with no period of time during which reproductive forms of both demes coexist, then interbreeding can occur between the members of the respective populations.

Annual Subspecies

If the members of one distinctive population or aggregates of population within a given species mature only during different year from those of another population of the same species, then the respective populations might be termed *annual subspecies*, if temporal isolation between their reproductive forms is complete.

Geological Subspecies

Populations which function during different geological times, respectively, have absolutely no chances of becoming synchronic. Hence the test of how freely and normally their members could interbreed is purely conjectural.

Ecological Subspecies

Distinctive, different, macrogeographically sympatric, infra-specific, populations or aggregates of populations which are isolated microgeographically, but whose members would crossbreed rather freely and normally if the populations were to become microgeographically sympatric under natural conditions. These occur in different niches, biotopes, or populations of biotopes, which cannot be indicated successfully on ordinary read-maps but require the use of topographic maps and usually of special faunistic maps to indicate the exact microgeographic habit of the ecological subspecies.

Polytopic Subspecies

That which is composed of widely separated populations, or is a geographically heterogeneous subspecies. When subspecies of a species differ in a single diagnostic character like colour, size, or pattern, it may happen that several different and somewhat widely separated populations independently acquire an identical phenotype population. Although such visually identical populations are different genetically, yet since the subspecies is not an evolutionary concept, these are combined by taxonomists into a single subspecific taxon called *polytopic subspecies* (Mayr, 1969).

OTHER INFRASPECIFIC GROUPS

Deme

Morphologically homogeneous group of organisms which are either from a single locality or from a single kind of habitat and is the starting point in all classifications. Mayr (1953, 1969) called them *Phena* and Simpson (1961) *Demes*. Both these names are generally accepted. The term 'Phenon' was used by Camp and Gilly (1943) to describe a particular kind of taxon of specific rank. The term 'Deme' was proposed by Gilmore and Gregor (1939) as a neutral root. It includes 'groups of organisms in the various special purposeful genecological and genetical categories'. It is a group of individual animals of one species or subspecies so localised that they are in easy and more or less frequent contact with each other, the unispecific members of a single community in the most limited sense (Simpson, 1961). Deme and Phenon have no nomenclatural status.

Variety

It was proposed by Linnaeus and was in common use for many years. This has been the most controversial and abused term in zoological taxonomy. Under the typological principles each species had a fixed pattern and anything that did not fit under this pattern was named a *variety*. Early evolutionists including Darwin failed to recognise it either as an individual variant, or a group of such variants or morphs conceptually associated by the variation alone and not forming a population, or a distinguishable population within a species analogous to or perhaps identical with a species. But such distinctions occur in nature. Now this term has been abandoned from the zoological nomenclature and any such name used after 1960 is rejected outright.

Form or Morphotype

These terms have either been applied to varieties in the second sense or to the aggregate of populations rather than to all varying organisms forming a population. Edwards

(1955) defines morphs as "distinguishable sympatric" inter-breeding population of a single species". They, too, have no nomenclatural status.

Cline

This term was used by Huxley in 1938. It is also available to variation within a species. It was used for a character gradient. It was defined as "a gradation in measurable characters". It is also not a taxonomic category and a single population may be composed of as many different clines as it has characters. It is formed by a series of continuous populations in which a given character changes gradually. *Isophenes* are lines representing points of equal expression of the character (of equal phenotype) at right angles to the cline. Clines may be of morphological, physiological, ecological, and other characters and also of the percentage frequencies of polymorphic characters. The geocline, ecocline and chronocline are for geographic, ecological and successional clines, respectively. The geoclines are frequently used. But no cline is a taxonomic category.

Formenkreis and Rassenkreis

Kleinschmidt (1926) was the first to distinguish geoclines from the linnaean morphological and monotypic species. He used the term *Formenkreis* (form—cycle or array of forms—implicitly geographical), for the geographic series. He thought it "a collective category of allopatric subspecies or species". In palaeontology, it is defined as "a group of related species or variants". No doctrine would support the concept that the Formenkreis is identical with the species of systematics. In contrast species is and remains the fundamental unit of the system, from which all research of infraspecific variation must begin. Similarly, *Rassenkreis* applied by Rensch (1929) to polytypic species is not used now. Rensch later preferred *Artenkreis* for Rassenkreis. A Rassenkreis is a genetic species with a series of intergrading but local populations, occasionally so different that two terminal populations cannot interbreed directly even though still exchanging genes through intermediate populations. An *Artenkreis* is a genetical species which does not break up into geographic races, thus nearly resembling the typological concept of species. Allopatric populations are often so distinct from each other that there is little doubt about their having reached species level (Mayr, 1969).

SUPERSPECIES

Mayr (1931) introduced this term in place of Artenkreis of Rensch (1929). It is defined as "a monophyletic group of closely related and largely or entirely allopatric species" (Simpson, 1961). The doubtful populations which could not be kept in species or subspecies were kept under a new term *'semispecies'* to mark their intermediate nature. Others suggest inclusion of not only the members of the superspecies but also all borderline cases in speciation. This category can be intercalated as usual in the hierarchy between species and subgenus. Simpson (1961) explains superspecies as "groups of populations that seem on other grounds (morphology, ecology, etc.) to have passed beyond the point of potential interbreeding and to have acquired separate evolutionary roles, but that are not demonstrated to have done so by the more conclusive evidence of remaining separate when sympatric. It is to be presumed that they are still near the critical point of speciation, that of definite isolation, and it cannot be quite certain whether they are really past that point and are not just below it. They are nascent species, when survive, shall collectively form a subgenus or eventually a genus but have hardly yet reached that degree of divergence and expansion".

Some workers recognise two types of superspecies— geographical and chronological.

Taxonomic Collection—
Identification—
Description and Publication

Taxonomy deals with the collection of individual organisms, their classification, identification, proper preservation, publication, etc. Collection of specimens is the foremost but difficult task for a taxonomist. Sometimes it is very frustrating when satisfactory collection, is not made. The collection can either be borrowed from the museums, institutions, and individuals or collected by the specialist himself. The latter method is the best as it gives ample opportunity to the collector to know its characteristics in natural environment.

Our country has a very rich flora and fauna. A total of 126,656 species of living organisms have been described from this subcontinent; probably another 400,000 are waiting to be described. India is also one of the global centres of diversity of crops and livestock contributing around 8% of global biodiversity. It is home to 77,729 described animal species representing 6.4% of the world's fauna. Out of this diverse animal wealth nearly 62% is endemic. Due to rapid growth in population leading to extensive urbanisation (like construction of roads, depletion of green cover due to expanding agriculture and developmental work, in biologically rich forests) and illegal trade together with natural catastrophes, natural abodes of animals are being rapidly destroyed. During last some years biodiversity hubs are seriously affected by wars; nearly 80% of the world's major armed conflicts from 1950-2000 occurred in regions identified as the most biological diverse and threatened" places on Earth. Such bio-hotspots contain the entire populations of more than half of all plant species and at least 42% of all vertebrates and are highly threatened. Many animal species have become extinct and quite a good number of them are following suit. Scientists fear that in the next 30 years, one-quarter of all species could be lost forever. Many may die out even before we know of their existence, taking with them their potential value of medicine and agriculture. According to recent study in India, 81 species of mammals, 47 birds, 15 reptiles, three amphibians and a large number of butterflies, moths and beetles are on the verge of extinction. Twenty animal species have not been sighted during the last 100 years and they are categorised as positively extinct". Moreover, we have already been witness to the extinction of the Indian cheetah, the lesser Indian rhino, the pink-headed duck and the mountain quail. As per latest report from Worldwatch Institute, 233 species of primates, the "closest relatives of humans are the most imperilled group of mammals on the planet due to their habitat destruction". In India, "Hanuman Monkeys" (regarded as sacred) are still in long term decline. Even highly esteemed species like langurs will be gradually reduced to a collection of isolated population on temple ground and little patches of forests. The nightingale made immortal in an ode by famous British poet John Keats, could become extinct within 30 years. Ninety per cent of its population has fallen in the last 40 years. Though the elimination of species is a part of the continuous process of evolution, the alarming thing is the speed at which this is now occurring—due to human pressures. Harvard biologist Edward Wilson (27 May 1992, Indian Express) estimated that nearly 140 species become extinct every day. From the estimated number of Asian elephants left in the wild across the world including India, Sri Lanka, Nepal and Indonesia (which is presently 50,000), a good number, i.e., 10,700 elephants were lost in the forest across India between 1980 and 2002. International Conservation Union (IUCN) warned that more than half of the world's wide ranging ocean sharks are threatened with extinction because of being targeted for their fin (for "shark fin soup") and meat.

The scientists at a Washington based Climate Institute predict that some 700 square kilometres of Indian coastal area face the threat of inundation and 7.1 million people of displacement in the next decades. Over the coming 50 years, the Institute predicts global warmings could result in the destruction of fisheries, increased storm damage, and displacement of millions of people. Much damage is predicted by the year 2015 and the rest by 2070, when the effects of global warming from greenhouse gas emissions will have done their worst. Another interesting fact which has been brought to light is the threat being posed to the global population of amphibians like frogs, toads and salamanders by the rising level of ultraviolet light from the sun. The sun's ultraviolet b-radiation is killing the eggs of some amphibian species that are already declining. The capacity of these species to repair the damage to their genes when exposed to natural levels of radiation is rather limited. It has also been established that rising temperatures enhance cloud cover on tropical mountains, leading to cooler days and warmer nights, both of which favour a particular Chytrid Fungus to grow and reproduces on many of our frog species. The fungus has already killed over 100 species of frogs in the cool highlands and the disease has begun to flourish in warm years, putting more frogs living at mid-altitudes at great risk. In our concrete jungle the amphibians (e.g. frogs and toads) control the population of mosquitoes, cockroaches and several such disease spreading vectors.

Under these circumstances it should be our urgent need to collect as many of the world species as possible. The extensive and quick collecting of fauna is the only way to compare the samples of populations before and after the destructive activities of man and natural catastrophes. The process of describing new species is considered to be equivalent to generating a new hypothesis in other branches of biology. These hypotheses are falsified when descriptions are found to apply to already describe species- thus forming "synonyms". Presently above 85% of existing terrestrial species and above 90% of marine species still await discovery With the clock of extinction now ticking faster for many species, discovering more and more new species merits high scientific and societal priority. The researchers at Indiana University in USA after analyzing largest- ever microbial data of microbial, plant and animal community obtained from government, academic and citizen science sources, estimated that Earth contain one lakh crore species of which 99.999% still remain undiscovered (Times Trend, Times of India, April, 2016). Many times the species found missing are arbitrarily classified as extinct. The highest rate of rediscovery may come from searching for species that have gone missing during the twentieth century and have relatively large ranges threatened by habitat loss. Whichever way the methods are deployed to rediscover the missing species, the fact remains that there is a huge gap when compared to enormous number of species which are getting extinct. Due to efforts made in this direction, scientists have found a Bornean rainbow toad, last seen in 1924, in the mountains of Borneo. Similarly, a hatching of a rare reptile with lineage dating back of the dinosaur age, has been found in New Zealand mainland for the first time in about 200 years. A species of frog (yellow-spotted bell frog), thought to have been extinct for 30 years, has been rediscovered in Australia.

There are many standard museums in the world that contain a samping of many areas which are inaccessible due to their remoteness or due to political reasons. Zoological collections represent an irreplaceable and valuable resource material and information on the unique biota. The millions of specimens in the collection of various museums provide accumulated knowledge on life on Earth in the past millions of years. These collections also serve as a guarantee of scientific rigor both for experts as well as amateurs. They are used heavily in the teaching of many undergraduate and graduate courses in zoology. There are other valuable collections like **holotypes** which form the basis of the published work. The other identified specimens serve as great reference material and are most valuable for identification from comparison. The Natural History Museum, London alone has 22 million specimens: 250,000 primary types, and 440,000 different species from the world. The museum is the world- renowned centre of research, specializing in taxonomy, identification and conservation. The specimens in this museum have great history and scientific value, such as specimens collected by Darwin. The museum is particularly famous for its exhibition of dinosaur skeletons. Recently, curator of the Natural History Museum, discovered a box containing beetle specimens collected by David Livingstone from Africa from 1858-1864. These were supposed to have been lost since 150 years. Due to the recent rapid change in the species concept, strong emphasis is laid on sufficient collection of material of a species to know its range of distribution and variation which is basic to speciation. Out of about 78,000 described species of animals, specimens of only 51,000 species are located in the Indian Museum, (Kolkata). Specimens of an additional 1000 animal species may be available with other institutions; that leaves about 26,000 species for which the specimens reside only abroad. In case of animal groups like aphids, India, holds specimens of only about 10% of the described species. Foreign agencies holding Indian material are charging heavy fees of thousands of rupees for help in identifying a single specimen.

Animal Species so far described from India

Protista	2577
Mollusca	5050
Crustacea	2972
Insecta	50,920
Other invertebrates (including Hemichordates)	11,254
Protochordata	116
Fishes	2549
Amphibia	206
Reptiles	485
Birds	1228
Mammals	372
Total	77,729

The mission of institution, especially museum housing animal collection, is to collect, document and store safely in readily accessible specimens of the local, regional and national or global diversity of living organisms. These specimens are record of genetic and morphological variation, past and recent geographical distribution and other biological information. Often they are the only remaining material of extinct species or the only record of the species seen only once in the world. The information from natural history collections about the diversity, taxonomy and historical distribution of species worldwide is becoming increasingly available over the Internet (Guerra-Garcia, 2008). Computerization of collections and development of electronic catalogue are providing new capabilities for curating collections (Graham et al., 2004). Unfortunately, both in developing and developed countries such institutions are poorly funded, especially for upkeep, leave aside adding more collection. This is against the aim and objective of Convention of Biodiversity.

SPECIES REGISTRY

Unfortunately even two centuries after Linnaeus developed his binomial (binominal for ICZN) classification system that brought order to the naming of species, there is still no universally accepted authority for registering the names of life forms on earth. Nearly 1.8 million species have been described and named till today. New discoveries are being made regularly. Names can be published virtually anywhere. Any officially accepted central registry of species would be a welcome boon for biologists. Presently, some organizations have been set up to attempt to list the so far described species. The important ones are given below:

i. **Species 2000:** It is a private non-profit organization, established in 1996. It works in partnership with Integrated Taxonomic Information System (ITIS) of North America with close working relation with USA, Canada and Mexico agencies. It receives funds from Global Biodiversity Information Facility (GBIF), Denmark. Its objective is to provide Indexing and Links for all known species as baseline dataset for studies of global diversity. The species 2000 in collaboration with ITIS (see page 23 in the book) started the **'Catalogue of Life Consortium".** The 2002 Catalogue of Life now lists 26,0000 species on CD-ROM and on the web (Bisby et al., 2002).

ii. **All Species Foundation:** It is also a non-profit organization dedicated to the complete inventory of all species of life on Earth. It seeks to catalogue all earth species by 2025. Thus, it has one of the greatest scientific goals of the new century. It began in 2001 as a spin-off of the "LONG NOW FOUNDATION". It has its office located at San Francisco, USA.

iii. **Tree of Life:** The tree of life project is another ambitious project from a phylogenetic perspective (see page 229).

iv. **Other Projects:** There are other projects which are restricted to geographical areas rather than globally.

These are funded by European Commission. These are: a, **Fauna Europea**—It was started in 2000 with the objective of making a web-based checklist of all European land animals; b, **European Register of Marine species (ERMS)**—It started in 1997 to prepare a checklist of marine organisms and its first edition is complete and available both on web and in printed format.

Besides, there are many other projects in several countries, funded separately, which could prove great asset in increasing the biodiversity census. In 1998, the project **"Fauna Iberia"** was launched in Iberian Peninsula to produce an inventory of the animal diversity in this area; 72 monographs on animal groups belonging to 11 phyla are already edited under this project. But its completion is going to take years. **Swedish Taxonomy Initiative (STI)** is another one started in 2002 with the aim to prepare a complete inventory of Swedish fauna and flora of multicellular organisms within 20 years.

v. **Zoobank:** It is an initiative of International Commission of Zoological Nomenclature and would be indispensable tool for taxonomic and biodiversity researchers.

It is a mandatory open-access registry for animals and taxonomic act in zoology. The proposal has already received wide support from taxonomists, conservationists, etc., the world over. The main advantage of this mandatory registration of animal names will be an increased "visibility" of taxonomic works together with a check for Code-compliance. It will also include species description and links to images via **Morph Bank** (is an open web repository of images serving the biological research community; established in 1998 by a consortium of systematic entomologists and currently housed at the School of Computational Sciences at the Fioride State University, USA) and gene sequence data via Gene Bank (is an comprehensive database that contains publicly available DNA sequences for more than 165,000 named organisms; established in 2005 with National Institute of Health, USA).

Descriptions of new animal species and their nomenclatural acts remain "hidden" in thousands of specialized journals and other publications (monographs and CDs).

Zoobank is now the official registry of zoological nomenclature as per International Code of Zoological Nomenclature. This zoobank will overcome the obstacle to access widely distributed taxonomic information from widely distributed scientific journals. This will be quite useful in our need for more rapid description and cataloguing of out fast disappearing biodiversity.

Zoobank comprises 30 members from 14 countries. Twenty members are ICZN commissioners and nine are external experts. Currently, Zoobank accommodates the registration of four different kinds of data objects:

1. Nomenclatural Acts: Published usages of scientific names of animals, which represent nomenclatural acts as governed by the International Commission of Zoological Nomenclature (ICZN). Most of these acts are "Original Descriptions" of new scientific names for animals, but other acts may also include emendations, lectotypifications, and other acts as governed by the Commission.

2. Publications: Publications that contain Nomenclatural Acts, as defined above.

3. Authors: Anyone who is an author of one or more publications (as defined above).

4. Type specimens: Type specimens for scientific names of animals. The registration of Type Specimens is considered provisional and is not yet fully implemented.

As the beginning of the century of extinctions, science has only discovered a very small proportion of living species of the globe. So zootaxonomy needs efficient, rigorous and automatic Nomenclatural Rules at the present- rather than investing time, energy and money in remaining millions of already named taxa in order to follow alternative nomenclatural systems, e.g., "Phyllogenetic" ones, which do not theoretical superior to the current Linnaean-Stricklandian one (Dubois, 2011). The current Code, result of 250-year improvement process, is based on very sound and healthy Rules, being theory-free regarding taxonomy, relying on nomina to taxa by a system of ostension using onomatophore, and so an objective basis Principal, priority, for recognizing the valid nomen of taxon in case of synonymy or homonymy (Dubois, 2011). He further added that the Nomenclatural System need to be modified in such a way that it rightfully appear to many zoologists as an complete set of rules dealing with all situations and needs of zoological nomenclature. In case of incompleteness, there will be fear amongst us that the zoologists might be forced towards "Phyllocode". So the Code is Zoology's Supreme Court for names.

COLLECTING WAYS

There are various ways to collect animals round the year. These methods may be from picking up insects flying towards a torch light in the evening to trawling or dredging for deep-sea animals. The latter method requires elaborate and specifically designed equipment operated by a crew of sailors and scientists on an ocean-going vessel. Now collecting is a highly specialised procedure in which a collector concentrates only on a group of organisms. Faunal surveys are carefully planned areawise in a particular geographical area. One must be in the knowledge of all possible geographic information of the place to be surveyed including the distribution of various types of vegetation, altitudes, seasons, means of transport, lodging, etc. It is also necessary to examine the previous collections of the concerned group to know the various localities of the already collected material. Some persons develop collecting as a hobby. Various techniques are employed for collecting animals Piechooki (1966) gave a good account of various techniques for the collection of macroscopic specimens. New techniques are continually being developed for all sorts of animals, like mist nets for bird collecting (Fig. 39), ultraviolet lamps for insect collecting, etc. Different kinds of attractants or baits are also successfully employed for collecting specific types of animals. Sometimes even other animals are employed for collecting specific animals. The artificial changes brought about in the environment cause the animals to come out of their hiding. Field surveys or extensive expeditions involve many hurdles both political and technical. The technical difficulties are under the control of man and can be made easier or removed by using different aids but the main hurdles come in the form of political restrictions on areas within a country or from country to country. There are some areas in every country which are completely banned for ordinary citizens and so cannot be surveyed to sample the fauna. Some areas are permitted to be surveyed only with special permission of the particular government and involves many frustrating procedures. Some countries have completely banned the collection of animals by visitors and thus remain unexplored even by the native due to unspecialised personnel and lack of funds. Such obstructions have resulted in almost complete halt in the dissemination of knowledge of fauna of various countries.

Fig. 39. A part of mist net with house sparrows (*Passer domesticus Linn.*) entangled in it.

Insects form the most dominant group of animals due to their enormous number combined with diverse habits and habitats. There is hardly any place on the planet which is not invaded by these creatures. These are found in the air; in fresh or brackish water; on the foliage or stems of trees and shrubs, low-growing plants; or the ground or near the roots of low-growing plants; internally in plants, plant products, seeds, and fruits; among debris, in the nests or habitations of animals and man; in the soil; and on or in other insects or animals. By knowing their peculiar habits and habitats one can easily pick them up with fingers. However, if one needs to collect quickly and in large numbers, special collecting equipments together with the study of extensive field observations are necessary.

Insects can be collected by sweeping, using general purpose nets during sunny days: light traps for insects attracted to light; baits and bait traps; and different types of traps for flying insects, e.g., Malaise traps, etc. Sweeping and beating are the most productive methods of collecting large number of insects. The purpose of sweeping is to dislodge insects from the vegetation by means of insect net. It is done on herbage, the flowers, or the foliage or trees with rapid sideways strokes of the net. The contents of the net should be examined after every few sweeps and the desired specimens are removed before they are damaged due to continuous sweepings. The principle of beating is to hit a branch of a tree or shrub hard enough with a heavy stick in such a way that the insects and other arthropods fall on the tray or white sheet placed below from where they can be captured.

Different types of aspirators are also quite useful in picking up large number of small insects and other arthropods. Some insect pests of plants and trees are reared in the laboratory from their infested materials. Fleas, lice, certain parasitic flies, bugs and beetles attack various birds and mammals. For this they are trapped or sometimes even shot. Small insects and other small arthropods inhabiting soil, debris, etc., are collected through Berlese funnel or flotation methods. Aquatic insects and other arthropods are best collected by dip net (Fig. 40) or plankton net (Fig. 41). A general account of the habitats and collecting ways of insects and other invertebrates is given in Table 3.

Insect Net

It consists of a wire ring, a bag and a handle (Fig. 42). The ring should be 12 to 15 inches in diameter and made of about 3 mm iron or steel wire. The ends of the wire are straight which can fit into the grooves of the handle. The tips are bent inward in such a way that they can easily fit into the holes of the handle. A detachable ring is quite useful as the bag then can be easily replaced when it is torn, wet, dirty or to be changed for one of another material. A folding ring (Fig. 43) is quite easy to carry while travelling. The bag is either of muslin cloth, fine nylon net or any other material depending upon the method and purpose of collecting. Organdy is an excellent material

Table 3. Collection and preservation of Insects and other arthropods.

Animal group	Habitat	Collecting ways	Preservation
1	2	3	4
A Insects			
1) Thysanura	Dwellings, soil, leaf-litters, rotten wood, under stones, bark and nests of ants and termites.	Picking up with a small brush moistened in alcohol; aspirator; berlese funnel.	80% alcohol.
2) Diplura and Protura	Soil, debris, bark, fungi, on foliage, under stones and woodland litter.	—do— (except aspirator used rarely).	—do—
3) Collembola	Soil, debris, under stones, some in stored grains, snowfleas on snow, some as pests on certain crops.	—do—	—do—
4) Orthoptera	Diverse habitats such as on trees, shrubs, grass to swamps, under stones, logs, debris, on light, and buildings.	Sweeping' beating with fingers or forceps; or other specialised methods for specific group.	Insect boxes after pinning; smaller forms in 80% alcohol also.
5) Dermaptera	Loose marshy earth, under stones, logs of wood, dung hills, most debris or any humid place secluded from life, and on light; some (Hemimerina and Arixenina) ectoparasites of rats and bats.	Picking with forceps.	Insect boxes after pinning; ectoparasites in 80% alcohol.
6) Plecoptera	Margins of rivers, lakes, streams, generally resting on stores legs, grasses, shrubs, tree trunks or other objects, on snow or ice near streams or rivers.	Beating, sweeping; on light.	80% alcohol.

(Contd.)

Table 3. (Contd.)

1	2	3	4
7) Ephemeroptera	Immature aquatic; adults terrestrial; nymphs in lakes, ponds crawling freely on the bottom of submerged debris, sometimes making burrows in mud, under stones, pieces of wood, etc; adults near water.	Nymphs by hand or by dredging or raking the bottom of pond or stream; adults by sweeping or beating foliage: or individually with forceps; and light traps.	80% alcohol; sometimes pinned.
8) Odonata	Nymphs aquatic; adults flying in vicinity of water, few active at dusk or even after dusk.	Nymphs as above; adults by sweeping or netting individually.	Nymphs in 80% alcohol and adults pinned, may be kept in alcohol.
9) Isoptera	Underground in galleries, mounds above ground, vegetation as pests of various crops, wood, books, etc., organic debris; on light etc.	Picking with forceps and also berlese funnel.	80% alcohol; rarely pinned.
10) Embioptera	Debris, cracks in soil, under stone, on or under bark, or among epiphytic plants, mosses, or lichesn; on light.	Picking with forceps.	80% alcohol.
11) Psocoptera	Vegetation, on or under bark, in debris; nests of birds, rodents, wasps and ants; undisturbed books and papers.	Searching and picking with forceps; beating; sweeping; berlese funnel; aspirator.	80% alcohol sometimes mounted on points.
12) Phthiraptera	As ectoparasites of birds and mammals.	Searching and picking with forceps.	80% alcohol.
13) Hemiptera	Under loose bark, among vegetation at base of grasses, other plants, or in vegetation, debris, few parasites of buds and mammals; some on light.	Searching and picking with for ceps; sweeping; asporator; light traps.	Pinning; and 80% alcohol.

(Contd.)

Table 3. (*Contd.*)

1	2	3	4
14) Thysanoptera	All kinds of growing and dying vegetation, in flower, soil, on foliage, fruits, bark and decaying waste and fungi.	Picking with calem hair brush moistened in alcohol; beating; aspirator; berlese funnel.	80% alcohol; best in AGA solution (8 parts alcohol + 5parts water + 1 parts each of glyscerine and glacial acetic acid).
15) Neuroptera	Vegetation; attracted to light.	Searching and picking with hand or forceps; light traps; sweeping.	Pinning; immature forms in 80% alcohol.
16) Mecoptera	Damp wooded places, wingless on snow also.	Sweeping and picking with hand or forceps; light traps; sweeing,	80% alcohal; sometimes pinned.
17) Trichoptera	Vegetation near lakes, rivers, streams, attracted to light.	Sweeing during day time; with forceps; light trapps.	80% alcohal; sometime pinned.
18) Lepidoptera	Highly diverse habitats, flying over vegetation during day time; attracted to light; pests of many crops.	Sweeping; light traps; baiting or using natural attactants; rearing; netting individually.	Pinning.
19) Coleoptera	Diverse habitats, under stones, logs, loose bark, in stems, roots, etc., leaf mould, moss and other plant debris; in water; on light; dung hills; stored grains.	Searching and picking with hand or forceps; sweeing; beating; somealso through berlese funnel.	Pinning, carding very small formsand grubs in 80% alcohol.
20) Strepsiptera	Parasites on Hemiptera, Hymenoptera or Orthoptera.	Dissecting out female stylops from hosts, or rearing hosts till emergence of male.	80% alcohol.
21) Hymenoptera	Flying during day time over vegetation, parasitise insects, also pests of crops and forest trees, and other diverse habitats; social, live in colonies.	Sweeping; picking with camel hair brush moistened in alcohol; rearing on hosts; with forceps; aspirator; netting individually.	Pinning; small forms and larvae in 70% alcohol after first putting then in 1% formalin for 1 to 5 minutes.

(Contd.)

Table 3. (*Contd.*)

1	2	3	4
22) Diptera	All kinds of habitats, pests of plants, and parasites on birds and mammals, parasitoids, predators, flying over vegetations, some in water; attracted to light; soil; leaf litter, debris, etc.	Sweeping; rearing from hosts; picking up with camel hair brush moistened in alcohol; forceps; aspirator; light traps; baits; some through berlese funnel; netting individually.	Pinning; smaller forms in 80% alcohol.
23) Siphonaptera	Ectoparasites of mammals, few on birds.	Shooting or trapping the host animal; searching the nests or bedding of the host; from nests of birds and mammals by teasing the material on white enamelled pan; or placing it in berlese funnel.	80% alcohol.
B. Acari 1) Mites	Vegetation; soil and organic debris; ectoparasites of birds, mammals, reptiles, insects, arachnids, myriapods, etc., their nests or along their run ways; fresh water, salt water including bottom of ocean shallows and ocean depths.	Picking with camel hair brush; aspirators; berlese funnel or flotation method; trapping or shooitng of hosts and their removal from host's body.	80% alcohol with few drops of glycerine; ectoparasitic ones first put in 2% aqueous solution of chloral hydrate to make them soft and devoid of sucked-in blood.
2) Ticks	Mostly ectoparasites of mammals, few of birds; host's nests; .free in locations waiting for their victim.	Trapping or shooting of hosts and their removal from their body with forceps; teasing nest's material with forceps or placing it in berlese funnel.	—do—
C. Araneae (spiders)	Trees, shrubs, or herbs; litter; debris; buildings.	Sweeping; beating vegetation; searching and picking with forceps; aspirators; sometimes also through berlese funnel.	80% alcohol plus glyserine.

(*Contd.*)

Table 3. (*Contd.*)

1	2	3	4
D. Crustaceans	Primarily aquatic; few in soil rich in humus; copepods parasites of fish.	Plankton net, fish net; bottom grab; hand picking; berlese funnel.	80% alcohol plus glyserine.
E. Annelida 1) Earthworms	.In soil; various water bodies; gardens; fields, etc.	Handpicking or with forceps; damp soil inhabitants by a sieve or terylene nets; mud dwellers scooped with sieve and washed with water, worms thus left on sieve are picked up with forceps.	First washed in water, then in 5% formalin or 80% alcohol plus glycerine 9:1 ratio; terrestrial ones first asphyxiated in stoppered bottle entirely filled with water.
2) Leeches	Aquatic; parasites on vertebrate and invertebrate hosts, terrestrial.	Picking with forceps; dip net.	First inject 5% formalin in specimens and then preserve either in same preservative or 80% alcohol.
F. Molluscs	Aquatic; terrestrial, on vegetation.	Handpicking; dip net, soup strainer; fish net for large forms from seas, etc.	First kept in 40°C hot water until extended and preserved in 80% alcohol or formalin–alcohol mixture (1 part 40% formaldehyde + 4 parts of 70% alcohol + 1 part water).

because it is cheap, durable, does not bunch up or scratch the wings of Lepidoptera, dries quickly and can be used with some success even if it is wet. The only disadvantage with it is its low transparency as compared to other cloths and higher resistance to air. The meshes of the cloth should be as large as possible but small enough to hold the insects to be captured. The edge of the bag, where it is attached to the ring, should be made of strong cloth like canvas, heavy muslin, or linen. The strong cloth is folded to form a hem of about three inches so that the ring can pass through it. It also increases the strength of the bag around the rim which is the point of greatest wear during collecting operation. The handle of the net should be three-fourth metre long but light. The net is thus made in such a way that it should be as light as possible having the least possible air resistance but reasonably strong and durable.

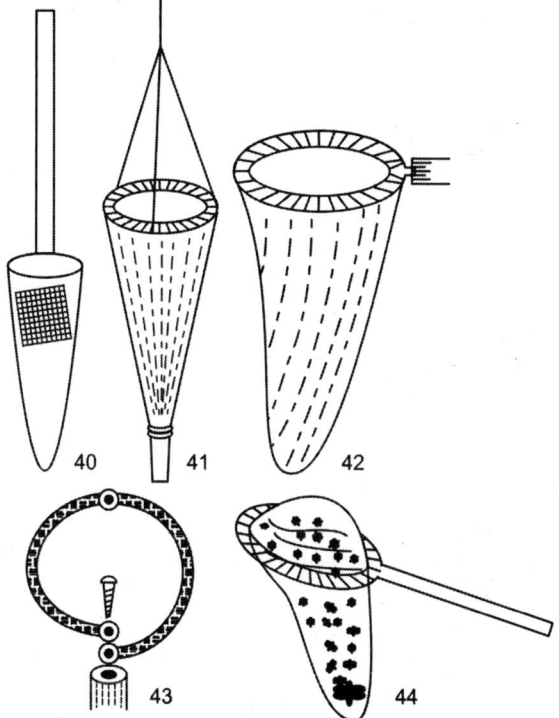

Fig. 40-44. Different kinds of collecting nets.

The small insects can be removed from the net with a killing bottle or aspirator. If the captured specimen is large, the handle is twisted quietly lapping the bag over the rim and the specimen is then enclosed in the bottom of the bag (Fig. 44). The bag is then grasped enclosing the specimen in a small fold of the cloth and with the other hand the open end of a killing bottle is inserted and pushed into the net upwards till the specimen is enclosed by it. The bottle is then withdrawn from the net and corked. In case of active or stinging specimens like bees and wasps, the fold of the net containing the insect is inserted into the killing bottle until the specimen is stupefied.

Do not use the net over wet vegetation because most of the insects are damaged, or on vegetation having stout thorns, or very close to barbed wire to avoid its wearing.

Aspirator

It is a simple suction apparatus used for collecting small insects and arachnids. These are of several designs but the most commonly used ones are discussed here (Figs. 45, 46). In both cases a vial of glass or preferably transparent plastic is used. In one case (Fig. 45) the vial is opened at one end only while in another case (Fig. 46) it is opened at both ends. The open ends are provided with tightly fitting rubber stoppers to avoid the crushing of small insects which otherwise may crawl between the stopper and the wall of the vial. Two glass tubes pass through the stopper. A rubber tube is attached to the outer end of the suction tube for sucking purpose. The other end of this tube, which remains in the vial, is covered with a piece of fine muslin cloth to prevent insects and arachnids from entering the tube. The second glass tube is opened at both ends. In the second case both the open ends of the vial are fitted with stoppers (Fig. 46) and one glass tube passes through each stopper.

In practice the outer end of the intake tube is placed near an insect or arachnid and suction is applied sharply through the suction tube. This suction creates a partial vacuum in the vial thereby drawing the arthropod up through the intake tube. When the aspirator is not in use, the outer end of the intake tube should be plugged with cotton to avoid the escape of arthropods once caught in the vial. Although the inner end of the suction tube is provided with a muslin cloth, more care is still needed to avoid contacting parasitic arthropods or other arthropod-borne pathogens while sucking the suction tube with mouth. In such cases it is always advisable to use some suction-producing apparatus. The insects or other arthropods collected in the aspirators may be killed by emptying them into an open killing bottle. The killing agent should not be kept in the aspirators, especially during its use.

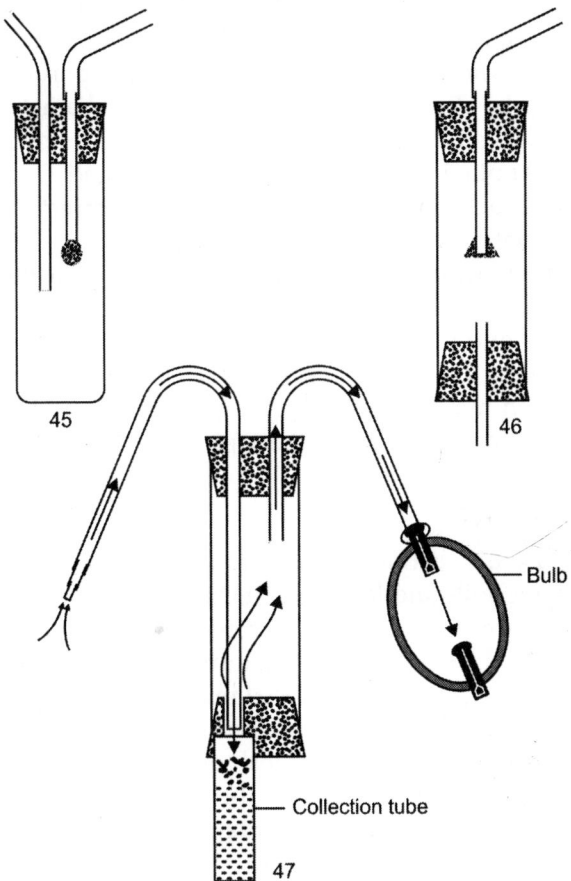

Figs. 45-47. Different types of aspirators for collection of small insects and arachnids.

Another type of aspirator is the Bulb Aspirator (Fig. 47). It is quite useful for the collection of mites, small insects, and spiders. The vial is of transparent plastic or heavy duty glass whose both ends are provided with tightly fitting stoppers.

There is a provision for fitting interchangeable nozzles of suitable diameters in the outer end of the intake tube depending upon the type of collection to be made. The other end of the intake tube runs through the entire length of the vial and placed in the narrow hole of the other stopper. This hole has smaller interior diameter and a larger exterior diameter. The narrower portion of this hole should always be slightly larger than the outer diameter of the intake tube for free movement of the air through the whole system. Similarly the wider portion of this hole should be slightly smaller than the outer diameter of the collecting vial so that it is tightly attached to the stopper. The suction is applied through the suction bulb and the desired specimen is sucked in through the intake tube directly into the collecting vial containing 70 per cent alcohol.

The only disadvantage with this type of aspirator is the presence of condensation inside the main tube and the intake tube due to the alcohol in collecting vial. This results in the sticking of the specimens along the intake tube. The condensation may be stopped to a great extent by keeping a piece of blotting paper inside the aspirator.

Berlese Funnel

It is quite useful in extracting insects and other small arthropods from organic soils and leaf litter. It can also be used to extract these organisms from loose bark, rotting wood, fungi, mosses, flowers, stored food products, manure and the material from nests of birds, mammals and social insects.

It is a simple apparatus consisting of a metal or plastic funnel (Fig. 48) having a wire mesh on its bottom for holding the sample. The narrow end of the funnel is received in a beaker or any other container containing 70 per cent alcohol with a few drops of glycerine to avoid desiccation in case the alcohol evaporates. The funnel is then covered with a lid having a hole in the middle for an electric bulb. As the upper part of the sample dries up due to the heat of the bulb, the organisms avoiding heat start migrating deeper and deeper into the funnel. They finally fall at the bottom of sample and are collected in a container kept below.

Flotation Method

This method is quite simple and also used to extract insects, mites and other arthropods from soil or matted vegetation. It is also good for collecting eggs and pupae of insects from such materials. The sample is broken up in a basin containing a mixture of magnesium sulphate in water in 1:3 ratio. It is then stirred gently. After some time the organisms start floating over the surface of water from where they are collected on a sieve or filter paper.

Killing Agents and Killing Bottles

Various kinds of killing agents are used but the best ones are those which kill the insects immediately without affecting their colours or unduly hardening them. The most commonly and widely used killing agent is the cyanide, especially potassium cyanide because it is least likely to deliquesce. Some commonly used killing bottles are as follows:

Cyanide Bottle

It consists of a wide mouthed bottle or jar of heavy-duty glass with a well fitted cork or lid (Fig. 49). A layer of granulated potassium cyanide of about 10 mm thickness is spread at the bottom of the bottle. Then the powdered dry plaster-of-paris is poured over it till it forms a layer of 1.5 to 2 cm thickness and then four or five drops of water are added to it. Later a thick paste of some more plaster-of-paris in water is poured over the previous mixture till it forms a 10 mm thick layer. The uncorked bottle is then kept open for 20 to 30 hours to allow enough time for the plaster-of-paris to dry. A blotting paper is then spread over it for absorbing moisture given out by the cyanide and to avoid direct contact of the specimens with the killing agent. The only disadvantage with the cyanide is that it makes the insects hard and brittle besides affecting their colours if left in the bottle for too long.

Since cyanide is deadly poisonous, the bottles must be labelled as POISON and stored with greatest care.

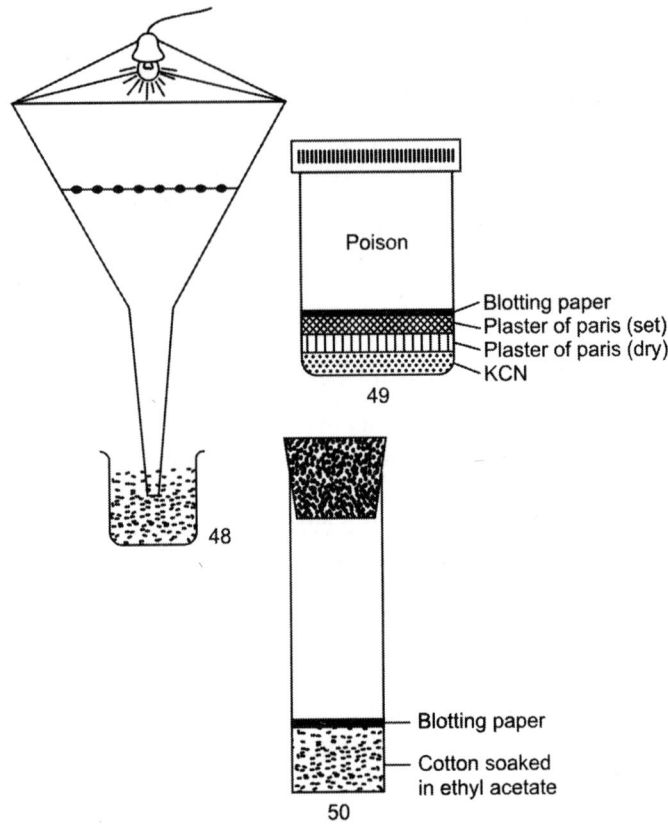

Figs. 48-50. Berlese funnel. **49, 50** killing boltles.

Ethyl Acetate Killing Bottle or Tube

Ethyl acetate is also an effective killing agent for insects, especially beetles, hymenopterans, etc. Any glass bottle can be taken to make the ethyl acetate killing bottle (Fig. 50). Cotton soaked in this killing agent is placed at the bottom which is then covered by a piece of blotting paper for preventing the direct contact of the specimens with the cotton. A few more drops of killing agent, are then poured over the blotting paper to make the bottle more effective. At the end of each day the insects collected in the bottle are taken out and preserved. If they are not required to be taken out, absorbent cotton soaked in a few drops of killing agent is placed over the collected material and then the bottle is corked tightly. The specimens remain in a relaxed condition for some days if kept in this way and are easily pinned.

Other killing agents used for killing insects are tetrachloroethane, carbon tetrachloride, ether, chloroform, benzene, ammonia, and ethyl dichloride. Although these are much safer to use and easy to handle than cyanide, they too require more precautions. The vapours of carbon tetrachloride, if inhaled in excess, can eventually damage the liver. Some of them take more time to stupefy the insects and excess of them may wet the specimens causing great damage to fragile scales, hairs or wings. Benzene, ethyl acetate, and ether are inflammable. Benzene and ammonia may affect the colour of the insects. Like cyanide, carbon tetrachloride and chloroform also make the specimens hard and brittle. Ethyl dichloride is good but it, too, affects the colouration of insects, especially orthopterans. The tetrachloroethane is the safest and most efficient killing agent.

Similar to cyanide bottle, all such killing bottles should also be labelled as POISON.

The insects should not be left for too long in the killing bottle. They should be removed before they get dirty or damaged. There should also be no overcrowding of specimens in the killing bottle. Separate small killing bottles may be preferred for collecting few specimens in each one of them. If all kinds of insects like tough and fragile or large and small are put in the same bottle, they are all likely to be damaged by one another. Some grasshoppers and beetles expel excretions which ruin other specimens in the bottle. Some scorpionflies emit a brown fluid which foul the bottle and its other specimens. It is, therefore, advisable to hold them for a minute till they emit a brown fluid before putting in the killing bottle. Lepidopterans should never be mixed with other specimens to avoid their spoilage by the scales and hair, even their killing bottles should not be used for other insects unless they are first properly cleaned.

The larvae of insects can be best killed by immersing them in boiling water before they are placed in liquid preservative to avoid their distortion.

DATA OF COLLECTION

Both the preserved and live animals must bear collection data. A specimen without such data is completely useless for a taxonomist. The collected preserved specimens, either dried or submersed in alcohol, etc., must not be without the actual data. The modem taxonomist needs still more information like films of courtship, displays and other aspects of behaviour, sound analysis and other characters like histological, cytological, chemical, etc. In this way he is fully aware of the diversity of nature which requires a far broader approach than was envisioned by taxonomists of yesterday. Thus every specimen in the collection must be labelled containing the following data:

> Geographic locality
> Stratiographic position (for fossils only)
> Data
> Stages (adult male, female or immature form)
> Altitude or depth (for aquatic organisms)
> Host
> Name of collector

Such data should also be recorded in a field notebook during the collection. During long expeditions, it is convenient to give numbering to specimens according to the place of collection and date. These numberings are then written in the notebook along with the pertinent collection data. Later, all the labels containing the pertinent collection data can be prepared in the laboratory and attached to every specimen.

PRESERVATION OF COLLECTED MATERIAL

The method of preservation varies from group to group. Taxonomists and others (who also deal with animals) must know that the specimens are preserved in such a way that they are not destroyed either through the action of insect pests, mould, oxidation or bleaching by sunlight, dyeing out, protein decay, etc. It is very essential for the collector to know in advance the various techniques already in use for his group. He may improve upon the old techniques or invent new ones for the better preservation of the specimens. Good preserved specimens are still in satisfactory condition even after more than 200 years of their collection. As more and more species are becoming extinct, the problem of permanent preservation is becoming increasingly important. The works of Ross (1941), Oman and Cushman (1948), Wagstaffe and Fidler (1955), Storey and Wilmowsky (1955). Myers (1956a, b), Duellman (1962), Levi (1966), Oldroyd (1970) and Uprety and Kapoor (1979) are useful in knowing the collecting and preservation techniques for various groups of animals. A general account of habitats, collecting ways and methods of preservation of some invertebrates is given in Table 1.

CURATING

Once the animals are collected, labelled and preserved, the next important job is of their safe custody, cataloguing, etc. This is the work of a curator. Curating is a very important task of a taxonomist, especially when one day he is likely to become the incharge of a museum to look after the whole collection. Curating is an extremely varied business that takes a lot of time of most taxonomists. A curator then needs to have an expert knowledge of his collection. He knows the function of his collection, the groups of animals in which collections are needed, areas which need to be urgently sampled, and the various policies with regard to the use of such collections. His primary responsibility is to preserve the collection, to accumulate, maintain and conserve a documented record of his collection. Quite useful and latest information on curating can be procured from the journal *'Curator* published by American Museum of Natural History, New York; the journal of *Museology* published by the University of Pennsylvania, Philadelphia; and the *Museum Journal* published by the Museum Association, 87 Charotte Street, London.

PREPARATION OF SPECIMENS

Due to great diversity of animal fauna, the methods of their preparation for storage and identification, too, are diversified. The main purpose of better preparation is that all structures of systematic importance are rendered clearly visible. There are many works on the techniques used for the preparation of different kinds of animals and these are required to be consulted. The skins of animals like birds and mammals are for direct study while the mammalian skulls need to be properly cleared before study. The protozoans can be better preserved if the techniques of Corliss (1963) are followed. In some cases microscopic slides are prepared of the whole animal or part of the organ for proper taxonomic study. Since the study of genitalia forms important taxonomic characters in many groups of insects, either their microscopic slides are prepared or it is studied in glycerine and then preserved in a tiny glass vial which is then placed along with the specimen in the box or separately carrying appropriate numbering.

Some Insects are better studied if preserved in alcohol or other liquid preservatives. Most of them are pinned and then stored dry in insect boxes. There are various methods of proper mounting of insects depending upon the size and group to which they belong. The proper mounting of insects is possible only when they are in a completely relaxed condition. The process of relaxation and various methods of their mounting are discussed below:

RELAXING INSECTS

For proper mounting the hard bodied insects must be in a relaxed condition. This helps in the proper display of their body parts of taxonomic importance. The insects are best mounted immediately after they are killed. If the specimens become quite dry and stiff, they must be relaxed in a relaxing box or any other type of similar apparatus. A relaxing box may be made from any large metal, glass or plastic container. At the bottom of this container a thin sheet of about 2 to 4 cm thickness of synthetic sponge or other porous material is placed. The sponge piece is then saturated with water. A cotton swab soaked in about 10 to 15 ml of phenol or ethyl acetate is placed in one corner of the container to prevent the formation of mould which otherwise develops on the specimens within two to three days. It is also necessary to place a sheet of blotting paper on the inside of the lid to avoid condensed water to fall over the specimens.

The insect specimens to be relaxed are then spread in a *petri dish* which is then kept in the relaxing box or the envelopes containing the specimens are kept as such in it. The time required for their relaxation depends on the size or type of specimens. Most specimens are satisfactorily relaxed by leaving them overnight in the box. To avoid damage and discolouration the specimens should not be left for too long in the relaxing box. Readymade relaxing boxes are easily available.

MOUNTING

The hard bodied insects are mounted on entomological pins. These are standard pins made of steel, enameled or japanned black and so do not rust. These are one to one and half inches long but the length differs from manufacturer to manufacturer or even within the same packet. The diameter of the pins also varies and numbered accordingly. Depending upon the length and diameter the pins are generally classified into three categories— English, Continental and Minuten nadeln. The English pins are 18 to 30 mm long and quite stout. These can be handled well with pinning forceps and preferred for staging or double mounting. The continental pins come in three sizes—35 mm long (numbers 000,00,0 and 1 to 7), 38 mm long (numbers 8 to 10) and 50 mm long (numbers 11 and 12). These are good for direct pinning. The Minuten nadeln are very fine, short (numbers 10,15, 20 or 22 mm) with or without heads and used for only double-mounting of very small insects.

Mounting can be done by either of the following methods depending upon the size, structure, and group of insects:

Direct Pinning

Freshly killed specimens are best pinned. The pin should be thick enough to pass through the insect body without damaging it. Since there are different ways to insert pins through the insect depending upon the group it belongs to, pins should pass through the correct points in the insect body (Table 4).

The pin is always inserted vertically through the body or sloping very slightly in such a way that the front part of the body is very slightly raised. The specimen is then pushed up

Table 4. Correct pinning of insects

	Insects	Points of Pinning
1.	Orthoptera	Insert pin through the right side of the pronotum near its posterior edge (Figs. 51, 52); long axis of the body should be nearly at right angles to the pin but with the head of the insect downward; set the abdomen drooping below the wings.
2.	Dermaptera	Insert pin through the anterior part of right elytron (Fig, 53).
3.	Phasmida	Insert pin through posterior-most part of the metanotum in the midline (Fig. 54).
4.	Odonata	Insert pin in the midline of the thorax between the bases of hindwings (Fig. 55).
5.	Hemiptera	Insert pin through scutellum near its anterior margin just to the right side of the midline for those specimens having large scutellum (Fig. 56); but in those with small scutellum or covered by the enlarged pronotum (e.g. Notonectide), it is inserted through pronotum, near its anterior margin just to the right of midline.
6.	Neuroptera	Insert pin vertically through the middle of the thorax (Fig. 57). Mecoptera and Trichoptera are also similarly pinned.
7.	Lepidoptera	Insert pin vertically through the middle of the thorax (Fig. 58).
8.	Coleoptera	Insert pin near base towards inner edge of right elytron (Fig. 59).
9.	Hymenoptera	Insert pin directly through thorax to the right of midline (Fig. 60).
10.	Diptera	Insert pin at the right of the midline of thorax near base of wings with its point emerging in front of right mid coxa (Fig. 61).

the pin until 1/4 part of pin is on top of specimen so that the pin can be easily grasped with fingers or pinning forceps without having any direct contact with the specimen. The labels bearing data, etc., are put below the specimen. If all the specimens are uniformly mounted, it greatly improves the appearance of the collection and makes the examination and comparison of the specimens with one another easier. Uniform pinning is easily done with the help of pinning blocks. A pinning block is made of hard-wood, metal or plastic in which holes of desired depths have been bored (Fig. 62).

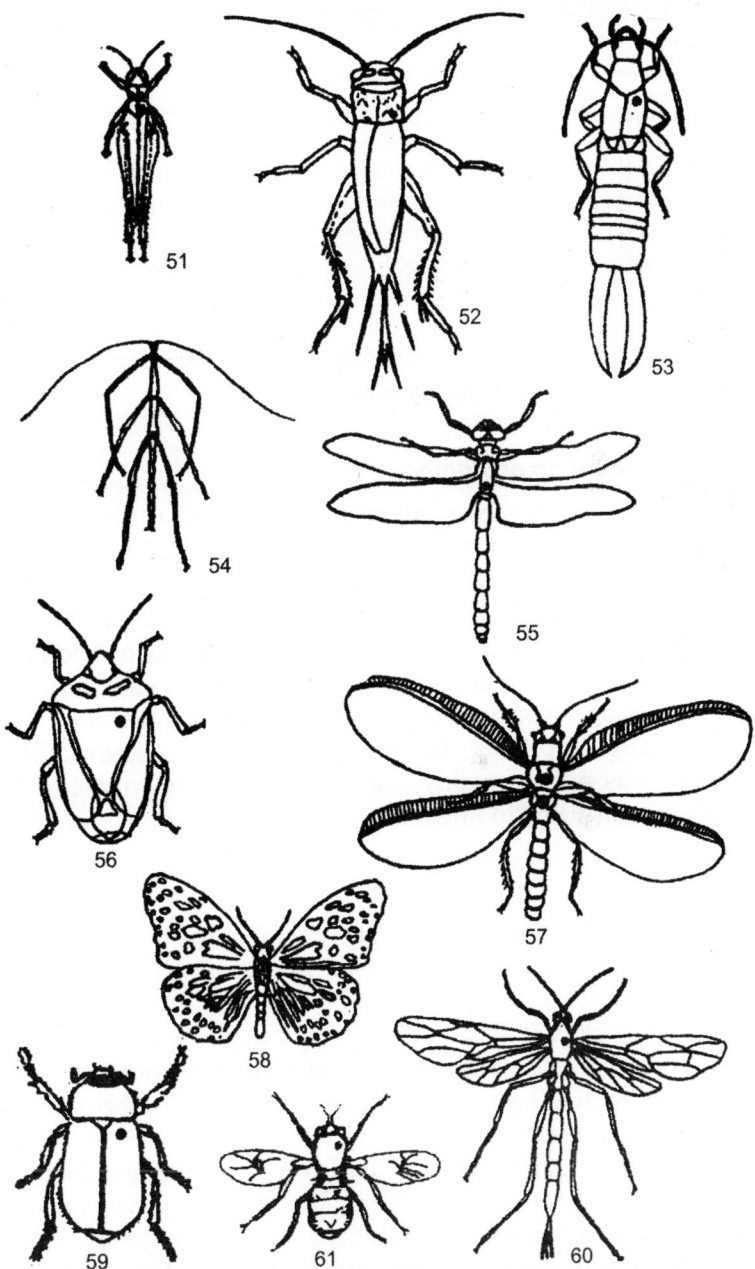

Fig. 51-61. Various ways of direct pinning of insects.

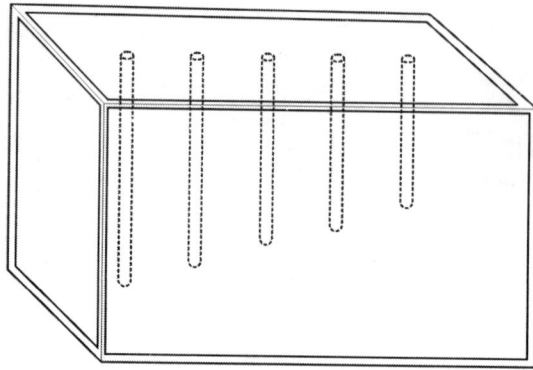

Fig. 62. Pinning block.

Double Mounting or Staging

This is quite useful for small insects. The insect is first pinned on a support or stage using headless Minuten nadeln. The other end of the stage is then supported on a large entomological pin which bears labels of data, etc, (Fig. 63). The stage may be of polyporous pith, strips of cork, pith or balsa wood. The best material for staging is the polyporous pith, a bracket fungus, due to its right consistency to hold pins well, and being pure white, also improves the appearance of the mounts. The polyporous pith is readily available in strips of suitable size with various firms in England.

Pointing

The smaller dried insects (except Diptera) are preferred for this type of mounting. It is either done by attaching the specimen with adhesive to the tip of a small triangle of thin card which is then supported on an entomological pin (Fig. 64) or by attaching it directly to the side of the slightly turned up tip of the point with adhesive (Fig. 65). The broad end of the point is then supported on an entomological pin.

The good points are prepared from heavy paper or thin but stiff and heavy card or two-ply Bristol board. These are cut into truncated triangle (6 mm long, 2 mm wide at base and 0.5 mm wide at its apex) but the size varies depending upon the size of the insect. Although several kinds of adhesives are used in attaching the specimens to the points, shellac gel is the most satisfactory for small and medium-sized insects. Even a minute quantity of it binds the insect firmly. In its preparation, 150 cc of white shellac is first boiled for about 20 minutes to which is then added 10 cc of 90 per cent ethyl alcohol. This mixture is then boiled again. It is poured in vials and allowed to cool in cold water. Finally, it is converted into a kind of petroleum jelly. Due to the evaporation of solvent the gel gradually thickens but a few drops of alcohol makes it thin again.

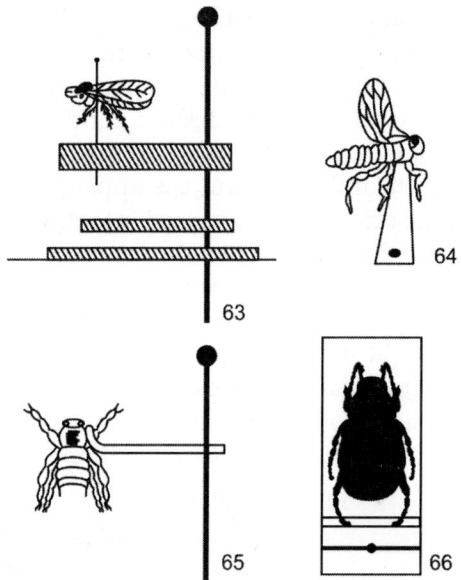

Figs. 63-64. Ways of indirect pinning of insects.

Whatever adhesive is used, it should be in the minimum quantity. Excess of it damages the specimen Although various parts of the insect body are used in attaching them to the points, side of thorax below the wings or margin of the tergum and above or between the bases of the legs is the best. It allows the study of all important taxonomic characters.

Carding

This method is commonly used in mounting some beetles or other hard bodied small insects. The insect is glued to the card from the ventral surface (Fig. 66). Here the specimens are better protected against damage than in pointing. In case the underside of the specimen is to be examined, it is first relaxed.

It is then removed from the card by soaking it in hot water or ammonia solution. It is, therefore, advisable to use water soluble glues.

Spreading

The best mounts of insects are those which make the taxonomic study of all important body parts quick and easy. The head, wings, legs and abdomen must be properly displayed with the help of pin or forceps. Some supports are always provided for keeping certain body parts in proper order till the specimens are dried and retained in the desired condition. This is best done in freshly killed specimens when their internal parts are soft enough to take the pin and the appendages are pliable. The wings of insects are better spread into desired position on a spreading board (Fig. 67) which is also used for setting other insects. The wings of small forms like microlepidopterans, dipterans, etc. are also spread on them which have small horizontal straight but narrow cavities (Fig. 68). Thin paper strips are used to press the wings on the spreading board and the specimens are allowed to remain in this condition till they become dried to retain their desired position.

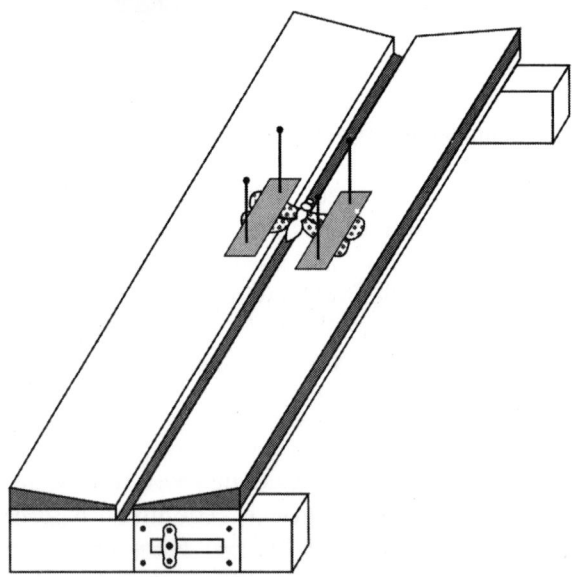

Fig. 67. Insect spreading board.

Care should be taken to keep the spreading board containing insects in empty insect boxes or other containers to avoid their destruction by ants, etc. The specimens must be dried before storage to avoid growth of mould over them.

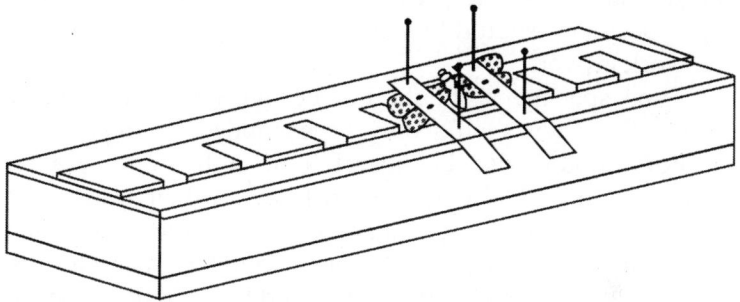

Fig. 68. Setting of small insects on spreading board.

STORAGE

Storage of collection has always been a difficult task of a taxonomist. The specimens are subjected to damage or destruction by various pests. They can also be ruined by dust, light or dampness. At a later stage there also occurs space problem due to the accumulation of more and more specimens. Rapid changes in temperature and humidity are also harmful to museum cases and specimens.

The nature and type of equipment necessary for safe storage differs from group to group. Care should be taken to get such storage equipments which guarantee the best maintenance of specimens in the long run. They should be both insect and dustproof. There are now many good scientific houses which manufacture reasonably good storage equipments. The collection should be stored in fireproof buildings that are also dustproof. Now it is seriously believed that such buildings should also be earthquake resistant. Presently more and more museums are relying on air-conditioned buildings, which provide uniform temperature throughout the year.

Special care is needed to curate the type specimens on which the names of the species are based. These specimens are virtually impossible to be replaced. Types serve as reference material to establish the identity of a doubtful nominal species. The curator is responsible to ensure its safety by all means. Such specimens should always be deposited with large standard museums which have special arrangements for their safety, Since this practice is not followed by many taxonomists, most type specimens have been damaged or lost by the authors who did not nave means to protect these specimens. Therefore, all type specimens should be deposited in a museum as soon as the work is over.

Type specimens should not be allowed to be handled frequently. They should only be examined by experts. The curators should also be liberal to loan them to qualified specialists. It should be practised to avoid their transport as far as possible. These specimens should be stored separately from the other general collection in order to facilitate their quick removal in case of emergency and to avoid their constant handling in a general study collection. They should be clearly labelled in distinct colours and numbered individually for quick reference. **Lectotypes** and **Neotypes**, too, are equally important and should be treated at par with the type specimens or **holotypes.** Some authors gift the type collection as well as others to large museums with the provision that the donor retains control and possession during his lifetime.

CATALOGUING OF SPECIMENS

The method of cataloguing is different from group to group. In some groups of animals, like higher vertebrates, each specimen is given a separate number and catalogued separately. This procedure is considered lime-consuming and it is now suggested that the methods employed for insects and molluscs be adopted for vertebrates as well. The entries usually followed in cataloguing vertebrates are---consecutive museum number, original field number, scientific name, locality, date, collector and remarks. The individual specimens of wet collection are usually not catalogued. Each collection as a whole is given a number, or each species in each collection. In case of insects this practice is usually not followed because of the large number of both species and specimens.

All the specimens collected from one locality or district by one expedition are catalogued together. The specimens are usually catalogued after they are identified, at least up to the genus level. This becomes quite useful for a permanent reference to the contents of the collection even after a long time after it is broken up and distributed in the systematic collection or even widely dispersed to other museums. Only the type specimens are almost unusually catalogued. In large recognised museums these type catalogues are usually bound in a book in which the types are serially numbered. The number of the type specimens refers back to the notes for complete detail.

There are various ways followed by different museums in using filing cards relating to the collections. Some museums have elaborate card filing system which permitted easy retrieval of all sorts of information regarding the specimens. This practice is being phased out slowly. Some of the large museums are experimenting to place all information on each specimen on a separate IBM card. This practice is now usually followed by some large museums, like Natural History Museum, London; U.S. National Museum, Washington, etc.

ARRANGEMENT OF THE SPECIMENS

The collected specimens should preferably be arranged according to the accepted classification. It is also necessary for the curators to see that the sequence of orders and families is standardised in many groups of animals. The unidentified collection should be kept separately. If genera or families are known, the specimens should be kept separately along with them.

MAINTAINING QUALITY OF COLLECTION

The primary responsibility of the curator is to see that his collection represents as much diversity as possible. He should be responsible for frequent sampling of the areas within his reach. After the collection is brought to the museum, it should be sorted out and handed over to the various sections made on the basis of major grouping of animals. He should be liberal to allow the outside specialists to come and work on the collection of their specialisations. This is a good way of getting more and more collections identified. He should also be aware of the recognised specialists in the world who can be contacted for getting the material identified. After obtaining the prior approval, he may lend the specimens to such qualified experts on requests. The material is returned to the museum as soon as the specialist has completed his studies. A specialist may retain the duplicates if he has entered in such agreement before accepting the collection. In case of anatomical studies, the dissection will partially or entirely destroy the specimens. This should be done with the permission of the donor and the specialist is bound by the obligation to preserve a pictorial record of the dissection.

Another way of raising quantity and quality of collection is through exchange. Exchanges are most desirable in groups of animals where series of unlimited size can be obtained and where the exchanges are in areas not readily accessible to the respective museums or institutions. Exchanges are sometimes necessary to make the identified collection complete. Many experts, who have excess of specimens, give them as 'open exchanges' without necessarily expecting any return. In any way the curator must assure the depositors that all necessary facilities are available with him for the safe storage of the collection. Any lapse on the part of curator may affect the reputation of his museum

IDENTIFICATION

Identification of the great diversity of living organisms around us is a challenging task that calls for a high level of professional training and infrastructure. The purpose of identification is basically to determine what kind of organism a given specimen is. The animals have become so much diversified and complex in their forms and activities that the identification of a given specimen is easily done only in a few well-known groups, like mammals, birds and butterflies and that too in certain parts of the world. The reliable identification is very difficult in the rest of the animal kingdom and more so in case of isolated specimens. It is quite possible for students and teachers of biology to acquire the competence to properly identify the taxa present in their own locality, given adequate support in terms of literature and given the motivation to acquire this knowledge.

The kind of identification may vary according to the stature of identifier. The professional worker would like to go up to the species level. The non-systematist requires the name for studying some fundamental aspect of an animal. Similarly an applied zoologist requires the name of the species in order to find out pertinent literature regarding basic information or appropriate control measures. All the information concerning the morphology, physiology, genetics, biochemistry, cytology, ecology, etc., of an organism would be essentially of no use or definitely less beneficial if it was not published as pertaining to a given species.

The systematist, in addition to using the species name as a key to the literature and a category for data storage, needs to identify an organism to species or at least to the point where he realises that the specimen (or specimens) he is attempting to identify, is an undescribed 'new species'. Further the identification of a specimen must necessarily precede the filing of this specimen into a scheme of classification. Thus, in taxonomic practice there is no sharp distinction between identification and classification. The classification always involves identification and the identification may lead to extension and improvement of the latter. For purposeful identification it is necessary to have knowledge of taxonomic methods in general, the characters and current terminology, the usual study techniques, pertinent literature, the natural history and comparative zoology of the group.

METHODS OF IDENTIFICATION

After the animals are collected they are first sorted out and at least tentatively identified up to order, family or if possible generic level. When this is done, the collection is ready to be studied by the specialist for precise identification. There are several different methods employed for identifying an organism. Surprisingly all the methods are based on comparison. The important methods used in identification are discussed below:

1. From literature
2. Keys
3. Pictures
4. Direct comparison
5. Combination of different methods

From Literature

The basic function of an identifier is to compare the specimens with the published descriptions of the species. This is a very tedious task involving hundreds and thousands of comparisons. There is need to skip over the great bulk of the species which are not closely similar to the specimen which is being identified. This is further facilitated if the keys are available in the group.

Keys

This is one of the most commonly used methods of identification. A key is essentially a printed information-retrieval system into which one puts information regarding a specimen-in-hand and from which one gets an identification of the specimen to whatever level the key is designed to reach. In the absence of the originally identified specimens of the species utilised for comparisons, the published description becomes the only tool. In larger groups with many species, it is the most tedious task to compare specimens with hundreds or thousands of published descriptions. This task can be solved if the keys to the main group are available. The main purpose of the key is to facilitate identification. Thus, the construction of the keys is an important job of a systematist. It is a tabular device designed for rapid identification, and based on the most convenient characters which are usually arranged dichotomously. It is thus a systematic framework for zoological classification with a sequence of classes at each level, of which the more restricted classes are formed by the overlap of two or more classes at the next higher level; its preparation requires a thorough analysis of the taxonomic characters and then the best possible stable characters are selected and arranged.

A good key is strictly dichotomous, not having more than two alternatives at any point. The language of the key is simple and telegraphic like that of taxonomic descriptions. The phrases in each alternative are usually separated by semicolons. Sometimes it becomes necessary to mention supplementary characters when primary characters raise confusion but used for completing the couplet. This makes the identification quite easy. Without the logic and simplicity of dichotomous keys, many good zoology students would not be able to follow the same and get lost. Memorizing a very large number of species around us in impossible and such keys need to be understandable for the identification of both higher and lower taxa.

Different keys are constructed for different purposes. Though keys are good and necessary, yet there are problems with regard to their use. A key made for adults shall not work for young ones. Moreover, even among adults if it is for males, it cannot work for females. This poses lots of problems when only male and female specimens are identified in the absence of the keys concerning either of the two. Even among adults, a key fails to give any clue for dwarfs, terata, or hybrids. Above all most keys are based on trivial characters and it becomes extremely difficult in some cases to run the specimen through the key up to family level, leaving aside identification at the lowest category, the species level. The problem with the dichotomous keys becomes most serious on inclusion of a wrong choice early in a lengthy key. This can make an inexperienced user very frustrating while running through couplets which never seem really to apply to the specimens at hand. If the user is attempting to identify a taxon which is not included in the key, it is possible to obtain an incorrect result with little difficulty. Most users of keys have had the experience of arriving at a terminal couplet in a long key only to find that neither choice applies.

There is still lot to be done in improving the earlier established keys, especially in the light of more and more taxa discovered continuously. Many of the above mentioned problems can be avoided in selecting key characters through which interpretation of character states becomes unambiguous on as many specimens as possible. It is also now understood that a dichotomous key which embodies hypotheses of taxonomic relationship will be more useful.

Sometimes the designing of identification keys for large taxa giving preference to the most conspicuous and least variable characters can certainly be done by the computers. Hall (1970), Morse (1971), Watson and Milne (1972), and Dallwitz (1974) have used computers for selecting the characters of organisms and organised them in dichotomous keys. Dallwitz even claims that the keys can be printed in the conventional bracketed style or in a tabular form which displays the structure of the key clearly and even the new species can be detected.

The idea of tabular key by Newell (1970, 1972) as an alternate to the dichotomous key has many advantages both to its designer and to the user. It is essentially a group of taxa by character matrix with character states coded for each taxon; e.g., even certain amount of missing data can also be overcome by the user provided that sufficient data are available to provide a unique combination of character states. When the group is too large to be represented in a single matrix, the initial matrix provides entry points to subsequent tables in which the subgroups are treated. Such a key may also exhibit the presumed phylogenetic structure of a group, so that an ambiguous or incorrect result may still provide some indication as to the relationships of the material. Wooley and Stone (1987) gave a good account of computer based programmes in the identification of organisms. Morse (1974) and Pankhurst (1978) presented overviews of several approaches to the development of computerised identification devices. Some of such approaches are briefly discussed below:

Monothetic Character Set Matching Method

Here the user essentially inputs a complete vector of character states for an unknown. The programme then uses a table look-up procedure to search for an exact match in a character state vector for a known taxon. Whatever logical possible combinations of characters could be drawn, all must be included in the table even though several character state vectors apply to a particular taxon. This method is *polythetic* in a way as no single character state vector is required for identification. This method is very efficient tor problems where a small number of characters are sufficient.

Polyclave Identification Procedures or Multiple Entry Keys

Here the programme has access to a data file containing known taxa and associated character state vectors. The user is free to input observed character states in any order. On every data entry the programme searches the data file and attempts to eliminate taxa with other than observed character states.

In addition, after each data entry the programme outputs to the user a list of possible taxa remaining. Such a device is a powerful tool tor an experienced user who desires to recognise rare or unusual character states in an unknown.

Online Key

it was proposed by Wilson and Partridge (1986) and is an extension of polyclave methods in which deliberate attempts are made by the programmer to optimise the sequence of character state input by the user. Here the characters are weighted in their programme using seven formulae which represent criteria for optimal discrimination or ease of use.

Online 3 Identification Programme

It was used by Pankhurst and Aitchinson (1975) and Pankhurst (1984). Here a single criterion (Gryllenberg's (1963) separation number) is used to provide optimised character selection in a polyclave format. Similar to other online keys, this programme provides the ability to check a presumed taxon against the database, using an algorithm to determine an optimised order of character state input. With the use of additional options, the user is able to retrieve the information already entered for an unknown and to determine the character states differentiating an unknown from a known taxon. The characters are used as per their merit (i.e. from its separating power) and also partly from its "weight" but finally the decision rests with the expert. In any case, the user is free to choose any character as per his choice in terms of merit.

XPER Programme

It is of recent origin and is similar to ONLINE 3 (Lebbe, 1984; Forget *et al.*, 1986). Here the user is made to enter more than one character state in response to a query. XPER is available to the public in France through MINITEL system and is now widely used for mushroom identification.

Both polyclave and online programmes seem to have many advantages over dichotomous keys. While the former would be the best for the experienced user for making informed choices for the order of character state entry for efficient discrimination, the latter shifts the burden of character choice to the programme by including algorithms for character selection. Thus, with the availability of database management software, any systematist (using a microcomputer) can construct monothetic key devices.

The computerised keys are still in their experimental stage and available only for a fraction of animal or plant kingdom. Thus, it is still considered better to confine the identification only with the help of traditional keys which are discussed below:

Types of Keys

These days different kinds of keys are used in taxonomic works. All are dichotomous and based on a series of choices. Some of the most important ones are :

   i. Indented key
  ii. Simple non-bracket key
 iii. Simple bracket key
 iv. Grouped type-key
   v. Combination key ⎤
 vi. Pictorial key   ⎟
 vii. Branching key   ⎬   For special purposes
viii. Circular key   ⎟
 ix. Box type key   ⎦

Indented Key .

The couplets in this key are indented from the left hand margin of the page, in such a way as to show their importance. Thus, the two or more members of the primary couplets are near the left hand margin, the secondary couplet is indented after leaving four or five spaces, the tertiary with equal number of spaces beyond the secondary and so on, to the end of the key. This type of key is advantageous in the sense that the relationship of various divisions is quite apparent to the eye and can be used in reverse also. When the key is short, it works alright. But in long keys, the alternatives get widely separated and take more space. Thus, it serves good purpose for keys to higher taxa or comparative keys. An example of such a type of key is given below:

A. Wings mostly hyaline
    B. Costal band dilated apically
        C. Scutellar bristles 1 pair.........*Bactrocera cucurbitae*
        CC. Scutellar bristles 2 pairs............................*B. tau*
    BB. Costal band not dilated apically
        C. Thorax with median yellow stripe........*B. diversa*
        CC. Thorax without middle stripe...........*B. dorsalis*
AA. Wings mostly opaque
    B. Wings with stripes
        C. Scutellum with 5 black spots...............*Carpomyia*
                                                    vesuviana
        CC. Scutellum with 4 black spots...........*C. zizyphae*
    BB. Wings reticulate
        C. Posterior margin of wings with 3 hyaline spots
                                 ....*Tephraciura xanthotricha*
        CC. Posterior margin of wings with 5 hyaline
             sports..................................*Spatulina acroleuca*

Simple Non-bracket Key

This key is also often used. The couplets are composed of alternatives for ready comparison. The specimens to be identified are run through this key forward. It can be used from backwards but gives slight inconvenience in tracing the backward run of the characters to be compared. An example of this type of key is given below:

1. Wings mostly hyaline...2
   Wings mostly opaque...5
2. Costal band dilated apically...3
   Costal band not dilated apically.....................................4
3. Scutellar bristles 1 pair..............................*B. cucurbitae*
   Scutellar bristles 2 pairs.. *B. tau*
4. Thorax with median yellow stripe .... *B. diversa*
   Thorax without median stripes...................... *B. dorsalis*
5. Wings with stripes...6
   Wings reticulate 7
6. Sculellum with 5 black spots...................... *C. vesuviana*
   Sculellum with 4 black spots...................... *C. zizyphae*

7. Posterior margin of wings with 3 hyaline spots ....
....*T. xanthotricha*
Posterior margin of wings with 5 hyaline spots ....
.... *S. acroleuca*

Simple Bracket Key

It is similar to the preceding one except that the numbers of
the couplets showing the continuation in the key are shown
in parentheses after the main numbers. When properly
constructed one can quickly and easily run through this key
both forward and backward. This is the best key in fulfilling
its diagnostic purpose. The main disadvantage in both these
types is that the relationship of the divisions is not apparent to
the eye like indented key. This is a commonly used key by the
present day taxonomists. A similar example of this type of key
is given below:

1.   Wings mostly hyaline...2
           Wings mostly opaque..5
2. (1) Costal band dilated apically..................................3
           Costal band not dilated apically............................4
3. (2) Scutellar bristles 1 pair........................*B. cucurbitae*
           Scutellar bristles 2 pairs............................. *B. tau*
4. (2) Thorax with median yellow stripe .... *B. diversa*
           Thorax without median stripe .... *B. dorsalis*
5. (1) Wings with stripes..6
           Wings reticulate 7
6. (5) Scutellum with 5 black spots............... *C. vesuviana*
           Scutellum with 4 black spots..................*C. zizyphae*
7. (5) Posterior margin of wings with 3 hyaline spots.....
           ....*T. xanthotricha*
           Posterior margin of wings with 5 hyaline spots.....
           .... *S. acroleuca*

Grouped-type Key

It is similar to the preceding one. In this the first member of
the primary couplet is set up, and the secondary, tertiary and
subsequent couplets are arranged beneath it and labelled 2, 3,
4 to the end of the first member of the primary couplet. The
second member of the primary couplet is then labelled 8, for

example, and the secondary, tertiary, and subsequent couplets are numbered in sequence under it. The first primary couplet is then labelled 1, with 8 in parentheses, to indicate the location of the second member of the couplet, the first secondary is labelled 2, with a 4 and 5 in parentheses, and the tertiary couplet is labelled 3 and 6 and 7 in parentheses. Many other schemes of numbering the couplets may be used.

Such a key has the advantage that the groups are conspicuous. It is economical of space as the entire printed page is used, and it is fairly easy to use in reverse. Its conspicuous difficulty is the fact that the members of the couplet are too far separated on the printed page to be easy to use and yet if one pays close attention to the numbers of the opposite members of the couplet this leads to no serious difficulty. In this kind of key it is possible to confuse the numbers used to designate the couplets and so one should be cautious about this.

1. (8)   Wings mostly hyaline
2. (4, 5)   Costal band dilated apically
3. (6, 7)   Costal band not dilated apically
4. Scutellar bristles 1 pair..................................*B. cucurbitae*
5. Scutellar bristle 2 pairs.. *B. tau*
6. Thorax with median yellow stripe...................*B. diverse*
7. Thorax without median stripe........................ *B. dorsalis*
8. (9)   Wings mostly opaque
9. (11,12)   Wings .... with stripes
10. (13,14)   Wings ....reticulate
11.   Scutellum with 5 black spots......... *C. vesuviana*
12.   Scutellum with 4 black spots..............*C. zizyhae*
13.   Posterior margin of wings with 3 hyaline spots ....*T. xanthotricha*
14.   Posterior margin of wings with 5 hyaline spots .... *S. acroleuca*

Combination Key

These keys contain the good points of both the indented key and those of either the key with couplets in simple non-bracket key or of the grouped-type key. If the combination key is long, the primary, secondary and tertiary couplets would be set as in the indented type; thus couplets A and AA would be set up near the left hand margin. Under A would be arranged, properly indented, couplets 8 and BB. Under B, if necessary, couplet C and CC, and so on for the other primary, secondary and tertiary couplets. Under C and CC, the couplets would be either in juxtaposition or grouped.

This key has the advantage of making the primary, secondary and tertiary groups conspicuous. If the couplets beyond those of tertiary rank are arranged in juxtaposition, it brings these minor differences close together so that they may be readily compared. Such a key has, of course, some of the difficulties and limitations of the indented key and also of key with the couplets in simple non-bracket key.

A. Wings mostly hyaline
    B. Costal band dilated apically
        1. Scutellar bristles 1 pair...................*B. cururbitae*
        1′. Scutellar bristles 2 pairs............................ *B. tau*
    BB. Costal band not dilated apically
        1. Thorax with median yellow stripe ..... *B. diversa*
        1′. Thorax without median stripe .... *B. dorsalis*
AA. Wings mostly opaque
    B. Wings with stripes
        1. Scutellum with 5 black spots .... *C. vesuviana*
        1′. Scutellum with 4 black spots.............*C. zizyphae*
    BB. Wings reticulate
        1. Posterior margin of wings with 3 hyaline spots..*T. xanthotricha*
        1′. Posterior margin of wings with 5 hyaline spots.. *S. acroleuca*

Pictorial Key

This key is also of significance. It is especially meant for field workers and non-taxonomists who can identify the commonly occurring species with the help of characters together with their figures in a comparative manner. There are several such keys for various groups of animals. Such a key for the separation of important tick genera is shown in Fig. 69.

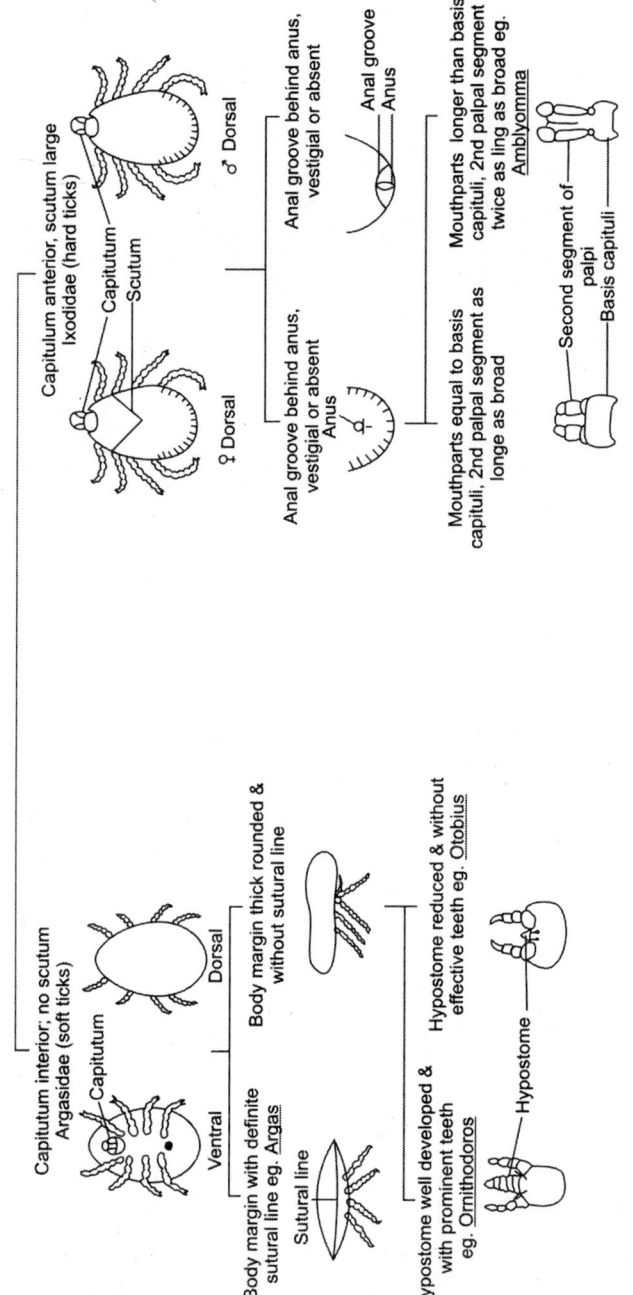

(contd.)

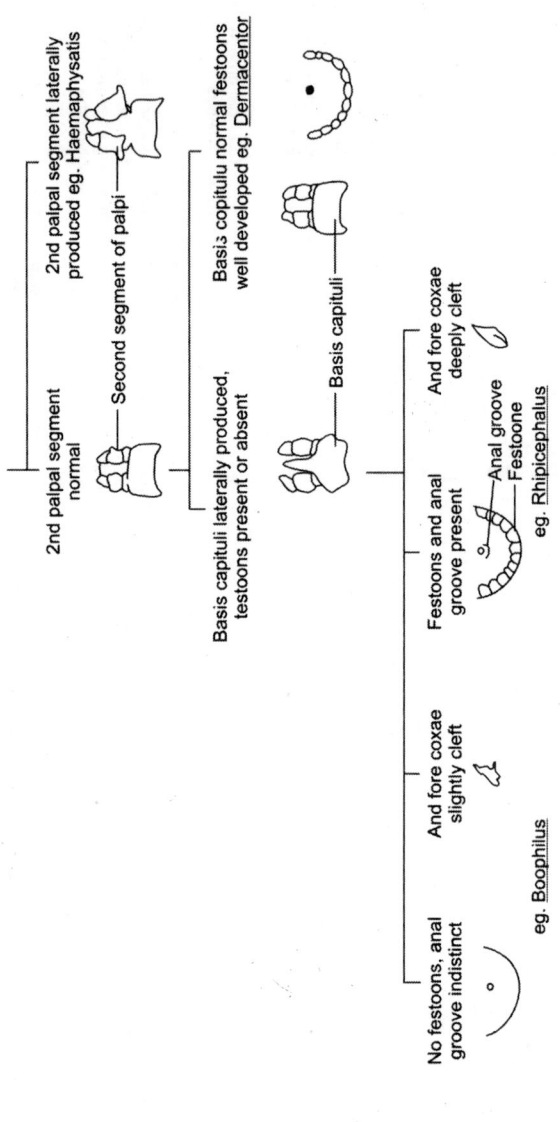

Fig.69. Pictorial key for the separation of important genera of Ticks (Modified from Cable, 1966).

Branching-type Key

This is also meant for easy and quick separation of the species as far as the group is small. It is quite useful for field workers. An example of this type is given in Fig. 70.

Circular Key

This key is also meant for small group. It is also quite useful for the non-specialists, especially field workers who want quick and immediate identification of commonly occurring important species. An example of this type is given in Fig. 71.

Box-type Key

Like the preceding two keys it is also meant for field workers and quite useful for quick identification of most common species. An example is given in Fig. 72.

Through Pictures

This is another easy method of identification in which pictures in the form of colour plates, black and white photographs, or line drawings are used. There are few animal groups in which nearly all the species can be identified from their pictures, especially if arranged in atlases showing all the forms. The pictures may represent the entire organisms or parts of organisms. These pictures are often used in combination with keys. The pictorial keys, as discussed above, consist of alternatives in the form of pictures (Fig. 69) with or without accompanying verbal alternatives. The pictures are especially of great use for those organisms which are highly patterned or characteristically coloured. The groups like mammals and birds can be identified by their coloured pictures. Among insects, butterflies are quite favourable for their identification through coloured pictures (Fig. 73).

Direct Comparison

This is one of the best means to identity the specimens. When the literature on a species is not available, the best way to identify a specimen is to compare it with that of the already identified one. This approach is useful to identify the

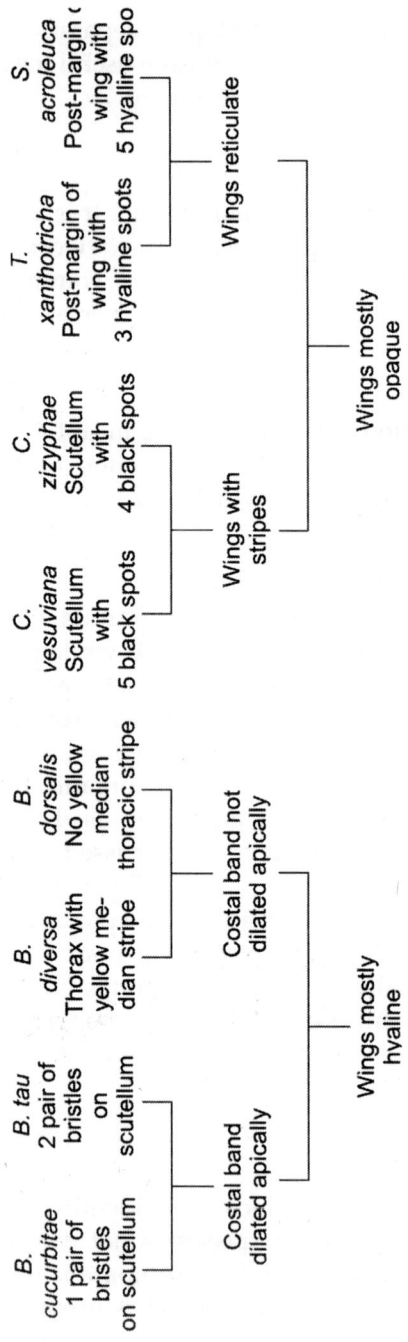

Fig. 70. Branching-type key.

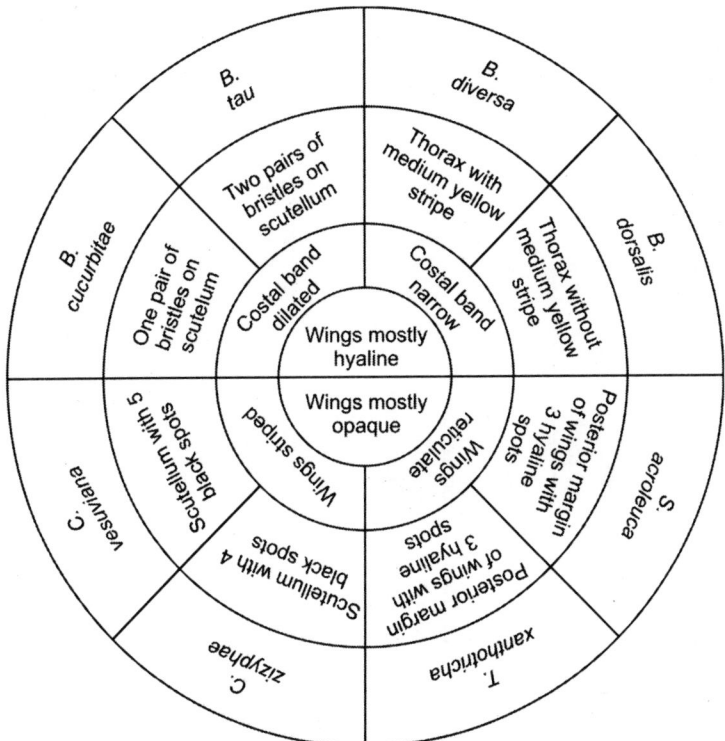

Fig. 71. Circular-type key.

specimens at any level. This method involves a highly technical task and requires considerable background of knowledge and preparation in the group. It is, therefore, not always advisable to rely on the identification solely based on comparisons with authentic collections without having the knowledge of the group in question. This is possible with a qualified expert.

The type specimens are the most authentic of all. These may be extremely useful in the identification of a species. Since these specimens are of great value, their use for routine identification should be avoided. The original verbal descriptions of species based on type specimens are the permanent records of the attributes of a given species. Such descriptions are particularly useful and constitute the only original record when type specimens are destroyed, lost or not available.

B. cucurbitae	B. tau	B. diversa	B. dorsalis	C. vesuviana	C. zizphae	T. xanthotricha	S. acroleuca
Scutellar bristles one pair	Scutellar bristles two median stripe	Thorax with yellow median stripe	Thorax without median stripe	Scutellum with five black spots	Scutellum with four black spots	Posterior margin of wing with three hyaline spots	Posterior margin of wing with five hyaline spots
Costal band dilated		Costal band narrow		Wings with stripes		Wings reticulate	
Wings mostly hyaline				Wings mostly opaque			

Fig. 72. Box-type key.

Fig. 73. Recognition of butterflies from their coloured pictures—1. *Danais melissa* (dark blue tiger); 2. *D. tytia* (chestnut tiger); 3. *D genutia* (common tiger); 4. *D. aglea* (glassy tiger); 5 *D. chrysippus* (plain tiger) (all Danaidae); 6. *Vanessa canace* (blue admiral); 7. *V. indica* (Indian red admiral); 8. *Cethosia biblis* (red lacewing, dorsal view); 9. Same in ventral view (all Nymphalidae).

Combination of Methods

The combination of all or many of these methods is very often utilised in the identification by both the specialists and non- specialists. The best and safest way of identification is to consult a specialist. Due to his strong background knowledge of the group in question, he is in the best position to make use of the highly specialised keys, original descriptions, and type specimens.

PROBLEMS ENCOUNTERED IN IDENTIFICATION

All these methods discussed above are not easily followed and worked out. One faces many problems when attempting to identify the specimens. Some of these are mentioned below:

  i. Although keys are useful for identification, these may have several drawbacks including failure to include the organisms being identified; failure to include the stage or sex of the organisms being identified; and failure to take into account any of a large number of possible variations in the morphology or other characters used.

 ii. Pictures, too, can be misleading if not clear and accurate. These may result in wrong identification if the species of the organisms being identified resemble the one depicted.

iii. Verbal descriptions of the species may be ambiguous, particularly when such characters as colour and texture are described. It is also necessary to have extensive knowledge of terminology in a particular group. The descriptions of type specimens are often in old, obscure journals which are not easily accessible and more so in a foreign language.

 iv. Direct comparison with the already identified specimens is also not easy. Such specimens of a group being identified are not always available, and even a good collection may not contain specimens of the same species or even genus or family as the specimen being identified.

  v. It is also very difficult or impossible to obtain type specimens for comparison; sometimes they are lost or destroyed.

vi. Difficulties are also faced in consulting the experts for the identification. Sometimes a specialist may no longer be alive or may never have existed. Even if he does, there is still the problem of getting the specimen(s) identified, since he may be residing in any part of the world. Although the specialist working in large museums are quite competent in the identification due to the enormous facilities at their disposal, they take sometimes up to two years in communicating the identifications even though it may be very urgently required. Moreover, there is presently extreme shortage of taxonomic specialists in many animal groups or in some no specialist is available at all.

vii. CAB, London (through CAB International Institute of Entomology) has even started charging heavily for identification service which was earlier free of cost.

SENDING INSECTS AND OTHER ARTHROPODS BY MAIL

Most of the identifications are done by specialists working in the institutions or museums. In either case they have to be sent by mail. Many times the insects are destroyed in transit due to faulty packing. It is, therefore, necessary to pin, mount and arrange them properly in a strong box to ensure their safe delivery to the specialists. Some big museums also distribute *pamphlets* containing instructions for proper mounting and packing of various animals before they are sent there for identification. Some of the important tips are mentioned below and which, if followed, will certainly minimise the risk of their damage in transit.

i. The insects should be correctly pinned, set and labelled. These should then be kept in wooden boxes (size according to requirement) (Fig. 74). Care should be taken that the cork sheet is thick enough to hold the pins deep and firmly attached to the bottom of the box. The larger specimens should be supported on either side with ordinary pins to avoid their possible damage due to swinging. The specimens on double mounts should have one additional pin passing through the pith, a little behind the specimen and the points should be supported by two pins, one on either side of the triangle, to prevent their damage due to circular motion.

Do not overcrowd the specimens in a box as it always increases the chances of their damage.

ii. A sheet of cardboard cut to fit the inside of the box should be kept over the top of the pinned specimens and the space between this and the lid of the box should be filled with cotton.

iii. At least 5 to 7 cm space should be left between the insect box and the outer container of thick cardboard. The space should be filled up with packing material like wood shavings, crumpled newspapers, or other similar material. The packing should be a little loose to minimise the risk of external shocks from being transmitted directly to the insect box.

iv. If there are two or more such boxes, they should be tied together and then placed in the outer container. This will avoid their damage due to jarring.

v. The address of the sender should be written both on the insect box as well as the outer container.

vi. Small insect specimens like jassids, small beetles, etc., and other arthropods can also be sent in dried condition in gelatin capsules of 2 to 3 mm length (easily available at various medicine shops) or plastic vials (Fig. 75) with empty space in between the specimens and the stopper filled with cotton. These are then placed in the cavities made in the thermocole piece (size according to need). The cavities should be such that the capsule and vial are firmly held in them. A similar piece of thermocole is then placed over the former and the two are then tied and finally packed in the outer container.

vii. Medium-sized insects can also be sent in large glass or plastic vials (Fig. 76). One or two insect specimens are first pinned on a thin strip of cork sheet which is then inserted inside a narrow place cut in the bottle cork. A narrow passage is made on one side of the cork which is plugged with cotton soaked in few drops of phenol to prevent growth of mould but allow aeration. A large pin should be inserted through the cork in such a way that it touches the opposite wall of the vial. This keeps the cork sheet in a fixed position.

Fig. 74. Correct way of pinning insect in an insect container.

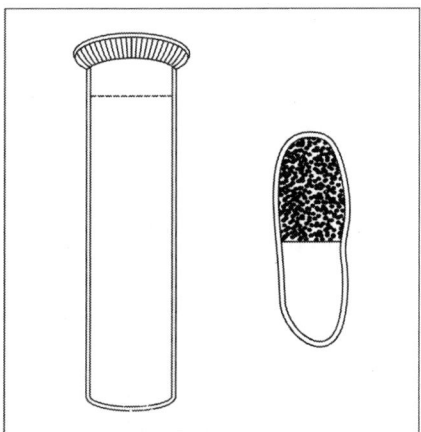

Fig.75. Plastic tube and gelatin capsule placed in thermocole piece.

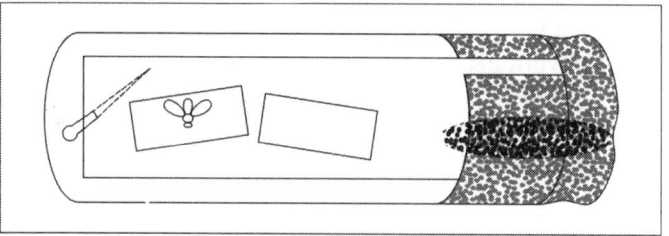

Fig. 76. Method of sending insects in glass vial.

viii. If there are more vials, these are then arranged in a row with enough packing material in between the vials and on the sides of the box to avoid the breakage. If there are more vials, a second row is similarly arranged with enough packing material. It is advisable to use plastic vials to minimise the hazards of breakage.

ix. Some insects like termites, aphids, dipterans, acarines, spiders and small crustaceans are preferably sent in small glass or plastic vials containing 70-80 per cent alcohol plus few drops of glycerine. The vials should be filled to the top and then tightly corked. These are then packed between layers of cotton in a box which is then packed in the outer container in the usual way. These vials can also be sent without corks but plugged with absorbent cotton and arranged in upright position in a strong jar containing 70 per cent alcohol or other preservative which is then tightly capped. The jar is then packed in a strong box in the usual way. Individual vials are safer in hollow wooden blocks or in cardboard mailing tubes, packed in absorbent cotton to prevent their shaking.

The small insects and other arthropods can also be sent in small polythene bags containing 70-80 per cent alcohol. The air is gently squeezed out of the bags which are then tightly tied with a knot around their necks to prevent leakage or drying.

x. The vials should not be placed along with pinned specimens. If this is absolutely necessary, the vials should be fixed with tape on the outside of the box.

xi. The box containing pinned insects must be fumigated.

The most effective fumigant is prepared by dissolving 10 gm of naphthalene powder in 50 mi of petrol to which is then added 0.1 ml of phenol to pevent growth of mould. A cotton swab soaked in little quantity of this fumigant is firmly fixed in one corner of the insect box.

Loose naphthalene crystals should not be kept in the insect box as they can damage the specimens on striking.

xii. Some insects like proturans, collemboians, diplurans, lice, etc., acarines and minute crustaceans are better studied in microscopic slides which can be packed in slide boxes (size according to need) available with various scientific firms. Enough packing material like cotton should be placed in between the slides and the lid of the box. The lid should be taped for more security. The slide box should then be packed in a strong outer container in the usual way.

xiii. The packing must be labelled as **fragile, handle with care** to minimise the risk of mishandling by the postal authorities.

xiv. When sending abroad by air or surface mail, the packet must also be labelled as **dried insects or animal for scientific study** on top of it. It is advisable to send valuable specimens in small packets by air mail, it should also bear another lable—**stamp gently.** By air mail the small packet will go with such similar packets and thus saved from bumps which would be more when sent by surface mail due to heavy weight parcels. To facilitate quick service from custom, the statement—**of no commercial value**—should be written on the top of the packet.

xv. Sometimes live insects are shipped from one country to another, This requires special handling and special permit from the concerned government. Eggs and pupae can be sent in tightly closed vials. It is advisable to send the pupae in cotton to avoid their damage in transit. Larvae should be sent in plastic bags containing enough food needed by them during transit. The loose end of the bag should be tied securely and all such similar bags are then packed in a box. Care should be taken in sending woodeating insects and their larvae. Larvae which do not feed during transit should be sent in damp sphagnum.

DESCRIPTION

Describing taxa is the principal means of making known the kinds of animals and their distinctive features. The description is thus one of the most essential steps in the taxonomic studies. Although both the non-taxonomist and taxonomist follow somewhat similar procedures in the identification, their main interests differ. The former does it to search the literature on the species of his own choice while the latter is concerned with either of the two things—First if the species identified by him is a known one, he will probably keep it for possible future study along with other specimens of the same species. Second, if he finds that the species identified by him has not been described (in being different from all published descriptions), he is sure that he has discovered a new species. He will proceed to describe it and place it in a group to which it appropriately belongs. His description will then serve as the basis to identify this new species for future workers.

The practice of describing taxa has been in use for over 200 years. The description of every taxon, especially genus and species, is of two types—general description and the diagnosis. The purpose of both these types of descriptions are different. The former includes a general picture of the described taxon and has a broader function. It also includes diagnostic as well as those characters by which it can be differentiated from yet to be discovered species. It may also include other characters like ecological, ethological, etc., which may be of interest to nontaxonomists. The *diagnosis* includes only those few characters by which the taxon in question can be easily separated from other similar or closely related taxa. The direct comparison of a species or other taxa with specifically mentioned species or other taxa is usually called *differential diagnosis*.

Since description forms the basis for the recognition of various taxa, it should be complete and **unambiguous**. Many species have been inadequately or badly described by Linnaeus and his followers. Some such examples of faulty descriptions are also seen in recent literature. This causes lots of problems in recognising the taxa. There is, thus, a great need for improving the descriptions of previously inadequately described species. Our main aim is not to name species but to know them. Thus, one *who provides genuine knowledge of species accomplishes far more than one who merely names them.*

SUBJECTS OF DESCRIPTION

Whatever ways the descriptions are obtained, the subjects are always the individual organisms. The descriptions may be based on one or several specimens; if the descriptions are based on more than one individual, they may have been obtained by a carefully planned and executed procedure so that they constitute a statistical sample of a population (i.e. theoretically they show the same distribution of traits as do the members of the entire population from which they were selected). Another more common alternative is that a group of specimens of supposedly the same species are obtained nearly in an erratic fashion, probably with several individuals taken from the same locality or with the different localities at different times and under different circumstances. A group of species thus obtained is called a series. There are many examples where species descriptions are either based on series and samples or on a single specimen, or on a series of a few individuals. For every species the description is based on the individual or group of type individuals. **Type** is that specimen on which the original description of the species is based. The locality from where the type is collected is called the **type locality.**

TAXONOMIC CHARACTERS

The entire zoological classification is based on the proper evaluation or weighing of taxonomic characters. It is defined as "any attribute of a member of a taxon by which it differs from a member of a different taxon". The Latin word *character* means a mark, a trait. Thus, the term better applied only to the attributes. When there are two or more attributes relating to the same basis of comparison and if they are ordered they will constitute a **morphocline.** If those attributes are not only ordered but also have their polarity established they will form a transformation series and thus will provide an evolutionary basis of comparison (Rodrigues, 1986). These characters may be positive or negative, including presence or absence of structures, qualitative or quantitative, structural, functional, behavioural or developmental. Still majority of the descriptions are based only on selected morphological characters. Although the modern trend is to include other sorts of characters like physiological, biochemical, serological, ecological, genetical and behavioural (dealt in Chapter 3), the morphological characters continue to be main focus of most descriptions. Moreover, we are still lacking knowledge in many of these new aspects.

In many instances not only sibling species are discovered but also closely related species are readily separated because of interspecific relationships as well as host-parasite/host-symbiont relationships. **Sibling species** have been found amongst termites on the basis of the presence of different staphylinid beetles in their colonies. The parasites are usually host specific. In some cases bird lice (Mallophaga), especially those present on bird's head are so specific that their accidental interspecific transfer is fatal to them. On the basis of the presence of two different merozoan parasites, *Dicymennea abelis* and *D. californica* on mollusc, *Octopus bimaculatus of* California, it was found to be a complex of two sibling species. Intracellular symbionts also offer useful taxonomic characters. Intestinal protozoan flagellates of termites have been proved useful in clearing confusion with regard to the identification of the species or even the classification of higher taxa of their hosts.

A taxonomic character is, thus, a feature which is present in all appropriate specimens at appropriate times. One should be very cautious in selecting the useful characters. The choice of characters is directly based on the accumulated experience of the taxonomist. He will select only those characters which are stable within the species or group, easily studied and quite distinctive in its easy separation from other closely related taxa. Thus a taxonomist has to search for good characters which are most effective in making the distinction and separation of the species or other taxa quick and easy. He has to keep in mind that these characters should not show wide variation among known specimens and intrinsic genetic variability. They are also not influenced readily by the environment but must be consistent and easily seen in the specimen. Sometimes variation in size is related to nutrition, colour to food preference, isolation or humidity. The patterning may also vary with age. The utility of such characters is clearly understood only with experience.

FUTURE OF DESCRIPTIVE TAXONOMY

It will not be justified to think that the job of descriptive taxonomists is close to being over, especially with the class insecta. Although about two-third of the total number of known animal species are insects, it is estimated that still about 8 to 10 million species of insects await discovery. It means that much less than half the species of the world are now known. Besides, in most cases the descriptions are quite inadequate or based only on adults. It becomes almost impossible to identify the taxa by running through such descriptions, or to identify the specimens of those stages or sexes on which there are no published descriptions. Thus, there is still great need for revisionary works or even the work at alpha level in insects and a few other groups. In some of the countries the fauna has not yet been satisfactorily discovered and to make the taxonomic studies complete and in line with the works done at other places there is a great need to discover and describe all such unknown forms.

TAXONOMIC PUBLICATIONS

No study in science is complete until it is made available to other scientists. The best way to make wide circulation of the work is through publication. It is very important for a taxonomist to know the urgency of type of the research most needed in his work, whether alpha, beta, or gamma levels of taxonomy. In taxonomy, the publications may be in the form of books, pamphlets, journal articles, symposium chapters, catalogues, check-lists, etc. A taxonomist should publish his work in appropriate journals so that the same is easily traced by fellow workers. He should not lower down the importance of new taxa by publishing in inappropriate works like those on population studies, biological studies, etc. Even if it is so, it should be mentioned in the title. Sometimes new taxa of one country are included in the faunal works of another country without any mention in the title. This is bound to remain out of knowledge of future workers. The major types of taxonomic publications are discussed as:

1. *Publications of new taxa:* These are ordinary descriptive papers of new subspecies, species, and genera. Such descriptions do not permit careful comparison of all related species. These descriptions may be justified when the names of taxa are required for biological or economic work They are also helpful in revisions of any work where the new species can be easily fitted in the classification. In addition, short papers concerning with new records of various taxa are also quite useful.

2. *Synopses and reviews:* These include brief summaries of current taxonomic knowledge of a group. These should not include new taxa. Such works actually bring together all scattered information at one place and thus very useful for future revisionary or monographic works. Examples of these types of works are given below:

   i. Sturtevant, A.H. and M.R. Wheeler. 1954. Synopses of Nearctic- Ephydridae (Dipt). *Trans. Am. Ent. Soc.,* 79 : 151-257.

   ii. Kapoor, V.C and Y.K. Malla. 1977. Dermaptera of Nepal and India (A Taxonomic Review). *J. Nt. Hist,* Nepal. 1 : 151-182.

   iii. Kapoor, V.C. 1970. Indian Tephritidae with their recorded hosts. *Oriental Ins.,* 4(2): 207-251.

3. *Taxonomic revisions:* These summarise and evaluate previous taxonomic works of a group incorporating new information. Some may be monographic in approach but not completely like monographs due to the inadequate material included in such works. Some are concerned with the arrangement of only new or previously established taxa. These may be done at species or generic level. The example of such work is given below:

   Foote, R.H. 1960. A revision of the genus *Trupanea* inAmerica North of Mexico (Diptera, Tephritidae). *Tech. Bull. 1214.* U.S.D.A., pp. 1-30.

4. *Monographs:* These are the most complete systematic publications involving full systematic treatment of all species, subspecies, and other taxonomic units and a thorough knowledge of the specialist on comparative anatomy of the group, the biology of the species and subspecies included, the immature stages in groups exhibiting metamorphosis, and detail distributional data (Mayr, 1969). It contains exhaustive bibliography as it assembles all the existing information on a group. The examples of such works are:

   i. Usinger, R.L 1966, Monograph of Cimicidae (Hemiptera- Heteroptera). Thomas Say Foundation Publications, 7:1- 585.

   ii. Hardy, D.E. 1973. The fruit flies (Tephritidae-Diptera) of Thailand and bordering countries. *Pac. Ins. Monog.,* 31 : 1-353.

   iii. Kapoor, V.C., J.S. Grewal and S.K.. Sharma. 1987. Indian Pipunculids (Diptera: Pipunculidae): A comprehensive monograph. Atlantic Publishers. New Delhi, pp. 204.

5. *Faunal studies:* These are detailed studies of the fauna of a single region. These are limited to a single group of animals and worked out only by experienced specialist. The faunas may cover all the animals of an area or of any major group. The examples are given below:

   i. Fauna of British India. Taylor and Francis, London. These run in many volumes in many groups of animats published since 1888 (now Fauna of India).

   ii. Bey-Bienko, G.Va. 1936, Insectes Dermapteres, *Faune de 1'* URSS, Moscow and Leningrad, 10: 1-212.

   iii. Faune de France, 1921-1967, Vote. 1-68. Office Centrale de Faunistique, Paris.

6. *Atlases:* These show comparative characters of animals in picture form. The purpose of atlases is purely taxonomic and includes semidiagrammatic drawings, full halftones or coloured plates. The examples of such works are:

    i.   Ferris, G.R. Atlas of the scale insects of North America. Stanford Univ. Press, Stanford Urtiv., Calif. 5 Vols. from 1937-1950.

    ii. Ross, E.S. and H.R., Roberts. 1943. Mosquito atlas. American Entomological Society, Philadelphia, 1 : 1-44, 2 : 1-44.

7. *Catalogue:* A catalogue is mainly an index to taxa arranged in such a way as to give a vivid picture of references for both zoological and nomenclatural purposes. According to Blackwelder (1967), a catalogue contains the following information:

    1. the original description reference,

    2. later references,

    3. synonyms with references,

    4. range,

    5. type locality' (also its repository),

    6. type of the genus,

    7-8. miscellaneous pertinent data (like biology, zoogeography, hosts, etc.).

The taxa are usually listed alphabetically. The merit of a catalogue depends upon the degree of its completeness and so its preparation is a very tedious task. A person must have great patience and thorough knowledge of the methods and sources of bibliography. The examples of a catalogue are:

    i.   Delfinado, M.D. and D.E. Hardy (Ed.). 1975 Catalogue of the Diptera of the Oriental Region. Vol. 1, Vol. II (Brachycera, Aschizan Cyclorrhapha); Vol. lll (Schizophoran Cyclorrhapha). The University of Hawaii Press, Honolulu, USA.

    ii. Metcalf, Z.P. 1964. Catalogue of the leafhoppers of the world. 16 vols (of various families of Homoptera).

8. *Checklist:* A checklist is an abbreviated synopsis and presents a list of names. These are more useful in

better known groups of animals, like birds, mammals, butterflies, etc. The examples of a checklist are:

i. McDunnough, J. 1938-1939. Checklist of the Lepidoptera of Canada and the United States of America. Part I. (Macrolepidoptera), pp. 1-275; Part ll (Microlepidoptera), pp. 1-171.

ii. Wood, A.M. 1989 (Rev. 1992). Insects of economic importance: A checklist of preferred names C.A.B. international, Wallingford, Oxon. OX10 8DE, UK. pp. 149.

9. *Field guide:* Some works are prepared by the taxonomists for the non-taxonomists to help them to identify the common animals in the field. These include clearcut and easily understood key characters. Among the field guides, well-known ones are: *Peterson Field Guide Series:* Putnam's *Nature Field Books* and Jaques Pictured Key Nature (Published by William C. Brown Company, London). Some field guides are specially designed in the form of pamphlets to the field workers for periodical check of the possible entry of new pests or for the identification of pests cropwise.

10. *Manual:* The manuals are also published in simple language including key characters for common species of animals. These are meant for students or layman who can easily identify the animals with the help of such works. Sometimes, these may be published as comprehensive works. The example is given below:

i. Pratt, H.S. 1951. A manual of the common invertebrate animals (exclusive of insects). McGraw-Hill Company, New York, 1-854 pp.

11. *Handbooks:* The term handbook is used either for field guides, manuals, or occasionally comprehensive volumes on a group of animals of relatively complete taxonomic treatment. The examples of such works are given below:

i. Wright, A H. and A.A. Wright. 1949. Handbook of frogs and toads. Cornell University Press (Comstock), Ithaca, N.Y., xii, 640 pages.

ii. Bishop, S. 1952. Handbook of salamanders. Comstock Publishing Associates, Cornell University Press, Ithaca, N.Y., 555 pages.

iii. Wright, A.H. and A.A. Wright 1957. Handbook of snakes. Comstock Publishing Associates, Cornell University Press, Ithaca, N.Y., (2 Vols.), 1105 pages.

12. *Treatise:* Handbooks are also sometimes named as *Treatise.* The examples of treatise are:

i. Grasse's *Traite de zoologie,* 1951, Vols. 1-17.

ii. Moore, R.C. (ed.). 1953. Treatise on invertebrate palaeontology. Geological Society of America and University of Kansas Press.

PREPARATION OF TAXONOMIC PUBLICATIONS

It is only in taxonomy that oldest works are considered valuable even today and universally consulted. The names and descriptions known since Linnaeus (1758) are still of same value as they were at that time. Thus a taxonomist must keep this in mind that the quality of his taxonomic publication should be such that it lives up to this timelessness.

Taxonomic publications have some basic requirements which must be followed by all the taxonomists. Some of these are discussed below:

1. *Description:* It is the main body of all published works. A description has a dual purpose. Firstly it serves to record and convey the data about the taxa described; and secondly it serves an essential basis for a new name.

2. *Keys:* It is also an important part of the publication. It denotes relationship amongst taxa with the help of important stable characteristics.

3. *Classification:* The latest and most commonly accepted form of classification should be followed.

4. *Synonymies:* This is also an important part of taxonomic publications this task is very tedious as it requires thorough screening of literature to find out the various names applied to each genus or species. In some cases

there are numerous such names and due to heavy cost on printing and paper, the editors of the journals do not welcome their inclusion except in monographic works. The chronological listing of such names gives a clearcut picture of the history of various names.

5. *Bibliography:* Every taxonomic article, small or large, always contains bibliographic references. The citation of these should be very selective than exhaustive due to the higher cost of printing. The designation "bibliography" means completeness of coverage of the subject and 'literature cited" indicates restriction of references, in both the cases arrangement is done alphabetically. Various journals have different ways of citing references. It is, therefore, advisable for the author to have a complete idea of the presentation of material in the paper of the journal in which he is to send it for publication,

6. *Nomenclatural parts:* The names should be written correctly. In zoological nomenclature all names, whether correct or misspelled ones, become permanent records.

   All such works written in any other language than English must contain a summary in English only. This is essential as the main requirement of a publication is that it should be read by more and more people.

7. *Illustrations:* The illustrations are keys to taxonomic descriptions in many cases. Taxonomic decriptions without illustrations become great liability on the taxonomists. It becomes extremely difficult to understand and compare the specimen in hand with the species description without illustrations. In most taxonomic works only diagrammatical line drawings are required and these give quite a substantial information of important characters of a taxon. It is advisable for a beginner to know "what and how to illustrate" by reading the following works:

   i. Blaker, A.A, 1965. Photography for scientific publication: A handbook. W.H. Freeman & Co., San Francisco, 158 pp.

ii. Cannon, H.C. 1936. A method of illustration for zoological papers. Association of British Zoologists, UK.

iii. Papp, C.S. 1963. An introduction to scientific illustration. Riverside, California, USA.

iv. Ridgway, J.L. 1938. Scientific illustrations. Stanford, California, Stanford University Press, USA.

v. Staniland, L.M. 1953. The principles of line illustrations. Burke Publ. Co., London, 224 pp.

vi. Zweifed, F.W. 1961. A handbook of biological illustrations. University of Chicago Press, Chicago, USA.

All taxonomic periodicals have their own style of presentation of material in a taxonomic paper right from the title to the bibliography. It is very necessary for a taxonomist to know the guidelines of that journal in which he desires to publish his taxonomic work. There are various general guides to the preparation of scientific papers. Few of them are given below:

i. Treloase, S.F. 1951. The scientific paper. How to prepare it. How to write. The Williams & Wilkins Company, Baltimore, USA, 163 pp.

ii. Style Manual for biological journals. Washington, D.C. American Institute of Biological Science, 1960.

iii. Woodsford, F. Peter (Ed.). 1968. Scientific writing for graduate students. The Rockefeller Univ, Press, N.Y., 192 pp.

In addition to these guides, it is also necessary for a taxonomist to acquire knowledge of proper use and spelling of technical terms through special dictionaries and glossaries like a few of the following ones:

i. Brown, R.W. 1954. Composition of Scientific Words. A manual of methods and a Lexicon of materials for the practice of Logotechnics. Washington, D.C., Pubd. by the author. Available only -with the Smithsonian Institution, Washington, D.C.

ii. Nybakken, O.E. 1959. Greek and Latin in Scientific

Terminology, Ames, Iowa, Iowa State College Press.

iii. Torre-Bueno, J.R. (3rd Print). A glossary of entomology. Brooklyn Entomological Society. Brooklyn, N.Y., pp. 1-329.

iv. Wood, R.S. 1966. An English classical dictionary for the use of taxonomists. Pomona College, xiv + 331 pp.

TAXONOMIC PAPER

Generally the body of the taxonomic paper would include the following:

i. Title

ii. Name of the Author(s) (Address may be given immediately below or as foot note or in the end, or according to the latest style of the journal, etc.).

iii. Abstract

iv. Introduction

v. Text (Main description including that of male, female and other variable paratypes if the number of specimens is large).

vi. *Remarks* or *Diagnosis* This is useful to define the taxa using most important characteristic and relationship with its close relatives.

vii. Acknowledgement

viii. References; Bibliography; or Literature cited, etc.

Most of the specialists in taxonomy know their fellow workers and immediately distribute the published works amongst them. They maintain a permanent mailing list for that purpose. The authors usually receive some number of copies of their published article as gratis for free distribution. The authors also order for more copies on concessional rate. These copies are often called as author's extras, separates or reprints Sometimes the term reprint is used to denote all the three. It is, therefore, necessary for us to know the differences in these three terms as explained by Blackwelaer (1967).

   i. *Author's extras:* These are pages removed unchanged from extra copies of the publication and so frequently contain parts of other papers. These may or may not contain the source and date of publication.

  ii. *Separates:* These are copies printed from the same type as the original minus the extraneous matter. The article shows the exact position of its beginning on the page. It is identical to the original in completeness, arrangement and pagination.

 iii. *Reprints:* These are the copies rearranged to fit most satisfactorily on a page and thus not printed at the same time as the original but from substantially the same type. Here the pages may be renumbered, the text rearranged with new setting of title, etc. Thus these are not the originals but the duplication of the original and are not of much advantage to taxonomists.

8

Reference Works in Taxonomy

There are various bibliographic works which are indispensable to both beginners and experienced taxonomists. The searching of literature is equally or rather more difficult than searching the material. It is quite useful to consult the "guide to the Literature of the zoological sciences" (7th ed.) by Smith and Painter (1967) for general zoological bibliographies and the work of Arnett (1970) for entomological literature only. Besides, electronic revolution during the last 15-20 years has made the world so small that no taxonomist/zoologist/scientist can remain aloof of the latest developments taking place in his sphere of knowledge. Some of the important bibliographies and literature-abstracting works are discussed below:

ZOOLOGICAL RECORD

The zoological record provides, in a series of annual volumes, a bibliography of zoological literature published throughout the world since 1864. This great reference work is indispensable for taxonomic work. Its publication was started in 1864 in England by a group of taxonomists like A.C.L.G. Gunther, A. Newton, E.C. Rye, W.S. Dallas and others under the name of 'The Record of Zoological Literature'. Up to 1869, it was a private venture and in 1870 the British Association of Advancement of Science supported this publication but under the new name 'The Zoological Record'. In 1871, an independent 'Zoological Record Association' continued its publication until 1886, when the Zoological Society of London took up the responsibility. This society continued its publication in cooperation with the

Natural History Museum, London and CAB International Institute of Entomology, London up to volume 114. On March 12, 1980 the Zoological Society signed an agreement with Biosciences Information Service, Philadelphia, Pennsylvania, USA for the publication of zoological record from volume 115 onwards. Its importance has been increased now due to the recommendation made in the fourth edition of the code that new taxa should be listed in the zoological record. This is the responsibility of the author who publishes any new name.

Each volume is divided into a number of sections corresponding to various groups of animal kingdom. During its history there have been various amalgamations and divisions of these sections as the concept of the various groups changed. Presently there are 27 separately issued sections and subsections of zoological record. The volume of literature indexed in the zoological record has grown considerably over the years since the first volume and now its volumes contain over 50,000 papers, originally published in over 6000 different journals, together with non-serial publications. It provides index to as much of the literature as possible though the works of popular nature are generally excluded. The publication of the record was on average five years late. The correct date of publication is taken as being December of the year following that to which the volume referred. The sections of the zoological record can be purchased singly (section-wise) or as an entire volume each year. These can be ordered from the Zoological Society, London, Regent's Park, London, N.W., 1. 4RY, England (except the Insecta section ordered from CAB International Institute of Entomology, 56, Queens Gate. London. S.W.7); the volumes prior to volume 103 i.e., 65-71 (1928-34), 85-102 (1864-1948) are available with Academic Press, 24-28, Oval Road, London NW 1 7DX, England. The volume 115 onwards can be ordered from Biosis, Philadelphia. USA-191003 which also publishes Biological Abstracts. It is now also available in CD Rom. Presently Zoological Record is published by Thompson Reuters, 3 Times Square in Manhattan, New York City, USA.

Of the 27 sections, 25 deal with different animal groups, one with general zoological literature and one lists the new generic and subgeneric names indexed in other sections. Within each section there are five indexes: the author index; the subject index; the geographical index; the palaeontological index; and systematic index.

Titles written in French, German, Italian, Latin, Portuguese or Spanish (incl. Catalan) are not translated. Titles in other languages which have been translated into English by the Zoological staff are enclosed in square brackets. More than a century of experience has made Zoological Record a respected resource of information from every field in animal biology, from biodiversity and environment to taxonomy and veterinary sciences. It has long acted as world's unofficial register of animal names. It covers 5000 serials plus many other sources of information including books, reports and meetings. Presently, 72,000 indexed records are added every year. The various sections and their titles are given below:

Section   Title

1. Comparative Zoology
2. Protozoa
3. Porifera and Archaeocyatha
4. Coelenterata and Ctenophora
5. Echinodermata
6A. Platyhelminthes and Nematoda together with Nemertinea, Mesozoa, Nematomorpha, Acanthocephala, Placozoa
6B. Annelida together with Rotifera, Chaetognatha, Echiura, Sipuncula, Gastrotricha, Kinorhyncha, Priapulida, Gnathostomulida, Pogonophora
6C. Conodonta and Fossil Miscellarea
7. Brachiopoda
8. Bryozoa (Polyzoa), Entoprocta
9. Mollusca
10. Crustacea
11. Trilobita
12. Arachnida together with Myriapoda, Tardigrada, Symphylida, Pauropoda, Onychophora, Arthropleurida, Pentastomida
13A. Insecta
Part A— General Insecta and smaller orders
13B. Insecta
Part B—Coleoptera
13C. Insecta
Part C—Diptera
13D. Insecta

Part D—Lepidoptera
13E. Insecta
Part E—Hymenoptera
13F. Insecta
Part F—Hemiptera
14. Protochordata
15. Pisces
16. Amphibia
17. Reptilia
18. Aves
19. Mammalia
20. List of new generic and subgeneric names

Today, **The Zoological Record** is produced by a team of 32 staff based in York. United Kingdom. Taxonomic indexing is carried out by graduate zoologists using highly sophisticated systems and data capture procedures developed in-house specifically for zoological record.

The best way to consult the zoological record is to start with the most recent volume and work back to the date of completion of the most recent catalogue or revision. If the exact title of the publication is to be seen, reference may be made to the bibliography of papers arranged alphabetically authorwise at the beginning of the section.

LITERATURE EARLIER TO ZOOLOGICAL RECORD

The most comprehensive review of the taxonomic literature before 1864 can be seen in the following:

   i. *Berichteuber wissenschaftlichen Leistungen.* It is published in different branches of zoology, including entomology and helminthology, in Archiv fur Naturgeschichte founded by A.F.A. Wiegmann in Berlin in 1835. Archiv fur Naturgeschichte is now published as Zeitschrift fur wissenschaftliche Zoologie.

  ii. *Bibliographic works of Englemann* (1846), Agassiz and Strickland (1848) and the catalogue of scientific papers published by the Royal Society (1800-1863).

 iii. *Index Animalia* (1758-1800), Sherborn, C.D. 1902, London, 1195 pp., 1801-1850, 1922-33, London, 28 pts,, 7056 pp.

These index Animalia give a complete list of generic and specific names proposed up to 1850. A gap of three years separates Sherborn's nomenclature from the first volume of the zoological record.

BIOLOGICAL ABSTRACTS

Since zoological record is always few years behind the schedule, there is always a need to consult other bibliographies for the most recent literature. Biological abstracts serve as an important source of recent literature. Its publication was first started by Biosciences information Service, 2100 Arch Street, Pennsylvania, USA in 1926 and since then it is regularly published. Presently it is published by Thompson Reuters through its subsidiary BIOSIS. It is published bimonthly and the taxonomic information is abstracted in 'Chordata, general and systematic zoology; and Invertebrata, general and systematic zoology parts'. In addition to the references it also contains a brief abstract of the paper and so is of great use in knowing the exact material of the paper which otherwise is not available immediately. Biological Abstract indexes over 350,000 new citations every year. It includes information from articles appeared in journals all round the world. Since January, 1970, the abstracts and citations of papers on mites and arachnids in Biological Abstracts and Bioresearch Index have been compiled into a separate publication, *Abstracts of Entomology*. One issue of it covers the matter of two issues of biological abstracts and one issue of bioresearch index.

The biological abstracts do not cover the taxonomic field completely and is no substitute for the zoological record. It is, therefore, necessary to consult other important abstracting journals like *Berichete uber die gesamte Biologie,* Abt, A; *Berichre uber die wissenschaftliche Biologie* (1925 to date); *Bulletin signaletique 16 Rem; Biologie et physiologice Animates* (1962) and *Gunther's* reviews of *systematic literature* 11956).

DISSERTATION ABSTRACTS

These are also useful and published monthly in the United States. This contains abstracts of all dissertations by contributing institutions in the United States arranged according to subject matter. If one needs more information, he may order a microfilm of an entire dissertation for a fee.

NOMENCLATOR ZOOLOGICUS

There are some works concerning the names of animals but the best and up-to-date work is of Neave. It is published by the Zoological Society of London. It includes a continuous record of the bibliographical origin of the name of every genus and subgenus in zoology, published since the tenth edition of Systems Naturae in 1758. It is indispensable for a zoological taxonomist. Volumes one to four of Neave's *Nomenclator Zoologicus* contain names published from 1758 to 1935: volume five (1936 to 1945), volume six (1946 to 1955), volume seven published in 1975 covering the period from 1956 to 1965, volume eight from 1966-1977 and volume nine published in 1996 for the period from 1978 to 1994 (Edwards, Hopwood, Tobias and Mauly, eds.) The volumes one to six can be obtained from the Academic Press inc. (London) white volumes seven to nine can be obtained from The Zoological Society, London. The names of genera and subgenera published every year are included in the Zoological Record. One compilation has been published including volumes 1-10 containing names of taxa from 1758 to 2004. There are an estimated 340,000 genera repesented in the text as well as approximately 3000 supplemental corrections.

THE CENTURY OF DICTIONARY

It is published by the Century Company, New York in six volumes (1889 to 1891). It is the most complete of all pronouncing guides for biological names. It is completely unknown to taxonomists and never listed as a source book. This dictionary fists thousands of zoological names, including not only names of common genera but also all the known names of more inclusive groups as well.

DIRECTORIES

These are quite useful in knowing the names of the taxonomists along with their specialisations. There are several such directories but the following are useful for consultation:

    i. *Directory of the Zoological Taxonomists of the World* by R.E. Blackwelder and R.M. Blackwelder, Carbondale, Illinois University Press, 1961.

ii. *Directory of the Zoological Taxonomists of India* by V.C. Kapoor, Kalyani Publishers, Ludhiana, India, 1974.

A list of some animal taxonomists, who are ready to identify particular animal group, can also be obtained from the net. Some do identification free of cost while others charge some fee.

ENTOMOLOGY ABSTRACTS

The publication of this abstract was first started in November, 1969 and is published monthly. Each volume contains some 8000 abstracts from 3000 primary journals. The abstracts are numbered consecutively. All the insects cited in the abstracts are classified up to families. It is also quite useful as it contains the important information of the research paper along with the complete bibliographic reference. Since the zoological record is always in arrears, it is useful in providing latest works in insect taxonomy which is separately sectioned. It is published by Information Retrieval Limited. 1 Falconberg Court, London, WlV 5FG, England.

HELMINTHOLOGY ABSTRACTS

It is a monthly journal compiled from CAB Abstracts database and produced by the Division of Animal Health and Medical Parasitology in Cooperation with the International Institute of Parasitology. It can be obtained from CAB International information Services. Wallingford, Oxon OX 10 8DE. United Kingdom, it was started in 1932 arid since then it is regularly published.

PROTOZOOLOGICAL ABSTRACTS

These are published monthly by Commonwealth Agricultural Bureaux, Farnham House, Farnham Royal, Slough, SL2 3BN, UK and first appeared In 1977. These also contain taxonomic and nomenclatural information.

REVIEW OF AGRICULTURAL ENTOMOLOGY

Review of Agricultural Entomology (Review of Applied Entomology—Ser. A—Agriculture till 1989) is an important monthly journal prepared from the Commonwealth Agricultural Bureaux (CAB) database produced by the Division of Crop Protection and Genetics in association with CAB International Institute of Entomology, London.

In addition to the applied aspects of insects, it is quite useful in knowing the latest works on the taxonomy of insects, especially those of agricultural importance. The abstracts contain summaries of the articles together with the complete reference and author's address. Occasional invited review articles in crop protection science with no rigid format are also published. It can be obtained from the same address mentioned for Helminthological Abstracts.

REVIEW OF MEDICAL AND VETERINARY ENTOMOLOGY

Similar to above one, it was earlier named as Review of Applied Entomology—Ser. B—Veterinary. It is also quite useful in knowing the latest works on the taxonomy of insects of medical and veterinary importance. It can also be obtained from the same address mentioned for Helminthological Abstracts.

AGRINDEX

It is a monthly publication produced by AGRIS, the International Information System for the Agricultural Sciences and Technology coordinated by the Food and Agriculture Organisation of the United Nations (FAO). AGRIS was started in January, 1975 and includes 127 national centres and 18 regional/international centres serving the needs of both developed and developing countries alike. All these centres provide bibliographic references of agricultural publications to its Coordinating Centres at FAO headquarters in Rome. An AGRIS Processing Unit is outposted at the International Atomic Energy Agency (IAEA) in Austria. Vienna, where processing and maintenance of the database are carried out under contractual arrangement. Taxonomic and nomenclatural information of animals of agricultural importance is given in sections on Plant Protection' and 'Animal Science Production and Protection.'

Agrindex is available in English, French and Spanish editions from FAO, via delie Tesme di Caracalla, 00100 — Rome, Italy or various AGRIS Centres.

BIBLIOGRAPHY OF AGRICULTURE WITH SUBJECT INDEX

It is published monthly by the Oryx Press, 3930 E, Camelback Road, Phoenix, AZ 85018, USA. The data for its publication is provided by National Agricultural Library, US Department of Agriculture. The magnetic tapes supplied by the library are computer processed, using several vocabulary control mechanisms. The resulting indexes provide access to the literature in considerably less time than would be required by traditional indexing techniques. It contains useful information on insects under section Taxonomic Entomology'. It also contains recent literature on insects. There are numerous other bibliographies dealing with zoology as a whole or with groups or subdivisions. It is advisable for a student of taxonomy to familiarise himself with the available bibliographic aids.

PUBLICATIONS OF INTERNATIONAL COMMISSION ON ZOOLOGICAL NOMENCLATURE

The International Commission on Zoological Nomenclature was set up on 18 September, 1895. It acts as an advisor and arbiter for zoological community by generating and disseminating information on the correct use of the scientific names of animals. Thus, it is dedicated to **"achieving stability and sense in the scientific naming** of animals"**. The Commission comprises 28 members from 20 countries, mainly practicing zoological taxonomists. The members are elected by **"Section of Zoological Nomenclature"** established by the International Union of Biological Sciences (IUBS). The regular term of a member is six years, reelection is possible various times (after 18 years, a three year absence is necessary before a person is reelected again) . The Commissioners cannot be over 75 years of age. If any member does not participate in the decisions and do not communicate for more than three years, the Commission has the right to terminate the membership. The work of the Commission is supported by a small secretariat based at Natural History Museum, London and funded by **"International Trust for Zoological Nomenclature"(ITZN),** a charitable organization. The Commission has the right to provide rulings on individual problems brought to its notice, as arbitration may be necessary in contentious cases, where strict adherence to the **"Code"** threatens stability in usage. These rulings are published in the **"Bulletin of Zoological Nomenclature".** It is responsible for some of the most important literature in zoological nomenclature and which a zoological taxonomist cannot afford to miss. These are mentioned below:

i. *International Code of Zoological Nomenclature:* It is a booklet edited by N.R. Stoll and ethers and published in 1964 as second edition. The third edition was published in 1985 and the fourth edition was published in 1999. It can be obtained from International Trust for Zoological Nomenclature housed in the Natural History Museum, Cromwell Road, London SW7 5BD, UK or American Association for Zoological Nomenclature, c/o USDA Systematic Entomology Laboratory, MRC-168, National Museum of Natural History, Washington D.C. 20508, USA. The cost is £19 or US$ 35.

ii. *Bulletin of Zoological Nomenclature:* The publication of long series of 'opinions' was started in 1907 and it continues even today. Up to 1996, 1859 opinions were issued, Francis Hemming edited a series of 27 volumes of opinions and declarations of the International Commission of Zoological Nomenclature. In 1943, it was thought to start a special bulletin, the *Bulletin of Zoological Nomenclature,* for the sole purpose of debating the nomenclatural problems. It is issued in smaller parts, "instructions to Authors" are published in each part of the Bulletin and it is necessary that these are followed otherwise the draft may be returned. The Bulletin is published four times a year for the International Commission on Zoological Nomenclature by the International Trust for Zoological Nomenclature, c/o Natural History Museum, London.

iii. *Lists and Indexes Containing the Rejected and Approved Names:*

a. The Official Lists and Indices of Names and Works in Zoology were published in 1987. It includes details of all names and works on which Commission has ruled since 1895 up to 1985, there are about 9900 entries.

b. In the five years (1986-1990), 946 names and five works were added to the Official Lists and Official Indices. It has been published as a supplement to the above one.

c. Between 2001 and 2005, a further 958 names and works have been placed in the Official Lists and Indexes, and 14 names already on the Lists and Indexes have been amended (see vols. 58-62 of the Bulletin of Zoological Nomenclature),

d. As per Art, 79.4 of the Code an available name must appear in the "List of Available Names in Zoology" published by International Commission of Zoological Nomenclature. In case of any controversy or confusion regarding usage of such names, the matter should be brought to the knowledge of the Commission (Art, 79.5).

iv. *Towards Stability in the Names of Animals:* To Commemorate its centenary (1895-1995), the Commission has published a history of the development of nomenclaure since the 18th Century in the form of *Towards Stability in Names of Animals—A History of the International Commission of Zoological Nomenclaure 1895-1995,* It contains 104 pages with 18 full page illustrations. It costs £30 or US$50 (including surface postage) from Commission's Office.

GUIDES TO JOURNALS

Since all zoological and entomological journals (except the specialised ones) contain articles on systematic zoology, it is very important to know them. The best guides to know such journals are—*World List of Scientific Periodicals*, 4th edition, Butterworth's, Washington 1963; and *Ulrich's Periodicals International Directory*, 16th edition, 1971-76 (a classification guide to current periodicals, foreign and domestic—a Bowker

Serial Bibliography) published by R.R. Bowker Company, 1180. Avenue of the Americas, New York 10036, USA.

A few important journals are mentioned below:

   i. *Systematic Zoology:* It is an extremely useful journal as it contains good articles on animal systematics. It was first started by the Society of Systematic Zoology in Washington in 1951 and since then it is regularly published. It usually contains papers dealing with cytological or chemical attributes of species and higher taxa, distribution patterns, behaviour, taxometrics, variability, endemism, extinction, serology, geographic distribution, speciation, rates of evolution climate rules, allometry, etc. Since 1998 its name has been changed to **Systematic Biology** and published by Taylor and Francis, London for the Society of Systematic Biologists, London.

   ii. *Publications of the Systematic Association, London:* Some quite useful publications, occasional symposia together with others has been a routine of this association since 1953.

   iii. *Taxon:* It also contains articles of general interest and was first published by International Association for Plant Taxonomy, Utrecht in 1951. It is useful for zoological taxonomists as well.

BOOKS

There are some quite useful books on various aspects of animal taxonomy. Some of them deal with general aspects while others include special theoretical aspects of animal taxonomy. Some of these are by Ridgway (1912), Ferris (1928), Thompson (1937), Park *et al.* (1939), Huxley (1940), Simpson (1945, 1961), Maerz and Paul (1950), Chamberlin (1952), Mayr *et al.* (1953), Keen and Muller (1956), Cain (1963), Sokal and Sneath (1963), Alston and Turner (1963), Gier (1965), Leone (1964), Solbrig (1966), Blackwelder (1967), Hawkes (1968), Papp (1968), Mayr (1969), Jardine and Sibson (1971), Wright (1974), Ross (1974), Jeffery (1977), Pankhurst (1978, 1991), Ferguson (1979), Nelson and Platnick (1981), Scott Ram (1990),' Panchen (1992), Minelli (1993), Quicke (1993) and Stevens (1994).

LATIN AND GREEK TERMINOLOGY IN TAXONOMY

The Latin and Greek words commonly used in naming various taxa of animals are not clearly known to many. The binominal nomenclature used for animals by Linnaeus is derived from Latin and Greek words. Although Latin is now largely unused, it can still be used in naming animal taxa. It is still quite useful to understand the source of scientific names. While the Latin names do not always correspond to present English common names, they are often related and if their meanings are understood, they are easy to remember.

LATIN ABBREVIATIONS

There are several different abbreviations and words of Latin language used in various taxonomic works. It is very necessary for a taxonomist to know them. The most commonly used ones are explained below:

aff., affinis—akin to; having affinity with.

al., alli, alorum—others, of others but not identical with. *apud or ap.*—with, in the work of; see in. *auct, auctorum*—of authors.

auct, non., auctorum non—not (of); used to indicate misapplied name.

ca., circa—about (with reference to dates). *cf., cfr., confer*—compare (with). *cit, citatus*—cited.

comb, nove., combination nova—new combination.

descr., descripto—description.

e.g., exampli grata—for example.

emend., emendatus—altered.

emen., emendavit—emended.

e.p., ex paste—in part.

et al., et alia—and others.

ex.—from, according to (used to connect the names of two persons, the first of which published a name proposed but not published).

excl. exclusus—excluded.

excl. gen., exclususo genere—with the genus excluded.

excl. spec, exclususo speceii—with the species excluded.

excl. spec., excl. specim., exclusis speciminibus—with the specimens excluded.

f., fig, figura—figure.

gen. nov., genus novum—new genus.

ib., ibid., ibidem—the same in the same place.

i.e., id est—that is in the same reference (book or article).

ined., ineditus (-a, -um)—unpublished.

in litt., in litteris—in correspondence.

l.c., loc. cit., loco citato—in the place cited, used to avoid repetition of a bibliographic reference already given (publication and page).

m., mihi—to me, after, used after a name to indicate the writer's responsibility for its publication.

n., nob., nobis—to us.

nec—not.

nom. ambig., nomen ambiguum—ambiguous name.

nom. conf., nomen confusus—confused name.

nom. cons., non. conserv, nomen conservadum—a name to be conserved.

nom. dub., nomen dubium—dubious name. *nom. non rite public, nomen non publicatum*—name not properly published.

nom., nov., nomen novum—new name.

n.n., nom. nud., nomen nudum—naked name.

nom. rejic, nomen rejiciendum—a name to be rejected; a rejected name.

nom. oblit, nomen oblitum—forgotten name.

op. cit, opera citato—in the work cited or article previously cited for this author; of a bibliographic reference already given (no page cited).

pp., pro parte—in part.

q.v., quod vide—which see.

sched, scheda—label (of a specimen).

sec, secundum—according to.

sens str., s.s., sense stricto—in the strict sense, in the narrow sense of a taxon, in the sense of the.

sens, lat, s.l., sensu lato—in the broad sense; i.e., of a taxon, including all its subordinate taxa and/or other taxa sometimes considered as distinct.

sp., specie—species singular.

specie indeterminata—undetermined species.

spp., specie—species plural.

ssp.—subspecies singular.

ssp. nov.—new subspecies.

sspp.—subspecies plural.

stat. nov., status novus—new status.

syn., synonymum—synonym.

typ. cons, typus conservandus—a type to be conserved.

viz., videlicet—namely

LATIN WORDS

ex nomine—by or under that name.

fide—on the authority of, or with reference to publication, to a cited published statement.

incertae sedis—of uncertain position in a list or scheme.

lopsum calami—a slip of the pen.

nomen dubium—a name not certainly applicable to any known taxon.

nomen inquirendum—a name of doubtful assignment.

nomen oblitum—a name that has not been used in the primary zoological literature for more than 50 years.

nomen triviale—the second part of the name of a species or the third part of the name of a subspecies.

seu.—either, or.

sic—*thus. teste*—according to. vere—the true. *vide*—see.

LINNAEAN SIGNS

♂ and ♀ *are* Linnaean signs for male and female, respectively used in taxonomic literature. A good account of their origin is given by Minst (1957).

TAXONOMY AND BIODIVERSITY ON THE WEB

It is the time of computers. Computers have been very popular to access any type of information through internet. When a vast amount of information is available, at this point of time, in other scientific fields through various websites, taxonomy and biodiversity are yet to come up at par to these fields. These disciplines are made for the web because of having quite rich information. It is also an opportune time for the taxonomists and conservationists not only to popularize their fields but also to get wide support to make available as much data as possible on the web. The websites serve as great via media to make available to the readers the research findings on these fields, essentially, free of cost the world over. Recently, Godfray (2002) has proposed that all taxonomic information can be made available for computers in the form of **'Unitary Taxonomy.'** Similarly, Sereno (2005) proposed the database as **'Taxon Search'**, allowing rapid recall of taxonomic and definitional information.

Internet taxonomy and biodiversity is becoming more and more popular. Presently, many websites are available concerning with various aspects of taxonomic and biodiversity information. These are quite useful in the broad sense when dealing with subject data or in narrow sense when dealing with individual scientific name or sometimes common name. It is not our aim here to enumerate all of them; but for the convenience of readers few are mentioned below:

i. **All Species**—See page 146.

ii. **BioNet International**—See page 19.

iii. **Biosis**—Systematic, Taxonomy and Nomenclature—Offers links to information on these subjects. Biosis information solutions for the global life sciences community are now provided by Thompson Scientific, USA. These databases are most complete. **BIOSIS PREVIEWS** is also provided by **Biosis;** this database includes citations from Biological Abstracts and Biological Abstracts/Reports, Reviews and Meeting (formerly Bioresearch Index), major publications of Biosis.

iv. **Description Language for Taxonomy (DELTA)**—A method for encoding taxonomic descriptions for computer processing that can be used to produce natural-language descriptions, conventional or interactive keys, cladistic or phenetic classifications and information-- retrieval systems.

v. **Expert Center for Taxonomic Identification (ETI)**—Develops and produces scientific and educational computer-aided information system. Also hosts the World Taxonomist Database link and World Biodiversity Database.

vi. **Global Biodiversity Information Facility (GBIF)**— See page 51—A way to access biodiversity data he'd in natural history museum collections, libraries and data banks around the world.

vii. **International Commission on Zoological Nomenclature (ICZN)**—Provides access to the system for ensuring that every animal has a unique and universally accepted scientific name.

viii. **Integrated Taxonomic Information System (HIS)**— See page 20.

ix. **National Center for Biotechnology Information (NCBI) Browser Taxonomy**—To retrieve the complete lineage of your species. This database does allow searches by common name and is very complete. If you do not find your species, you can try a variation on the name.

x. **National Science Foundation (NSF) of USA**— Directorate of Biological Sciences—Systematic and Population Biology.

xi. **Species 2000**—See page 146.

xii. **The Systematic Association**—The Systematic Association, Natural History Museum, London, is committed to furthering all aspects of systematic biology. Its publications are on systematics or systematics related to environment protection.

xiii. **Society for Systematic Biology**—It publishes the renowned journal 'Systematic Biology since 1998 both in print as well as online.

xiv. **Taxonomy Lab. "The Nuts and Bolts" of Taxonomy and Classification**—Illustrates creativity involved in taxonomy, ancestral traits, derived characters play in generating schemes-excellent and thorough overview of taxonomic classification.

xv. **Taxonomic Data Working Group (TDWG)**—Provides an international forum for biological data projects and includes new conference, information about standards and discussion forums.

xvi. **Tree of Life**—The Tree of Life web project (TOL) is a collaborative effort of biologists and nature enthusiasts from around the world. On more than 10,000 World Wide Web pages, the project provides information about biodiversity, the characteristics of different groups of organisms and their evolutionary history. Its goal is to contain a page with pictures, text and other information for every species and for each group of organisms, living or extinct.

xvii. **Wikispecies**—Wikimedia Foundation project that aims to catalogue all species presently known. It is an open and free directory of species covering groups like Animalia, Plantae, Fungi, Bacteria, Archaea, Protista and all other forms of life.

xviii. **World Diversity Databases**—A growing taxonomic database and information system that aims at documenting all presently known species. Taxonomic trees, descriptions, synonymies, literature citations and molecular database queries can also be accessed,

xix. **Zootaxa**—An online journal that publishes papers on animal taxonomy. It is a truly mega-journal making the difference for taxonomic-research. Accepted papers are published as soon as they are ready. It started in 2001 under the responsibility of National Science Foundation (NSF). USA. Both the printed and online editions are published on the same day.

xx. **Digital Taxonomy**—Recently. the word 'Digital Taxonomy' or Cyber Taxonomy has been in use on website. It is becoming quite useful for viewers. **Digital Taxonomy Home** page presents wide-ranging resource information for biodiversity data management in the **World Wide Web,** thus promoting the effective use of computers for handling biological software development projects. It provides a range of links on software, hardware, methodologies, standards, data sources and projects related to biodiversity data management, covering DELTA (Description-language for taxonomy), taxonomic databases, ecology, morphometries and phylogenetic analysis software tools, computer techniques and internet addresses of developers and distributers of free bioinformatics software. The pages are quite useful for exchange of useful tools for the developers of biodiversity software.

xxi. **Name Bank**—Patterson et al. (2008) suggested the use of this term to meet the purpose of **Taxonomic Indexing** where it is required to have a registry of all names in all forms for all organisms. Taxonomic Indexing is useful to accommodate special problems of using names of organisms to index biological material. It can link alternate names for the same entity and distinguishes between uses of the same name for different entities. The names are then placed within a indefinite number of hierarchical schemes.

Godfray (2002) proposed integration of all taxonomy within a single peer-reviewed web portal, to replace our current system of species descriptions scattered across hundreds of hard-to-find printed journals. International Code of Bacterial Nomenclature already has a unitary taxonomy. Here all valid names are now published in a single journal with a new starting date for nomenclature of 1980. Universal web taxonomy (Wilson, 2003) does not depend on changing taxonomy rules. It requires a stable well funded and user friendly information source available worldwide. The proposal made for bacterial nomenclature is not acceptable to both the two Codes (Zoology and Botany).

9

Zoological Nomenclature

Nomenclature (*nomen*—name; *calore*—to call) means allocation of names to the taxa. Naming animals is the first and foremost task of every taxonomist. The scientific names should be unambiguous, unique, universal and stable as they serve the only means of reference thereby avoiding the need of continuous usage of cumbersome descriptive phrases. An animal's name is the key to its literature. Proper naming of an animal is thus the fundamental principle of nomenclature. **It seems pertinent to mention here the Chinese Proverb that "The beginning of wisdom is to call things by their names".**

The vernacular names exist in all languages. Even the most primitive tribes in Africa, South America or New Guinea had vernacular names for commonly occurring local animals e.g., elk, robin, or slicker. The common and vernacular names are useful for everyday communication but are limited by location and language. Each culture and language system give their own names to the plants and animals occurring around them. This traditional age old practice of naming species may be called **"Folk Taxonomy"**. Folk taxonomy gradually evolved into a formal system for organizing the names of organisms and which is then called scientific **"Taxonomy"**. Since local people are apt to recognize a vast number of organisms, their cooperation is of immense help in organizing collection survey in understanding the biodiversity in both tropical and subtropical environments. The local names pose great problems not only in zoological nomenclature but also in biological nomenclature as a whole. Even within a single language the same name is used in different senses to denote different kinds of organisms, or the same kind of organism is called by more than one name. Thus, the naming system must conform to a set of rules, reducing confusion that may arise from using local names and providing a uniform language to ensure accurate communication throughout the world. The zoological nomenclature tries to avoid such defects through following of sets of rules called *Code of Nomenclature*. Thus, taxonomy and nomenclature are interlinked and a taxonomist must know the accepted rules and procedures that govern the application of names. The formation and use of scientific names of animals are governed by the International Code of Zoological Nomenclature (ICZN) (fourth edition, 1999).

ORIGIN OF THE CODE

Linnaeus was the first to mention a set of rules of nomenclature in *Critica Botanica* (1737) and *Philosophies Botanica* (1751). The binomial (now binominal) system for species discovered by Linnaeus in 1758 becomes the foundation in naming the present day modern names. The binary form of species names were called as "Trivial Names" for both plants and animals. These names were for fieldwork and education but not as replacement to earlier "Phrase Names". The phrase names included species description for separation from other species of the genus. The simplicity of providing trivial names by Linnaeus revolutionized nomenclature and his binary nomenclature replaced the phrased names. Linnaeus published the 10th Edition of **"Systema Naturae"** in 1758 and here he consistently used binomial (Binominal) nomenclature. Fabricius (1778) improved upon the Linnaeus' rules of nomenclature in *Philosophica Entomological* for naming insects. In 1801, Rudolphi, a parasitologist, suggested a set of rules which were so much confusing that no one followed them at that time. During the 19th century more and more new and local sets of rules originated in different countries. The binominal nomenclature was mainly adopted by naturalists but with no acceptable rules governing naming of animals and this created lot of confusion by the 19th century. To overcome this confusion the zoologists in various countries were concerned about the seriousness of this problem. The

British Association for the Advancement of Science thought that the zoologists must control the growth and application of animal nomenclature by a code of laws centrally agreed upon. In June, 1843 the Association adopted the 'Stricklandian Code" (also called British Association Code) which was framed by a committee under the chairmanship of H.E. Strickland. Its official title was "Series of Propositions for Rendering the Nomenclature of Zoology Uniform and Permanent".

This code was widely accepted and translated into French and Italian languages and approved by the Padua Scientific Congress in 1843. It was later amended and republished in 1863 and 1865. This code continued to be followed by American and British zoologists for a number of years and formed the basis for other codes approved subsequently.

The American Association for the Advancement of Science made another code, Dall's Code, under the chairmanship of W.H. Dall in 1877. Later more and more codes were added, like French Code in 1881 (Chapter, 1881), Douville Code (for naming fossils) in the same year, American Ornithologist's Union (A.O.U.) Code based on Stricklandian Code in 1885 (Anonymous, 1886) and the German Code in 1893 (Anonymous, 1893). By the end of 19th century it became apparent that more and more problems had been created and so it became essential to have an International Code for *International Zoology*. In view of this Professor Blanchard prepared a set of rules which were presented at the First International Congress of Zoology held in Paris in 1889 but accepted only in the Second Congress in Moscow in 1892. Later at the Third Congress in Leiden in 1895 Blanchard's Code was labelled as a French product and a commission was appointed to provide and maintain a uniform system of zoological nomenclature. In 1928, a British National Committee prepared rules for Entomological Nomenclature. Later, a number of discussions were held in subsequent congresses to revise and re-revise the rules and finally a draft was prepared and circulated to the International Commission of Zoological Nomenclature (formed in 1910) on 11 January, 1961. It was then approved by the Commission and formally accepted by voting as 'The International Code of Zoological Nomenclature' adopted by the XV International Congress of Zoology, London, in July, 1958, in English and French and published by International Trust for Zoological

Nomenclature, London in 1961. Even this Code was subjected to minor revision in view of the amendments made by the XVI International Congress of Zoology held in Washington in 1963. The revised Code was published in 1964 in which Article 11d was amended; Article 31 was reduced to recommendation; Articles 31a, 39a and 60a (i) were deleted.

Even this code did not satisfy a number of workers. The 17th International Congress of Zoology (Monaco, 1972) decided to amend the 1961 Code and also transferred the responsibility for future Codes to the International Union of Biological Sciences (IUBS). The revised Code prepared under the Presidentship of Dr. C.W. Sabrosky was finally approved by the Commission, with the authority of IUBS in later 1983 and published in 1985 (3rd edition) by the International Trust of Zoological Nomenclature, c/o Natural History Museum, London. Even this edition of the Code could not settle some of the issues which remained to be decided in the future course of time. Moreover evolutionary concept was yet to penetrate the core of nomenclature. The Code still reflected traditional Linnaean framework developed more than 100 years before the widespread acceptance of evolutionary ideas. It provided no rules for the recognition of taxa and categories. Yet the available rules significantly restricted taxonomic practice by tying taxon names to categorical ranks—thus interfering with the Code's own avowed goals of providing explicit, universal and stable names for taxonomy because a taxon must not be changed every time it is placed in a different category (de Queiroz and Gauthier, 1994; Christoffersen, 1995). The mandatory categories, the names of genera as parts of binominals, typification and synonymy tied to Linnaean categories and redundant taxonomic names (Christoffersen; 1987; de Queiroz and Gauthier, 1992) all conflict with evolutionary approaches. The code therefore still perpetuated essentialistic taxonomic traditions rather than promote conceptual evolutionary innovations. In 1988, the International Commission on Zoological Nomenclature set up an Editorial Committee for the preparation of fourth edition and also published an invitation to all Zoologists to submit further recommendations. The Commission held open meetings in 1988 (Canberra), 1990 (Maryland) and 1991 (Amsterdam) for preliminary discussion of proposed improvements to the Code. The Committee met in Hamburg in 1993 and reviewed each Article of the existing Code in the light of the above discussions and all the comments/ suggestions received from the Zoologists all over the globe. The Discussion Draft for the fourth edition was distributed to all the interested persons for comments/suggestions. In return almost 800 pages of comments from some 500 sources were received; a number of these were also published in the *Bulletin of Zoological Nomenclature*. The revised draft was accepted by the commission by postal vote in 1997 with minor amendment and was published in 1999. Its use is effective from January 1, 2000. After the amendments made to the Code due to finally acceptance of Electronic Publications as valid, preparation for 5th Edition of the Code became inevitable. There are now several issues confronting the zoologists all over the world which are to be taken care of in the 5th edition of the Code. Foremost among them are Electronic Publication, registration of names and typification of newly discovered rare and endangered animal species.

The International Code of Zoological Nomenclature intends ICZNwiki to be a tool for developing the 5th Edition, with goals like:

1. To make the process more transparent, by allowing broader participation of world zoologists;

2. To preserve a record of the deliberations/discussions leading to the adoption of final text;

3. To develop more extemporary and fully explanatory parts/sections, leading to easy and understandable Code.

It is proposed to bring our new (5th) Edition of the Code in the year 2018.

BIOCODE

The question of having a Code of bionomenclature **(BIOCODE)** across all organisms was felt to overcome issues of homonymy between kingdoms and continuing disruption and confusion caused by **'legacy names'** together with complicated nature of the current codes (i.e. **International Code of Zoological Nomenclature (ICZN); International Code of Nomenclature for Algae, Fungi and Plants (ICN), rules are also supplemented** from International Code of Nomenclature for Cultivated Plants (lCNCP); and International Code of Nomenclature of Bacteria (ICNB) that have *evolved over the time*. The International Union of Biological Sciences (IUBN) has taken it seriously and appointed a team leader on "BIOCODE" initiation for the purpose of modernizing the Nomenclatural Codes to meet the present and future aspiration of the scientific community. Even the revised "DRAFT BIOCODE" prepared in 2011 (Greuter et al., 2011), has not been able to arrive at some possible conclusion of fulfilling the purpose that *"biology requires a precise, coherent and simple system for naming of organisms used internationally"*.

This new 'Biocode is quite tough and very hard to grasp. The positive aspect of this draft code is that it is much shorter and less complicated than other Codes mentioned above. The main background in the initiation of having a "BIOCODE" was felt due to: i. common concern that numerous organisms are becoming extinct before they are described; ii. too few taxonomists and complicated procedures involved in naming new species.

Biocode may help avoiding further fragmentation of the scientific communities working with different organisms into producing a potentially ever increasing numbers of special Codes, or different names for the same taxon Moreover, modernization of the "Draft Biocode" to include electronic and molecular phylogenetic era will also help in providing a system of internationally accepted rules which will facilitate the more efficient working for all taxonomists. This will further help in accelerating the process of describing more and more Earth's Biodiversity which is the need of the present hour.

Still a lot is yet to be done by biological taxonomists internationally to arrive at some positive conclusion whether to have **Biocode** or to continue with the existing practice of working with the already available different Codes. But one thing seems to be sure that it is still not easy in the near future to arrive at some most acceptable biological code or **"Biocode"**.

PHYLOCODE

The rules of naming organisms are governed by different Codes mentioned above. Over the last twenty years, some systematic workers have claimed that these Codes do not fully help in achieving name stability from the point view of evolution, non-typological and biological paradigm. In view of this phylogenetecisis proposed another Code, named as **"Phylocode" or "International Code of Phylogenetic Nomenclature"** in 1998; later promoted by Cantino, 2004; Cantino and Queiroz, 2004; de Queiroz, 2006; Harlin, 2005; Sereno, 2005; and Laurin and Cantino, 2007. In this Code, the use of **Ranks** (like species, genus, etc.) are not required. The Rank-based Codes define taxa using a rank and in many cases a **'Type Specimen"**. in Phylocode the content of taxa using rank are delimited using a definition that is based on phylogeny (i.e. ancestry and descent) and uses specifiers (e.g. species, specimens, apomorphic, means organisms sharing the same ancestor) to indicate actual organisms. The defined taxon is thus an ancestor and all other its descenders. The phylocode will only allow the naming of clades' instead of genus, family etc, not of paraphyletic or polyphyletic groups; Latin name of a species shall no longer be made up by its traditional two parts (i.e. binominal nomenclature) but will consists of only one word. Thus, in Phylocode a scientific name for a group of species is to be defined in relation to its position on a phylogenetic tree. It thus reflects a philosophical shift from naming species and subsequently classifying them (i.e. into higher taxa) to naming species and clades. Only species and clades should have names and all ranks above species are excluded from nomenclature.

Phylocode is still a draft and controversial. Phylocode in its present form seems to be almost totally opposite to those Codes which advocate the use of binominal nomenclature which has stood the test of time since almost last three centuries. Although there are some weak points in these Codes but these are amended from time to time. To replace these Codes with Phylocode is not acceptable by most of the taxonomists although there is a small group of phylogenetecists who are strongly supporting phylocode. If one follows the Rules of Phylocode, not only the number of taxon names would dramatically increase (as they are rightly said as Geometric Constructs) (Benton, 2007) but even all valid names now in use would have to be redefined and registered, in this way the world would change under the phylocode and to follow its new regulation and new activities will be wastage of precious time affecting the discovery of new biodiversity and their conservation. Moreover, the phylogenetic relationships for many taxa are not yet clear and the use of this code will only create more confusion instead of clarification. In any case it would be useful if the taxonomists consider phylogeny in the descriptions of taxa in defining the monophyletic groups.

INTERNATIONAL CODE OF ZOOLOGICAL NOMENCLATURE

International Code of Zoological Nomenclature includes a set of rules in zoology with only one fundamental aim. i.e., providing the maximum universality and continuity in classifying all animals in accordance to taxonomic judgement. The Code is a guiding force in naming animals and allows zoologists some degree of freedom in naming and classifying new species. Any dispute pertaining to nomenclature is decided by applying the Code directly, not by referring any precedent.

This Code consists of three main parts—the Code proper. Appendices, and Glossary. The Code proper includes 'Preamble' followed by 90 consecutively numbered 'Articles' grouped in 18 chapters. Each article is composed of one or more mandatory provisions which are sometime accompanied by Recommendations or illustrative examples. The use of Recommendations is not mandatory but lays down the best procedure for cases not strictly covered by the application of rules. These are designated by the number of articles with which they are associated, followed by appropriate capital letters, e.g., Recommendations 10A, 72B, 73C, 74D, etc.

There are three Appendices. The first two have the status of recommendations and the third is the Constitution of the Commission. The terms used in the text are clearly defined in the Glossary.

Preamble

"The international Code of Zoological Nomenclature is the system of rules and recommendations adopted by the International Congress of Zoology; since 1973, by the Division of Zoology of the International Union of Biological Sciences, and since 1982, by the General Assembly of that Union. The objective of the Code is to promote stability and universality in the scientific names of animals, and to ensure that each name is unique and distinct. All its provisions and recommendations are subservient to those ends and none restricts the freedom of taxonomic thought or actions. Priority of publication is the basic principle of zoological nomenclature. Its application, however, under conditions specified in the Code, may be modified to conserve a long-accepted name in meaning. When stability of nomenclature is threatened in an individual case, the strict application of the Code may under specified conditions be suspended by the international Commission on Zoological Nomenclature. Precision and consistency in the use of terms are essential to a Code of nomenclature. The meanings given to terms used in this Code are given in the Glossary which is an integral part of the Code. The International Commission on Zoological Nomenclature is the author of this Code".

RULES OF NOMENCLATURE

It is not the sole purpose here to mention all the 90 Articles as these are already published in the form of a booklet. But since it is difficult to understand the technical part of these articles a brief but important account of the rules is discussed here. The Zoological Code recognizes three main types of names: Family-group names (Super- family, Family, sub-family and Tribe); Genus-group names (Genus and Subgenus); and Species-group Names (Species and Subspecies). The other ranks are not covered under the Code. Zoologists sometimes use few names above super-family.

Uni-, Bi-, and Tri-nominalism

The zoological nomenclature is independent of other systems of nomenclature and the name of an animal is not rejected merely on the ground that it is identical with the name of a taxon that is not animal (Art. 1.4). At the same time it is good to avoid the introduction of those generic names in zoology which are already in use in botany. It is, therefore, advisable to consult the *Index Nominum Genericorum (Plantorum)* and *Approved List of* Bacterial Names before establishing new genus-group names (Art. 1 A).

The scientific names of animals from subgenera and above are uninominal (Art. 4.1, 4.2). These are plural nouns (or adjectives used as nouns) for names above genus and singular nouns for genus and subgenus. They begin with an upper-case letter. The zoological code stipulates standardised endings (Art. 29.2) for the names of superfamily (-oidea), family (-idae). subfamily (-inae), tribe (-ini), and subtribe (-ina). The names of species are binominal (Art. 5.1) and those of subspecies trinominal (Art. 5.2). For exampie :

Genus	Subgenus	Species	Subspecies	Author
Bactrocera	*(Afrodacus)*	*aberrans*	*nigritus*	*(Hardy)* 1955

The first (genus) and the third (speciəs) words formed the binomen (Binominal nomenclature—correct spelling of binomial). The second word in parenthesis is the subgenus and the fourth indicates the subspecies which forms the trinomen (Trinominal nomenclature). The presence of subgenus does not affect the status of nomenclature. The subspecific name always begins with a lower-case letter. Hardy is the original author of this taxon published in 1955. Beginning of binominal nomenclature, consisting of a group name (generic name) and a specifying name (specific name) are also sometimes found not only among the pre-Linnaean authors but even among primitive people. But it was only Linnaeus who made it the basis of a consistent system of nomenclature, like binominal or trinominal nomenclature The precise significance of binominal nomenclature was not entirely clear to Linnaeus nor to most of his successors and even to many of the contemporaries. The names of binomen have actually opposite functions— the specific name expresses distinctness while the generic name relationship. In botany nomenclature, the equivalent to **'binominal nomenclature' is 'binary nomenclature'** or **'binomial nomenclature.'** The fourth edition now rules that a new name published after 1999 is made available only if it is clearly indicated as being new (preferably be the term as "sp. nov"., "gen. nov"., "fam. nov"., "nom. nov". or by the directly equivalent term in the language in which the scientific paper is written). The published new taxa should be brought to the notice of Zoological Record for their inclusion in it (Art. 8.5; 8A).

The zoological Code does not recognise the names of the taxa of infrasubspecific rank i.e., below subspecies. A scientific name proposed expressly as the name of a "variety" or "form" after 1960 is infrasubspecific and excluded from zoological nomenclature (Articles 1.1.1; 1.3.4; 45.6.3) but if before 1961, these are regarded as subspecific names (Article 45.6.4).

The names of subtribes to superfamilies are *family-group names;* genera and subgenera are *genus-group names;* and species and subspecies are *species-group names.* The basic language of the scientific names is Latin and names from other languages are Latinised or treated as such. The principal area of conflict concerns gender concord, i.e., the requirement of agreement in gender between an adjective specific name (epithet) and the generic name with which it is at any time combined. In the recent years, gender concord has been challenged from various quarters and it is presently one of the serious questions in the future of zoological nomenclature. Even the fourth edition could not solve this problem in species-group names. However, some changes have been made in the fourth edition for the simplification of the identification of gender in genus-group names and the formation of stems for family-group names. It is now believed that this will satisfy those having little or no knowledge of Latin. A name proposed for a **collective group** will be of genus-group name (Art. 42.2.1) and a name proposed for **ichnotaxon** can be of family-group name or genus-group name or species-group name as per its usage (Art. 10.3).

The use of author's name is optional (Art. 51) but it is desirable to at least mention the same once in each work (Rec. 51A). When 3 or more joint authors are responsible for a name, the term **"et al"**. should be mentioned after the name of first author provided all the names are once mentioned somewhere in the same work (Rec. 51C). If the new name is established anonymously, the term *"Anon"*. may be used in place of name of the author (Rec. 51D). If name of the taxon is established by other person than the author of the work or by one in publication of joint authors, author is mentioned as B in A or B in A & B (Rec. 51E).

Uniparentally reproducing animals are usually given conventional species names (Art. 17). They are quietly listed in catalogues, or keyed out in monographs just like other named species. This uniform taxonomic treatment becomes extremely dangerous in view of the fact that such 'species' (in some groups at least) are sometimes called **Agamospecies'** or **Microspecies,** rather than species (Minelli, 1993). Some other workers (Dobzhansky, 1937; Mayr, 1969; Hull, 1980; Ghiselin, 1987) do not treat them as **'true species.'** In that case how can these be named as 'true species' (Minelli, 1995).

Another problem is with those animals which do not obviously belong to species, e.g., *hybrids* (Art. 1.3.3). Until recently, natural hybrids were regarded as peculiarity of the plant world, their very rare occurrence among animals being so exceptional as to be better ignored from the view point of nomenclature (Minelli, 1995). However, the biochemical and molecular studies have brought about revolutionary change in our concept with regard to natural animal hybrids. There are presently many naturally occurring hybrid forms which at least are as stable and well circumscribed as many conventional species. But these are still denoted by formulae (where a generic name is followed by an accession number or a locality name) rather than by Linnaean names from the point of view of nomenclature. But still there is no universality of attitude towards them (Minelli, 1995). Echelle (1990a, b) for instance believes that **non-Mendelian species'** of hybridogenetic fishes and reptiles should be treated similar to usual **'Mendelian species.'**

Name Changes and Instability in Nomenclature

One of the main drawbacks with binominal system is its instability. Genera are split or lumped. The species are frequently shifted from one genus to the other. Such changes not only annoy the persons who are affected by them but also reduce the efficiency of zoological nomenclature as a reference system. Though the zoological code has got provisions to govern such changes yet this continues to be another important problem.

The recent code also rules that no name can be replaced if it has been used as valid by at least 10 authors in 25 publications during the past 50 years and encompassing a span of not less than ten years, by an earlier **synonym** or **homonym** which has not been used as valid since 1899. Even grammatically incorrect but established names have been protected by the Commission. The Code also gives freedom to a taxonomist either to continue to use the original type species for genus-group taxon or taxonomic species actually involved as type in cases of misidentifications without referring to the Commission to avoid delay. The Commission has also safe-guarded names by adopting lists of names in major taxonomic fields. Names not typed in such adopted lists (Art. 10.7) will be treated as unavailable names.

Sometimes a name is spelled in more than one way in the same work. Under such cases the correct spelling is that which is first selected and used by the **First Revisor** or original author (Art. 24.2). It is also necessary to get printed all components of a scientific name in full on first mention and later on it is abbreviated. In case abbreviated word creates confusion, it may be used in a little different manner e.g., mosquitoes like *Anopheles,* use *An.* and for *Aedes,* use *Ae.* Similarly, for other such names different abbreviations may be erected (Rec. 25A).

Michener (1964) favoured replacement of binominalism by uninominalism or mononominalism. He proposed the freezing of the original scientific names of a new species for all times by connecting generic and specific names with a hyphen. This proposal has superficial advantage in terms of avoiding the problems of homonymy, change of generic and specific combinations and change of specific endings to agree with the gender of generic name. The fourth edition of the Code has simplified the identification of gender in genus-group names and the formation of stems from family-group names. But it creates confusion in cases where the species have been transferred from one genus to the other, more so belonging to altogether different family. For example, the bee species *jenseni* was earlier published under the genus *Nomia* in the family Halictidae but presently belongs to the genus *Leioproctus* of the family Colletide. According to Michener's proposal this species shall be continued to be called as *Nomia jenseni,* thereby creating wider confusion than proposing any improvement in its nomenclature. On the other hand, the binominal nomenclature shows classification of the taxon which is one of the most important steps in the whole hierarchy. It clearly shows that one taxon, the species, is a member of a next higher taxon, a genus. When the classification of a group is well advanced and the genera are truly natural units, the binominal can be the basis of prediction and of phylogenetic, evolutionary and zoogeographical speculation. The mononominal would destroy this. Thus, it would not be good to discard the well established and time tested binominal system for an untested system that might profit us in certain mechanical ways but would lose for us in fundamental ways.

Use of Parentheses

The author's name is put in parentheses when the species is transferred from one genus to the other retaining its original author and date (Art. 51.3). It is called *New combination,* for example, *Hemilea bipars* (Walker, 1862) Hardy, 1959. This means that Walker described the fruit fly species *bipars* originally in some other genus but Hardy in 1959 transferred it to its correct genus *Hemilea.* This example gives the names of both the original author as well as the one who transferred it although the use of this type of *double-citation* is purely optional under the Zoological Code. *Sophira bipars* Walker, the original name of the species becomes the **basionym** of *Hemilea bipars.* Similarly, the date in full or a part of it is included in parentheses if it is not specified, but demonstrated by evidence from the work itself.

Thus, **"new combination"** is defined as the first post-original combination of any generic name with a previously published species-group name. It excludes rank changes of species-group names within any nominal genus to which they have formerly been referred. Smith and Perez-Higarda (1986) suggested the term **"onymorph"** (Greek, *onyma,* name and morphe, form) for every unique name association for species and subspecies. Thus, any new association of species-group names with each other or with any nominal genus is properly regarded as a **"new onymorph"** (name-form) while only a new association of any given species-group name with a nominal genus is regarded a "new combination".

Use of Square Bracket in Taxonomic Works

The use of the citation of author's name or the date in full (or part of it) in square brackets indicates that the name has been taken from indirect source (or other than the original work). The use of the square brackets denotes the original anonymity, in citing synonyms, square brackets are also used to include statement of misidentification.

Use of Punctuation Marks

Comma is always used in between the author's name and the year (Art. 22A.2.1), for example *Cranopygia bhallai* Kapoor, 1966. There should not be any punctuation mark in between the name of the species and the name of the author. The above example indicates that Kapoor is the original author of this species. When other authors in other literature make use of such already published names and if it is necessary to refer such subsequent users of the name, they should be cited in a way that should not give the impression that they are authors of junior names (i.e., junior homonyms or repetition of already published ones). The Zoological Code provides provision for denoting such subsequent usages. They should be written after the original citation but be separated from the latter by a colon or semi colon but never by a comma or full stop, e.g., *Phagocarpus immsi* Bezzi, 1913 and *Phagocarpus immsi:* or; Munro, 1935. This shows that Munro in 1935 made use of the name validly published for the first time by Bezzi in 1913.

No diacritic mark, apostrophe, diaeresis, hyphen or ligature of the letter *a* and *e (ae)* or *o* and *e (oe)* should be used in a zoological name. A name (whether old or new) with any of such marks should be corrected by the deletion of the mark concerned and any resulting parts are to be united excepting one specified use of the hyphen and except that, when, in a German word, the umlaut sign is deleted from a vowel, the letter 'e' is to be inserted after that vowel. For example, the name *Terrae-novae* is to be corrected to *terraenovae*, *d'urvillei* to *durvillei* and *n'unezi* to *nunezi*; but *mulleri* becomes *muelleri* and is not a homonym of *mulleri* (Art. 32.5.2).

If the first element of a compound species-group name is a Latin letter to denote a character of the taxon, it is connected to the remainder of the name by a hyphen, e.g., *c-album* (Art. 32.5.2.4.3).

Numericals in Compound Names

The use of numericals is also forbidden by the Code and if a number of numerical adjective or adverb forming a part of a compound name is discovered, it is to be written in full as a word and united with the remainder of the name, e.g., *decimlineata* not *10-lineata* (Art. 32.5.2.7).

Use of in, ex, sensu, stricto, sensu lato, non or nec, proparte and partim in taxonomic works

The Latin preposition '*in*' is used to connect the names of two persons, the second of which was the editor (or the main author) of a work in which the first author was actually responsible for validity (making available a name), e.g., *Dacus indicus* Agarwal in Kapoor denotes that Agarwal was responsible for the publication of the name *D. indicus* in a work edited by Kapoor or otherwise written.

The Latin preposition '*ex*' means 'from' or 'according to.' It is used to connect the names of two persons, the first author of which validly published a name which was actually proposed (but not published) by the second author. For example, *Forcipula indica* Brindle *ex* Burr means that Brindle was responsible for the valid publication of the name, *F. indica*, originally proposed by Burr which he himself never validly published. There is no necessity to denote the preposition '*in*' or '*ex*' but their citation helps others to know any extra detail required by the readers.

Sometimes different names are used in more than one sense, i.e., different names have been applied by different authors to taxonomically different, although superficially similiar, animals. This problem arises mainly due to originally 'published very brief and vague descriptions which do not allow correct typification (identification)'. Under these circumstances one must denote actual original taxa by the Latin word '*sensu stricto*' *(s.s., sens, str.)*. It denotes the use of the taxon in the strict sense or in the narrow sense, i.e., of a taxon, in the sense of the type of its name; or in the sense of its circumscription by its original describer; or in the sense of its nominal subordinate taxon when two or more subordinate taxa are there in a taxon; or with the exclusion of similar taxa sometimes united with it. For example, subgenus *Hillara s. str.* (Empididae, Diptera) means in the strict sense of the subgenus erected by its original author Meigen 1822. When taxa which include all its subordinate taxa and/or other taxa, sometimes considered as distinct, the Latin word '*sensu lato*' *(s.l., sens, lat.)* meaning in broad sense, is used. For example, *Empis s. lat.* (Empidae, Diptera) means that doubtful subgenera have been assigned to some species of the genus *Empis* in literature which need to be broken into correct subgenera on world basis. Here the subgenus shows to include all such heterogeneous groupings.

The Latin words *'non'* or *'nec'* means not. They are used to denote the actual authority in nomenclature. For example, *Forcipula indica* Brindle for *Forcipula pugnax* Burr (nec Kirby, 1891) means that Brindle described *indica* as new species for *pugnax* Burr (but not of Kirby which remains a valid species). Such citations also arise due to misidentification.

Another situation of misidentification is denoted by the Latin word *'partim'* or *'proparte' (p.p.)* which means that only a part of a taxon as circumscribed by previous author is being referred to by the writer. For example, *F. indica* new species Brindle for *F. pugnax* Burr (nec Kirby, 1891), *proparte* or in part, means that, part of the *pugnax* species of Burr (i.e., some specimens only) belongs to the new species *indica* erected by Brindle.

The authority of a taxon does not change if diagnostic characters or circumscription are altered. In case the author for bringing about such alteration in the taxon is to be mentioned; he should be mentioned after the original author in the following manner with the use of the Latin word *partim:*

Taenia solium Linnaeus, partim Goeze, means that Linnaeus is the original author while Goeze made alteration in its circumscription.

Priority

This is a very controversial part of the zoological nomenclature *but a basic law of International Code to promote stability.* Whenever two names belonging to the same taxon are discovered, the problem of the validity of one is decided by the *Law of Priority.* It means the valid name of a taxon is the oldest (taken from the date first published) available name with certain exceptions (name conserved by the commission), **nomen oblitum** (forgotten name) and limitations associated with the change of rank. The authority of a name in the family, genus, or species-group names is not changed on its elevation or reduction in rank within the group. A species-group name established for an animal but later found to be a hybrid (Article 17.2) is not a valid name even if it has priority over all other available names, but it may enter into homonymy. The only drawback in Zoological Code is that due to this principle even the names, poorly and incompletely described originally, become valid in being oldest while in other sciences such works of incompetency are merely ignored.

The most significant operational change made in the fourth edition of the Code is the introduction of automatic courses of action in cases which previously were required to be referred to the Commission. This automatic departure from the Principle of Priority in some cases is of significance where existing usage of names or spellings is threatened by the revival of unused names prepared before 1900 or in cases of misidentification of type species. A zoologist is free to fix as the type species either actually nominated species of original author or nominal species in conformity with name in use. Actually such cases comprise the most common ones in zoological nomenclature.

Starting-point Date

The starting-point date for zoological nomenclature is January 1, 1758 and the tenth edition of Linnaeus' *Systema Naturae* has been regarded to have been published on that date (Art. 3). Any work published in 1758 or before is treated to have been published after that date, except spider nomenclature which is considered to have started in 1757 [Clerck, C. 1757. Aranei suecici (Svenska Spindier, Stockholmiae)] (Ars. 3.1, 3 2); in botany the starting point will often be 1753 while in bacteriology it is quite new i.e.. 1980, but maintaining original authors and dates of publication. Today, as per Codes of Nomenclature for Zoology and Botany, every animal or plant name published before 1758 or 1753. respectively, is called **prelinnaean'** and thus not valid (even the early names published by Linnaean himself are **"prelinnaean"**).

Use of suffixes 'i', 'orum', 'ae', and 'arum' (Art. 31.1.2)

These suffixes are used for species-group names formed from modern personal names. A species-group name, if a noun formed from a modern personal name, should usually end in-

(i) 'i' if the personal name is that of a man, e.g., *smithi* from Smith, *bhallai* from Bhalla; if the name is based on a Latin or Latinised name of a modern name, the nominative singular ending —"us" should be replaced by the genitive singular ending 'i', e.g., *fabricii* from Fabricius, *aurivillii* from Aurivillius; if a name is based on a compound personal name, preference should be given to the better known component, e.g., *taken* from Bethune Baker; *guerini* from Guerin-Meneville; Greek and Latin declension should be followed in basing a name on the fore name of a modern person, e.g., *caroli* from Charles; if based on personal names bearing prefixes like 'Mac,' 'Mc,' it should be spelled as 'mac' and united as in *maccooki* from Mc' Cook; *macooyi* from 'M' Coy; if the prefix is 'O', it is united as *obrieni* from O' Brien; if prefixes consist of an article like le, la, l', les, el, il, lo, or containing an article like du, de, da, des, del, della, it should be united as *leclerci* from LeClerc, *logatoi* from Lo Gato, *dubuyssoni* from Du Buysson, *lafarinai* from La Farina; if prefix, abbreviated or not consisting of a nobiliary particle or indicating Christian sainthood, it should be omitted, as *chellisi* from De Chellis, *remyi* from St. Remy; a German or Dutch prefix, if united with a name, it should be included as in *vonhauseni* from Vonhausen, *vanderhoecki* from Vanderhoeck or if not united, should be omitted, as in *iheringi* from von Ihering, *wulpi* from van der Wulp; (ii) "orum" if men or man (men) and woman (women) together, e.g., *smithorum* from Smith; (iii) "ae" if of a woman, e.g., *guptae* from Gupta; *iosephineae* or *josephinae* from Josephine; (iv) 'arum' if of women e.g., *smitharum* from Smith.

Use of suffix 'ensis' or 'iensis' in Taxonomic Works

These are important suffixes used for species-group names based on a geographical name, e.g., *ludhianensis* from Ludhiana; *bangalorensis* from Bangalore; and *siciliensis* from Sicily.

Kinds of Names

A name which is published according to the requirements made in Articles 8 to 12 of the Code is the **"available name"** or **"legitimate name"** or **"valid name"**. A name which is just mentioned in any text or written on a label of an animal and does not contain description or does not satisfy the conditions laid down by articles 8-12, it is called a ***nomen nudum*** or naked name and has no status in nomenclature. A work (of new taxa) not printed on paper issued after 1999 in numerous identical, durable and unalterable copies may be treated as published if it clearly mentions that copies of the same are deposited in at least five major publicly accessible libraries mentioned in the work itself (Art. 8.6). Some kinds of materials are treated unpublished as per Art. 9, e.g., (a) abstract of papers, posters, etc., issued to participants at congresses, symposia and other meetings but not otherwise published; and (b) offprints or separates distributed after 1999 in advance of the date of publication specified in the work from which offprint is taken out. There are other names sometimes used in nomenclatural works, e.g.,

Nomen dubium: A name of uncertain application either due to non-availability of the original type and sufficient information, or because it is not possible to find out the taxon to which its type should be referred to.

Nomen oblitum: Latin, meaning forgotten names applied to a name after January 1, 2000 but unused since 1899.

Nomen novum or replacement name: A new name; i.e., a name expressly proposed and published to replace an earlier name that cannot be used for some reason, in case it is a junior homonym. Sometimes it is also called 'Substitute name.'

Nomen rejicundum: Names which are listed as 'officially rejected' are to be permanently rejected.

Nomen conservadum: A name the use of which has been officially sanctioned in spite of its contravention of one or more provisions of a Code. The procedure of giving sanction is called Conservation.

Nomen triviale: An expression used by Linnaeus and others for the specific name; applied by some authors in the same sense as 'vernacular name.'

Nomen hybridum: Name of hybrid, hybrids are normally individuals, not populations. Thus, such names have no status in nomenclature.

Vernacular name: The name of a taxon in any language other than the language of zoological nomenclature.

Preoccupied name: One that is a junior homonym; a name already in use for another taxon with a different type.

Pre-Linnaean name: Names which are published prior to January 1, 1758. These, too, have no status in nomenclature.

Incertae cedis: Names of a taxon of uncertain taxonomic position.

Species inquirenda: A doubtfully identified species needing further investigation.

Tautonyms

A *tautonym* is a name of a species or subspecies in which the second or even the third component of the name repeats the generic name, e.g., *Apus apus apus*. The use of such names is permitted by the zoological code (Art. 18).

Synonymy

Two or more names belonging to the same taxon are called **synonyms**. When there are a number of synonyms in a name the selection of a proper name applicable to the taxon is done through Law of Priority, i.e., the oldest one is taken as the proper name. It is called the **senior synonym** while the rest are the **Junior synonyms** (Art. 10.6, 11.6) and are placed as the *synonyms* below the accepted name of the taxon. Rate of synonymy differs from group to group. In some groups the observed rate of synonymy (i.e., the ratio of the total number of synonyms to the total number of names) can exceed 50% (Gaston and Mound, 1993). Solow *et al.* (1995) have given an excellent account to estimate the rate of synonymy through statistical model.

The term synonym is also used in reference to all *occurrences* of any name or set of names (usually synonyms) in the literature, or in given segments, for which synonyms do not provide proper sense. To overcome this Smith and Smith (1972) suggested another term, the **Chresonymy** (from Greek Chresis = use) for name-usage for secondary meaning of synonymy, namely a summary of occurrences or usages of any given scientific name or set of names.

The synonyms pose great problems for the taxonomists. These are created due to the lack of knowledge of the existing literature or not understanding the amount of variation which a species may possess. A careful study of the phenomena of individual variation in general and more particularly in a group with which the taxonomist is concerned is an indispensable prerequisite of all taxonomic works. It is estimated that more than half of all synonyms are due to underestimation of individual variation. Due to this, simple variants or races of one species are given independent specific status, for example 251 species of the genus *Anodonta*, a fresh water mussel, from France alone were later found the habitat variants of only two valid species of this genus while the rest were synonymised with them. To some extent such long lists of synonyms are reflections of increased knowledge, as well as indications of the primitive views of early workers and the weaknesses of taxonomists in general. These long lists of synonyms do sometimes become the point of criticism when these should not be of any concern to the general biologist. He may as well take the last name given, without quibbling. If it is a must for him to search the older literature for some points, then he may use the synonymic list as a guide. In other words lists of synonyms are none of the general biologist's business and concern matters he does not usually understand, nor which would yield him any biological information.

Thus, synonymy is of great significance as it provides considerable amount of information available in the literature under one or more of these names. The presentation of a synonymy is a key to information contained in the taxon. It is, therefore, very necessary that the synonymies should be correctly established and only by the specialists who have sufficient background of the concerned group.

The synonymy is of two types—*subjective synonymy* and *objective synonymy.* The *subjective synonyms*, also called *taxonomic synonyms* or *heterotypic synonyms* are those which are based upon different types, and remain synonyms only as long as their respective types are considered to belong to the same taxon. Thus, this type of synonymy is not absolute (Art. 61.3.1). It may be indicated by the mathematical sign of equality (=). The *objective synonyms*, also called *nomenclatural synonyms, obligate synonyms,* or *homotypic synonyms,* are those which are based upon the same type. Such synonymies are always absolute. It may be indicated by the use of the mathematical sign of congruence (≡) (Art. 61.3.2-4). Such a type exists when the names have the species as type; one name is proposed as a replacement for junior homonym, thereby automatically having the same type as the other, whenever the type is fixed for either of them; and one name is an intentional respelling (amendation) of another (erroneous and illegal replacement). An example is given under secondary homonym below.

Homonymy

The names which are spelt in an identical manner (identity in spelling) or in a manner so similar as to be considered identical under the provisions of the Code but based on different types are called *homonyms* (Arts. 52-60). The Zoological Code rules that if two or more *homonyms are* found, only the oldest is used *(senior homonym)* (Art. 52.2) while the rest *(junior homonyms) are* excluded from use (Arts. 23.3.5, 23.9.5, 39, 55, 60). This may occur when identical names based on different types are used at the same rank, e.g., subspecies in a species, species in a genus, and so on. The family-group names differing only in suffix are also considered as homonyms. The Code also rules that identical species-group names placed in different genera having identical names (homonymous) are not to be considered homonyms, e.g., *Noctua variegata* (insecta) and *Noctua variegata* (Aves). The existence of two or more identical names based on different types is called *Homonymy.*

The *Law of Homonymy* rules that any name that is a junior homonym of an available name must be rejected and replaced. If the rejected homonym has one or more available synonym(s), the oldest of these must be adopted as valid name, with its own authorship and date (Art. 53.3). The Code also rules that two or more species-group names of the same origin and meaning and cited in the same nominal genus or collective group are to be considered homonyms if the only difference in spelling consists of any of the following (Art. 58):

i. The use of *ae, oe,* or *e* (e.g., *caeruleus, coeruleus, ceruleus).*

ii. The use of *ei, i,* or *y* (e.g., *cheiropus, chiropus, chyropus).*

iii. The use of *c* or *k* (e.g., *microdon, mikrodon).*

iv. The aspiration or non-aspiration of a consonant (e.g., *oxyrhynchus, oxyrynchus).*

v. The presence or absence of c before *t* (e.g., *auctumnalis, autumnalis).*

vi. The use of a single or double consonant (e.g., *litoralis littoralis*).

vii. The use of *f* or *ph* (e.g., *sulfureus, sulphureus*).

viii. The use of different connecting vowels in compound words (e.g., *nigricintus, nigrocintus*).

ix. The transcription of the semi-vowel / as *y, ei, ej,* or *ij*)'.

x. The termination of -*i* or -*ii* in a patronymic genitive (e.g., *smithi, smithii*).

xi. The suffix -*ensis* or -*iensis* in a geographical name (e:g., *timorensis, timoriensis*) xii) Three pairs of names treated as special cases; e.g., *saghalinensis* and *sakhalinensis, sibericus* and *sibiricus, tianschanicus* and *tianshanicus,* otherwise even one letter difference is sufficient to prevent homonymy e.g., *Raphidia londinensis* and *Raphidia londonensis* (derived from Londinium and London, words of the same origin and spelling) are not homonyms. Such names, so similar in orthography (based on a different type) likely to create confusion, are also called *paranyms.*

Homonyms are of two kinds—**primary homonyms** and **secondary homonyms.** Primary homonym includes any of two or more identical species-group names referred to the same nominal genus when first published, e.g., *Rana tigrina* Linnaeus, 1758 and *Rana tigrina* Fabricius, 1795. The latter name is a junior homonym of the former and must permanently be rejected and replaced by a new name.

Secondary homonym includes any of two or more identical species group names referred to the same nominal genus as a result of subsequent transference, reclassification or combination of one or more taxa from another genus or other genera (Art. 52.3; 57). The junior secondary homonym, too, must be rejected and replaced by a new name e.g., *Forficula riparia* Dufour, 1805 is believed to be congeneric with *Labidura riparia* Pallas, 1790, and when transferred to *Labidura,* it becomes a junior secondary homonym, and is renamed as *Labidura indiana* Bolivar, 1897, being Bolivar was the original author of discovering and correcting this case of nomenclature. If subsequently *Forficula riparia* Dufour, 1805, is later no longer believed to be congeneric with *Labidura riparia* Pallas, 1790, the former specific names again becomes a valid name. *Labidura indiana* Bolivar, 1897, then becomes a junior objective synonym of *Forficula riparia* Dufour, 1805.

Typification, Type and its Kinds

The designation of a *nomenclatural type* is called **"typification"**. It is the means by which names are allocated to taxa. The type method is the only way to determine objectively and unequivocally the correct application of names to various taxa.

A **'type'** is a zoological object on which the original published description of a name is based. It is the objective basis to which a given zoological name is permanently linked. In other words it is the nucleus of a taxon and foundation of its name. Once designated the type cannot be changed, not even by the original author except by exercise of the plenary powers of the Commission (Article 78) through the designation of a Neotype (Article 75). The type of a nominal species is a specimen, that of a nominal genus is a nominal species and that of a nominal family is a nominal genus. A name-bearing type of species-group taxon proposed after 1999 and if based on preserved specimen or specimens, its depositroy should be mentioned.

During the past over 130 years the introduction of type method has brought about major conceptual changes in the theory of taxonomy and nomenclature. Early taxonomic works were dominated by Aristotelian concept of types. According to this concept all specimens conformed to the taxonomist's concept of the type of a taxon were considered typical. There were as many *types* as there were typical specimens which formed the basis of species description. This concept was followed by Linnaeus and his contemporaries.

Linnaeus never designated any specimen as type, not even under the circumstances when his description of a species was based on a single specimen. He even went on to the extent of substituting the old specimens by better new ones. 'Linnaean types' were never regarded reliable. The same practice was followed by others in Europe during the first half of the 19th century. Moreover, another practice frequently followed by the workers of this period was the transference of labels from one specimen to the other. This resulted in much more confusion in tracing out the type specimens. Until about the end of 19th century the impressions of Aristotelian type-concept remained visible. The typification of the higher taxa was done by the process of 'elimination' of a typical element and even the 1901 Code did not ratify this method for old cases but ruled for the designation of 'types' for the future works. The International Congress of Zoology held in Boston in 1947 for the first time provided provisions for designating types of the genera but for types of species even later. The Article 72 in its various subsections (i-x) on "Type series" now defines the true status of type on modern lines.

In terms of the range of variation occurring in a taxon, the type is not necessarily regarded typical of that taxon. Under these circumstances the purpose of a type is to provide a fixed point associated with a scientific name in the range of variations of organisms so that the application of the name can be objectively and unequivocally determined without taking into consideration the occurrence of discontinuities and boundaries between taxa.

Boisduval was perhaps the first to label all specimens from which the description was drawn as type (Holland, 1929). Then the workers started selecting one specimen as the 'type' and the rest were labelled **'cotype.'** Lord Walsingham objected to the use of the term 'cotype' and instead proposed the term **Paratype** in addition to other compound words like **homotype** or **homoeotype,** etc. The process of inventing names relating with the term type continued. This resulted in numerous different names which led us to believe that earlier workers were more concerned with the invention of new names for replacing type or expanding it rather than defining the exact role of the type. Horn (1929) mentioned 119 such words compounded with the word type. Waterston (1929) preferred the use of only one term, the type. Frizzell (1933) listed as many as 233 such names. Fernald (1940) listed 108, grouped in three categories —

    i. *Primary types or Proterotypes* (e.g., Allotype, Chierotype, Holotype, Paratype, etc.); ii) *supplementary types* (e.g., Neotype, Heautotype, Plesiotype, etc.); are the described or figured specimens used by any author to supplement or correct knowledge of a previously-defined species; iii) *Icotypes* (e.g., Homoeotype, Ideotype, Metatype and Topotype), are typical specimens that have been used in published descriptions or figures but consist of material which the authors have worked on or such as have been collected at the original locality. Fischer (1966) gave a list of principal kinds of types.

Blackwelder (1967) grouped such names of the types into the following seven categories:

i. Primary types (i.e., the single nomenclatural type e.g., Holotype, Lectotype, Neotype).

ii. Secondary types (i.e., the specimens from which the primary type must be selected, e.g., Syntypes, Paralectotypes).

iii. Tertiary types (i.e., other specimens originally set aside as of special taxonomic interest to supplement the primary type; e.g., Paratype, Allotype).

iv. Specimens identified as of special origin, e.g., Topotypes.

v. Specimens identified as to time or person of identification, e.g., Metatype, Homotypes or Homoeotypes, etc.

vi. Specimens identified as to special treatment or use, e.g., Plesiotypes, Hypotypes, etc.

vii. Replicas of type specimens, e.g., Plastotypes.

All these terms together with some more are very often or occasionally used in the zoological literature. All such terms are explained below:

1. Allotype—Muttkowski (1910) was the first to introduce this term. It is a specimen of the opposite sex to the type. It can be designated either at the time of typification or subsequently.

2. Apotype—A specimen, not the type, upon which a subsequent or supplementary description or figure is based; a *hypotype; a plesiotype.*

3. Autotype—Any specimen identified by the describer as an *illustration of his species and compared with the type; heautotype.*

4. Chirotype—A type specimen upon which a manuscript name is based.

5. Cotype—Any one of all specimens present before the describer at the time the description is drawn (not including type and allotype); a syntype; a paratype; an *isotype*. Cotype is now superseded and so not used (Rec. 73E).

6. Genoholotype—The species on which a genus is founded whether unique or one of a series, specifically named as generiotype by the author; essentially the same as the genotype; it is not used now.

7. Genolectotype—The species of a series selected as the type of the genus in which the describer of the genus placed it subsequently to the description; it is also not used now.

8. Genosyntype—One of a series of a species upon which a genus is founded, no one species being mentioned as type; it is also not used now.

9. Genritype—The type of the name of a genus; a generotype.

10. Genotype—The species which is designated as the type species of a genus, upon which it is based. Recommendation 67A of the 1999 Code (4th Edition) does not allow its use in referring to the type of a genus as the term creates confusion with its counterpart in use in genetics (where genotype means hereditary and genetic constitution of an individual). Only the use of the term type-species or a strictly equivalent form in another language is permitted in referring to the type of a genus. Blackwelder objects to this ruling and insists for its use in taxonomy because firstly, the word is in use in the literature of most groups of animals and its use in taxonomy preceeds its use in genetics by more than 100 years; secondly, although there are several alternate terms for genotype like type-species of the genus, type of the genus, generitype, genus-type, none of these lends itself to discussion of general aspects of the study on the basis of the genera; genotype is a useful word and it is not easily duplicated with any of the above expressions; and thirdly, the expression adopted in the Code, 'Type-species of the nominal

genus involves the user in the problem of whether the type is a nominal species, or the name of a species, as well as in the parallel problem of whether the thing typified is a genus, a nominal genus, or the name of a genus".

11. Geotype—A specimen from the type locality.

12. Hapantotype—One or more preparations of directly relaxed individuals representing different stages in the life cycle together forming a name-bearing type in an extant species of Protistans; it is a series that must not be restricted by lectotype selection but if a hapantotype is found to contain individuals of more than one species, components may be eliminated until it contains individuals of one species (Art. 73.3).

13. Haplotype—A generic type by a single reference (Only species).

14. Heautotype—See Autotype.

15. Holotype—The single specimen selected by the author of a species as its type, or the only specimen known at the time of description; a true type (Art. 73.1).

16. Homeotype or Homoeotype or Homotype—A specimen compared with the type by a person other than the describer and determined by him as conspecific with the type. This is a very valuable specimen when the type is damaged or lost.

17. Hypodigm—All the specimens personally known to a taxonomist at the time of describing a new species (Simpson, 1961); it is not acceptable to the taxonomists.

18. Hypotype—See Apotype.

19. Ichnotaxon—A taxon based on the fossilised work of an animal including trails, tracks and burrows (trace fossils), by an animal.

20. Icotype—A typical specimen which serves the purpose of identification but has not been used in literature.

21. Ideotype—A specimen named by the author after comparison with the type, but has not been from the type locality.

22. Isosyntype—A duplicate type of syntype, not cited in original description of the name.

23. Isotype—A duplicate of the type.

24. Lectotype—One of a series of syntypes which, subsequent to the publication of the original description, is selected and designated through publication to serve as "the type" (Art. 74).

25. Logotype—A type by subsequent designation.

26. Metatype—A specimen compared by the author of a species with the type and determined by him as conspecific with it.

27. Monotype—A holotype based on a single specimen; when holotype is correctly designated, it is synonymous with it.

28. Morphotype—A selected specimen of the second or later form of a dimorphic or polymorphic species; its use is permitted only by International Code of Nomenclature of Bacteria but not by Zoological Code.

29. Neoallotype—An allotype (opposite sex of the type) described after the publication of the original description.

30. Neotype—A specimen designated or selected subsequently to serve as the type of a name when all the original type specimens are destroyed or missing or believed to be so. (Art. 75).

31. Neoparatype—A paratype described after the publication of the original description. This resulted when at the time of description of new species there was no specimen other than holotype.

32. Onomatophore—A specimen on which the original name of the species is based; a 'name bearer", as equivalent to holotype (Simpson, 1961); it is also not acceptable to the taxonomists.

33. Ornatype—See Topotype.

34. Orthotype—A type species of the genus by original designation. It is also not in use now.

35. Paratype—A specimen cited in the original description other than the holotype (Rec. 73D).

36. Plastotype—A plaster cast of a type; used mainly in palaeozoology.

37. Plesiotype—A specimen upon which a subsequent or additional description or figure is based; any specimen identified with a described or named species by a person other than the describer.

38. Pseudotype—A type species of the genus by erroneous designation.

39. Topotype—A specimen collected in the exact locality from where the original type was obtained. It is very valuable if it is a homoeotype and when the original type has been lost.

40. Type—Same as Holotype.

41. Typotype—The type of type e.g., if the type of name, studied by an author, is a description or illustration previously published by an earlier author, then the specimen on which the earlier author's description or illustration was based, and which as such, the later author did not study, is the typotype of the later author's name.

42. Voucher specimen—Same as type; proposed as a replacement for type by Darlington (1971). It is actually nontaxonomic and cannot replace 'type' as the former represents only those individuals used or reported in a respective project.

Application of the Zoological Code to Such Names

The Zoological Code recognises only a few such terms like type, holotype syntype, paratype, lectotype, paralectotype and neotype. The invention and use of so many other terms as equivalent to 'type' or due to expansion of it is due to the fact that the authors still do not understand the true nomenclatural function of the type-specimen. Although all these pseudotypes have not been recognised by the code, they may serve useful purpose in the identification of the type of a specimen. As per Recommendation 72A of the Code, the term Allotype" may be used to designate among paratypes a specimen of opposite sex to the holotype and its use should be avoided for specimens other than paratypes. In most of the recent taxonomic literature, its use has been abandoned although it can be advantageous in adding the sexual characters of the opposite sex of the holotype.

Data on Holotype: According to Recommendation 73C. The holotype must be labelled containing the following data:

1. Size of the specimen

2. Locality, date and other relevant data

3. Sex

4. Development stage or form

5. Name of the host species in case of a parasite

6. Name of the collector

7. Collection in which it is kept and any collection or register number assigned to it

8. Altitude in metres for terrestrial species

9. Depth in metres for marine species to know the depth in metres below sea-level from where it was collected

10. In case of fossil species. Its geological age and stratiographic position, if possible in metres, above or below a well-established place.

Type Specimen and Live Specimen

International Code of Zoological Nomenclature states that "where the holotypes or syntypes are extant specimens, by a statement of intent that they will be (or are) deposited in a collection and a statement indicating the name and location of that collection (Article 16.4.2). It is debatable as to whether living specimens are allowed as types of those which have not been deposited in a museum (Dubois, 2009; Dubois and Nemesio, 2007; Donegan, 2008, 2009; Nemesio, 2009). Dubois (2009) pointed out that the wording of the Code is ambiguous and does not clearly prohibit the description of a species with living type specimens. It sends a very wrong message to conservationists in case of endangered species which are known by few specimens. Since the revised version of the Code is scheduled to be published in 2018, it can be looked into by the experts preparing the revised Code whether there is need to rephrase this Article to prohibit its strict use for endangered species.

Lectotype Designation: A *lectotype* (Art. 74) should be designated only by a specialist. If the description of a species is clearly based on one of the syntypes or one of the syntypes is properly illustrated, it should be named as *lectotype.* As far as possible a lectotype should be chosen from syntypes present in a public museum, preferably that museum which is having the largest number of syntypes or containing the collection on which the original author of the species had worked or the one containing most of the types of that original author. If the syntypes are from several localities, the selection of a lectotype be made from that locality which represents the maximum number of specimens of the locality mentioned by previous revision. For latest amendmend see page 267.

Neotype Designation: Article 75 of the Code regulates the designation of the neotypes when no holotype, lectotype or syntype exists either through loss or destruction. A neotype should be designated only in the revisionary work or under exceptional circumstances when neotype designation is necessary to solve complex zoological problems like confused or doubtful identities of closely similar species for one or more of which no holotype, lectotype or syntype exists. Otherwise, its designation is forbidden even if the original type is damaged or lost. If a neotype is lost, presumably another neotype can be selected under the conditions. If, after the designation of a neotype, original type—material is discovered the case is to be referred to the Commission (Article 75.4). After the designation of neotype, if previously lost holotype, syntype or lectotype of a species subsequently typified by a neotype. is rediscovered the original one will automatically displace the erected neotype (Art. 75.8). In case existing name bearing type of a species is indeterminate (i.e. nomen dubium) zoologist should contact the Commission (Art. 75.6).

Distinct Labels on Type Series: It is always desirable to use distinct labels for the holotype, allotype (if used), and paratype. Generally the holotypes bear red or pink coloured labels; the allotypes green; and paratypes yellow labels. This is done for their quick recognition.

Power of the Commission

Chapter 17 containing Articles 77 to 84 is devoted to this section of nomenclature, The international Commission on Zoological Nomenclature is a permanent body which derives all its powers from the International Congress of Zoology (Art. 77.1). The Commission has power to suspend the application of any provision of the Code (*plenary* powers, Art. 78.1) to suppress or validate any scientific name, and to annule or validate any typification, publication, or any published matter concerning nomenclature in order to promote stability and continuity in nomenclature (Art. 81). The decision of the Commission on any such particular case referred to it is termed an *Opinion* (Art. 80.2). The opinions are published in the Bulletin of Zoological Nomenclature by the International Trust for Zoological Nomenclature. Opinions become operative once they are published. Thus, the Commission acts on behalf of zoologists and has the power to waive or modify the provisions of their application to a particular case if it is causing any confusion. No individual zoologist, no matter how well respected is he in his field, can have exception in such cases. The Commission decides suitable action in response to proposals related to nomenclature submitted to it.

Any zoologist can submit cases involving nomenclatural problems to the Commission. Such applications are published with the object of providing an opportunity to interested specialists to submit comments to the Commission. For this purpose at least one year must elapse from the date of publication of amendments before the International Commission starts to vote on any given application. It is always desirable for an applicant to discuss the case with other workers in the field before submitting an application. This helps to ensure wider awareness and the likely reactions of others. While the application to the Commission is pending, he should continue to use the existing name until the decision of the Commission is published. Amendments take formal effect when they have been approved by the Commission and ratified at a General Assembly of the International Union of Biological Sciences. It is now clear from the rules of nomenclature that—firstly, well established names do not need to be changed for nomenclatural reasons; secondly, when a well-known and commonly used name is found to be invalid, the fact should be reported to the Commission for possible action under the Plenary Power; and thirdly, names which fit the criteria of usage are not to be changed but are to be conserved by notification in the Bulletin of Zoological Nomenclature.

The international Commission on Zoological Nomenclature also brings out declarations (Art. 80.1) the Commission is empowered to make between Congresses which are provisional modifications of the Code. After the publication of fourth edition of International Code of Zoological Nomenclature, only one official amendment has been made, i.e., to Article 74.7.3 for **LECTOTYPES** through a declaration appeared in Bulletin of Zoological Nomenclature in 2003. It is given below:

1. The wording of Article 74.7.3 is hereby amended to read 'contain an express statement of deliberate designation (merely citing a specimen "lectotype hereby designated", "lectotype" is insufficient").

2. An example is added directly below Article 74.7.3 to read 'Example statement such as "lectotype hereby designated", "lectotype by present designation", "I choose specimen X as lectotype" would fulfill this requirement, but "lectotype : specimen X" would not.'

3. The following recommendation is also added to read— **'Recommendation 74 G. Not merely for curatorial purposes.** The designation of lectotypes should be done as part of a revisionary or other taxonomic work to enhance the stability of nomenclature, and not for mere curatorial convenience.'

4. These amendments are backdates and apply to all works published after 31 December, 1999.

The official indexes of rejected and invalid names and works, and the official lists of validated names and approved ones (Art. 79), although published by the Trust separately, are considered to be integral parts of the Zoological Code. The Commission consists of a President, Vice-President, Secretary, Assistant Secretary and 25 members from 19 countries. Commissioners are zoologists or palaeontologists with nomenclatural experience but without any remuneration. They are expected to participate in all the commission's decisions and attend (as far as possible) meetings. The Commission meets triennially in conjunction with General Assemblies of the International Union of Biological Sciences. Nominations for election to the Commission may be sent to the Secretariat. Elections take place at Assemblies where the Commission meets. Votes are cast by members of the Commission and by other zoologists who are attending the assembly. The casual vacancies are filled by postal by-elections (Art. 79.2). One full chapter (chapter 18, Art. 85-90) is on regulations for governing the fourth edition of Code.

Modification of the Code

The Zoological Code may be modified only by International Congress of Zoology; or recognised equivalent like the Division of Zoology of the International Union of Biological Sciences; General Assembly of the International Union of the Biological Sciences to which the Congress has delegated such powers. These bodies act on a recommendation from the International Commission on Zoological Nomenclature, presented through and approved by the Section on Nomenclature of the Congress or another recognised body. The Code is drafted on behalf of the International Commission by an Editorial Committee appointed by the Congress or recognised equivalent. The proposals for the modifications of the Code are sent to the Secretary of the International Commission at least one year in advance of the next International Congress of Zoology or its recognised equivalent, as mentioned above. The Code may also be provisionally modified between congresses or their equivalents by means of *Declarations* of the International Committee on Zoological Nomenclature.

International Commission of Zoological Nomenclature is funded by a charitable trust (International Trust of Zoological Nomenclature). The Trust was down to its last pennies, the support base for ICZN was going to crumble. Due to foreseen acute fund crunch, there was going to be Nomenclature Chaos. This chaos was timely averted due to the intervention of Singapore. The National University of Singapore declared to fund the secretariat of the International Commission of Zoological Nomenclature for the next three years (i.e. 2013-

2016). The ICZN held their first meeting in Singapore from 17-20 November, 2013. The secretariat is run by an officer who coordinates all the activities around the world.

ELECTRONIC PUBLICATION

Following four years of highly charged debate, the ICZN voted in favour of a revised version of the amendment to the International Code of Zoological Nomenclature proposed in 2008. The purpose of the amendment (anonymous, 2012) is to expand and refine methods of publication allowed by the Code, particularly to permit electronic publication after 2011 with the requirements that the work be registered in ZOOBANK before it is published. The ICZN has thus amended Articles 8, 9, 10, 21 and 78 of the ICZN (Anonymous, 2012A,B).

Dubois *et. al.* (2013) pointed out that the publication of the amendment provides an opportunity to discuss some of the nomenclatural problems as given below:

1. All works published only online before are nomenclaturally unavailable;

2. Printed copies of the PDFs of works which do not have their ISSN or ISBN and which are not obtainable free of charge or by purchase, do not qualify as publication but must be seen as facsimiles of unavailable works and are unavailable to provide nomenclatural availability to any nomenclatural novelties they may contain;

3. Publications online of later released online public publication;

4. The publication dates of works for which online publications had been released are not of those prepublications and it is critical that the real release date of such works appear on the actual final electronic publication, but this is not currently the case in electronic periodicals that such online prepublications and which still indicate on their websites and PDFs, the date of releasing of prepublication as that of publication of the work;

5. Supplementary online materials and subsequent online formal corrections or either paper or electronic

publication distributed only online are nomenclturally unavailable;

6. Nomenclatural information provided on online websites that do not have a fixed content or format, with ISSN or ISBN, is unavailable.

In view of this these authors believe that, as long as, the problematic points linked to the new amendment and to electronic publication as a whole are not resolved, nomenclatural novelties continue to be published in paper-printed journals that have so far shown editorial competence regarding taxonomy and nomenclature, which is not the case of several recent electronic-only published journals. Mylon et.al (2014) agrees with Dubois *et. al.* but the problems will continue to come and so there will be solutions. Such acts are taken care of in a positive manner for the progress of nomenclatural activities.

References

Adams, R.P. 1970. Contour mapping and differential systematics of geographical variation. *Syst. Zool.*, **19:** 385-390.

Adanson, M. 1763. *Families des plantes*, Vol. 1. Preface, pp. cliv et seg., clxiii, clxiv. Vincent Paris, cccxxv + 190.

Agapow, P.M., O.R.P. Bininda-Edmonds, K.A. Crandali, J.L. Gittleman, G.M. Mace, J.C. Marshall, and A. Purvis, 2004. The impact of species concept on biodiversity studies. *Qt. Rev. Biol.* **79:** 161-79.

Albrech, P. 1993. Museum, collections and biodiversity inventories. Trends in Ecol. Evol., **8:** 372-375.

Alexander, R.D. 1962. The role of behavioural study in cricket classification. *Syst. Zool.*, **11:** 53-72.

Allah, H.H. 1940. Natural hybridization in relation to taxonomy. In Huxley, *The New Systematics*, pp. 515-528.

Allah, H.H. 1967. Acoustical communication in arthropods. *Ann. Rev. Ent.*, **12:** 495-526.

Alston, R.E. and B.L. Turner. 1963. *Biochemical Systematics*. Prentice-Hall Inc., Englewood Cliffs, N.J., 404 pp.

Amadon, D. 1S49. The seventy-five per cent rule for subspecies. *Condor*, **51:** 250-258.

Amadon, D. 1950. The species—then and now. *Auk*, **67:** 492-498.

Amadon, D. 1959. Behaviour and classification, some reflections. *Vischr. Naturf. Ges. Zurich.* **104:** 73-78.

Amadon, D. 1967. The superspecies concept. Syst. *Zool.*, **15:** 245-249.

Amadon, D. 1976. Treatment of subspecies approaching species status. *Syst. Zool.*, **25(2):** 161-167.

Anderson, R.M. 1948. Methods of collecting and preserving vertebrate animals (2nd ed.). *Bull. Nat. Mus. Canada,* Dept. Mines, No. 69, Biol, Ser., pp. 1-162.

Anderson, S. 1974. Some suggested concepts for improving taxonomic dialogue. *Syst. Zool.*, **23(1):** 58-70.

Andreev, S.V., B.K. Mortens and V.A. Molehanova. 1966. Various light traps and their application. *Zool. Zh.,* **45:** 850-857 (In Russian with English summary).

Anonymous. 1886. American Ornithologist's Union: The Code of Nomenclature and check-list of North American birds. New York, 392 pp.

Anonymous. 1893. Deutsche Zoologische Gesellschaft. Regies fur die wiss-enschftschaft Benennung der Thiere. *Verh. Oeutsch. Zool. Gesel. Dritte Keersaml., Gottingen,* 89-98 pp.

Anonymous. 1928. British National Committee on Entomological Nomenclature: Report on Rules of Entomological Nomenclature. *Proc. Ent. Soc. London.* 13 pp.

Anonymous. 1936. British Museum (N.H.) 1936, etc. Instructions for collectors. (Several animal groups and several editions).

Anonymous. 1961. The mathematical assessment of taxonomic similarity, including the use of computers. *Taxon,* **10:** 97-101.

Anonymous. 1969. *Systematic Biology: A survey of federal programmes and needs.* Office of Science and Technology, Washington.

Anonymous. 1970. *Systematics in support of biological research.* National Research Council, Canada.

Anonymous. 1971. *The systematic Biology Collections of the United States: An Essential Resource.* Pt. I: *The great collections: their nature, importance, condition and future.* The New York Botanical Garden, New York.

Anonymous. 1979. *Advances in Insect Taxonomy*—Workshop, Manali. Oriental Insects, Delhi, pp. 105.

Anonymous. 2012A. Amendments of Articles 8, 9, 10, 21 and 78 of the International Code of Zoological Nomenclature to expand and refine methods of publication. Zookeys. **219:** 1-10.

Anonymous. 2012B. The International Code of Zoological Nomenclature amendment of Articles 8, 9, 10, 21 and 78. Zoosystematica Rossica. **21(2):** 323-327.

Anthony, H.E. 1945. The capture and preservation of small mammals for study. *Ann. Mus. Nat. Hist. Sci. Guide,* No. **61:** 1-54 pp.

Arnett, R.H. 1970. *Entomological Information Storage and Retrieval,* Baltimore: Bio-Rand., 210 pp.

Ashlock, P.D. 1971. Monophyly and associated terms. *Syst. Zool.,* **20:** 63-69.

Ashlock, P.D. 1972. Monophyly again. *Syst. Zool.,* **21:** 430-438.

Atkins, D.E.. K.K. Droegeneier, S.L. Feldman, Klein Garcia-Molina, M.L. Messer-Schmitt, P. Messina, J.P. Ostriker and M.H. Wright. 2003. Revolutionising science and engineering through cyber infrastructure. *Report of the National Science Foundation Blue-Ribbon Advisory Panel on Cyber infrastructure,* pp. 1-52, NSF Washington, D.C.

Avise, J.C. 1974. Systematic value of electrophoretic data. *Syst. Zool.,* **23(4):** 465-481.

Ball, C.R. 1946. Why is taxonomy ill supported? *Science,* **103:** 7-13.

Ball, G.H. 1970. *Classification Analysis*. Standford Res. Inst., Menlo Park, Calif. Technical Note SRI Proj. 5533, 117 pp.

Ball, G.H. and E.W. Clark. 1953. Species differences in amino acids of *Culex* mosquitoes. *Syst. Zool.*, **2:** 138-141.

Ball, I.R. 1977a. A monograph of the genus *Spathula* (Platyhelminthes: Turbellaria: Tricladida). *Austr. J Zool.*, **47:** 1-43.

Ball, I.R. 1977b. La faune terrestre del' ile de Sainte Helene. 6. Vermes. 2. Turbellaria. *Ann. Mus. r. Afr Cent.* (Ser. 8° *Sci. Zool.*). No. **220:** 492-511.

Barber, H.S. 1951. North American fireflies of the genus *Photuris*. *Smithson. Misc. Coll.*, **117:** 1-58.

Barr, A.R. 1954. Punch-card taxonomy. *Syst. Zool.*, **3:** 143.

Basford, N.L., J.E. Butler, C.A. Leone and F.N. Rohlf. 1968. Immunological comparisons of selected Coleoptera with analysis of relationships using numerical taxonomic methods. *Syst. Zool.*, **17:** 388-406.

Basu Chaudhury, R.C. and J.B. Chatterjee. 1969. Ascorbic acid synthesis in birds. A phylogenetic trend. *Science*, **164:** 435-436.

Bates, M. 1940. The nomenclature and taxonomic status of the mosquitoes of the *Anopheles maculipennis* complex. *Ann. Ent. Soc. Am.*, **33:** 343-356.

Bather, F.A. 1927. Biological classification: past and future. *Quart. J. Geol. Soc. Lond.*, **83:** 62-104.

Baum, D. 1992. Phylogenetic species concepts. *Trends Ecol. Evol.*, **7:** 1-2.

Beandry, J.R. 1960. The species concept: its evolution and present status. *Rev. Canada Biol.*, **19:** 219-240.

Becak, W. and M.L. Becak. 1969. Cytotaxonomy and chromosomal evolution. *Cytogenetics*, **8:** 247.

Beckner, M. 1959. *The biological way of thought*. Columbia Univ. Press, N.Y., 200 pp.

Beer, J.R. de and E.F. Cook. 1958. A method for collecting ectoparasites from birds. *J. Parasit.*, **43:** 445.

Benazzi, M. 1973. *Cytotaxonomy and evolution; General remarks—vertebrate evolution*. Ed. A.B. Chiarelli and Campana. Academic Press, London & N.Y., pp. 1-3.

Benton, M.J. 2000. Stems, nodes, crown clades and rank-free lists: is Linnaeus dead? *Biol. Review*, **75:** 633-648.

Benton, M.J. 2007. The Phylocode: Beating a dead horse? Acta Pala. Polonica, **52**(3): 651-55.

Bergan, T. 1971. Survey of numerical techniques for grouping. *Bacteriol. Rev.*, **35:** 379-389.

Bergstrom, J. 1979. *Morphology of fossil arthropods as a guide to phylogenetic relationship*. In *Arthropod Phytogeny*, (ed.) A.P. Gupta. Van Nostrand Reinhold Co., pp. 1-58.

Bessey, C.E. 1908. The taxonomic aspect of species. *Amer. Nat*, **42:** 218-224.

Bier, M. 1959. *Electrophoresis: theory, methods and application*. N.Y. Academic Press, 563 pp.

Bigelow, R.S. 1957. Museum taxonomy and taxonomic research. *Ann. Soc. Ent. Quebec*, **3:** 68-74.

Bishy, F.A., J. Shimura, M. Ruggiero, J., Edwards and C, Haeuser. 2002. Taxonomy at the click of a mouse. Nature, **418**: 367.

Blackith, R.E. and R.A. Reyment. 1971. *Multivariate morphometries.* Academic Press, N.Y., 412 pp.

Blackwelder, R.E. 1940. Some aspects of modern taxonomy. *J.N.Y. Ent. Soc,* **48**: 245-257.

Blackwelder, R.E. 1941. The gender of scientific names in zoology. *J. Wash. Acad. Sci.,* **31**: 135-140.

Blackwelder, R.E. 1951. Systematics in zoology. *Science,* **113**: 3.

Blackwelder, R.E. 1955. The open session on taxonomists, *Syst. Zool.,* **3**: 177-181.

Blackwelder, R.E. 1960. The present status of systematic zoology. *Syst. Zool.,* 8: 69-75.

Blackwelder, R.E. 1962. Animal taxonomy and new systematics. In *Survey of Biological Progress,* (ed.) B. Glass, **4**: 1-57.

Blackwelder, R.E. 1964. Phyletic and phenetic versus omnispective classification. In *Phenetic and Phylogenetic classification,* (ed.) V.H. Heywood and J. McNeill. Syst. *Asson. Publ.* No. **6**: 17-28 pp.

Blackwelder, R.E. 1967. *Taxonomy.* John Wiley & Sons, Inc., N.Y, 698 pp.

Blackwelder, R.E, and A.A. Boyden. 1952. The nature of systematics. *Syst. Zool.,* **1**: 26-33.

Blaker, A.A. 1965. *Photography for scientific publication. A handbook.* W.H. Freeman & Co., San Francisco, 158 pp.

Blomback, B. and M. Blomback. 1968. Primary structure of animal proteins as a guide in taxonomic studies. In *Chemotaxonomy and Serotaxonomy* (ed.) Hawkes, pp. 3-20.

Bock, W.J. 1963. Evolution and phylogeny in morphogically uniform groups. *Am. Nat.,* **97**: 265-285.

Bogert, C.M. 1954. The indication of infraspecific variation. *Syst. Zool.,* **3**: 111-112.

Bolton, E.T. and B.J. McCarthy. 1962. A general method for the Isolation of DNA. *Proc. Natn. Acad. Sci.,* **48**: 1390-1397.

Bonde, N. 1975. Origins of higher group: viewpoint of early phylogenetic systematics. *Call. Int. CNRS,* **218**: 293-324.

Borgmeier, T. 1963. Basic questions of systematics. *Stud Ent.,* **6(1-4)**: 537-563.

Bossert, W. 1969. Computer techniques in systematics. In *Syst. Biol. Proc. Int. Conf, Ann. Arbor. Nat. Acad Sci. Publ.,* **1692**: 595-605.

Boyden, A.A. 1943. Serology and animal systematics. *Am. Nat.,* **77**: 234-255.

Boyden, A.A. 1959. Serology as an aid to systematics. 15th *Intern. Cong. Zool,* pp. 120-122.

Brand, J.M., M.S. Blum and H.H. Ross. 1973. Biochemical evolution in fireant venoms. *Insect Biochem.,* **3**: 45-51.

Britton-Davidian, J. 2001. How do chromosomal changes fit in? *J.Evol.,* **14**: 872-873.

Brown, F.M. and H.M. Heffron. 1928. Serum diagnosis and Rhopalocera. *J.N.Y. Ent. Soc,* **36**: 165-170.

Brown, R.W. 1954. *Compositions of the scientific words*. Wash. D.C. pub. by the author (distributed by the Smith. Inst., Wash. D.C), 882 pp.

Brown, W.J. 1959. Taxonomic problems with closely related species. *Ann. Rev. Ent.*, **4:** 77-98.

Brown, W.L. Jr. and E.O. Wilson. 1954, The case against the trinomen. *Syst. Zool.*, **3:** 174-176.

Brues, C.T. 1929. Present trends in systematic entomology. *Psyche,* **36:** 13-20.

Bryson, V. and H.I. Vogel (eds.). 1965. *Evolving genes and proteins.* Academic Press, 629 pp.

Burma, B.H. 1949. The species concept: a semantic review. *Evolution,* **3:** 369-370.

Burma, B.H. 1954. Reality, existence and classification: a discussion of the species problem. *Madrono,* **12:** 193-209.

Burt, W.H. 1954. The subspecies category in mammals. *Syst. Zool.*, **12:** 99- 104.

Butler, J.E. 1968. Determination of immunological correspondence for taxonomic studies by sensitometric scanning of antigen-antibody precipitates in agar gel. *J. Comp. Biochem. Physiol.*, **25:** 417-426.

Butler, J.E. and C.A. Leone. 1967. Immunological investigations of the Coleoptera. *Syst. Zool.*, 16(1): 56-63.

Butlin, R. 1987. Speciation by reinforcement. *Trends Ecol. Evol.*, **2:** 8-13.

Buzzati-Traverso, A.A. and A.B. Rechnitzer. 1953. Paper partition chromatography in taxonomic studies. *Science,* **117:** 58-59.

Cable, R.M. 1966. *An illustrated laboratory manual of Parasitology.* Burgress Pub. Co., Minneapolis, U.S.A., 165 pp.

Cain, A.J. 1954. The superspecies. *Syst. Zool.*, **3:** 145-146.

Cain, A.J. 1958. Chromosomes and taxonomic importance. *Proc. Linn. Soc.* London, **169:** 125-128.

Cain, A.J. 1959a. Taxonomic concepts. *Ibis.*, **101:** 302-318.

Cain, A.J. 1959b. Deductive and inductive methods in post-Linnaean taxonomy. *Proc. Linn. Soc.* London, **170:** 185-217.

Cain, A.J. (ed.) 1959c. Function and taxonomic importance. *Syst. Assoc.* Publ., London, no. **3.**

Cain, A.J. 1963. *Animal species and their evolution.* Hutchinson Univ. Press, London, 190 pp.

Cain, A.J. 1968. The assessment of new types of character in taxonomy. In *Chemotaxonomy and Serotaxonomy* (ed.) Hawkes, pp. 229-234.

Camp, W.H. 1951. Biosystematics. *Brittonia,* 7 113-127.

Camp, W.H. and C.L. Gilly. 1943. The structure and origin of species. *Brittonia,* **4:** 323-385.

Campbell, D.H., J.S. Garvey, N.E. Cramer and D.H. Sussdorf. 1964. *Method in immunology: a laboratory text for instruction and research.* W.A. Benjamin, Inc.. N.Y., 263 pp.

Candino, P.D. 2004. Classifying species versus naming clades. *Taxon,* **53:** 795-798.

Candino, P.D. and K. de Queiroz. 2004. Phylocade: A phylogenetic code of Biological Nomenclature. Avail, at http://www.ohio.edu/phylocode/.

Cazier, M.A. and A. Bacon. 1949. Introduction to quantitative systematics. *Bull. Am. Must. Nat. Hist.,* **93:** 347-388.

Chadwick, C.E. 1955. The practical entomologist and taxonomy. *J. Austr. Inst. Agric. Sci.,* **21:** 230-238.

Chamberlin, W.J. 1952. *Entomological Nomenclature and Literature.* 3rd ed. Dubuvue Iowa, William C. Brown Co.

Chaper, M. 1881. Regies applicables a la Nomenclature des Etres organises proposees per la Societe Zoologique de France, Paris, 1-37 pp.

Chen, P.S. 1962. Free amino acids in insects. In *Amino Acid Pools,* (ed.) J.T. Holden, pp. 115-138.

Chen, P.S. 1966. In *Advances in insect physiology* Vol. (eds.) J.W.L. Beament, J.E. Treherne and V.B. Wigglesworth. Academic Press, London, pp. 53-132.

Chiarelli, A.B. and E. Campanna (eds.). 1973. *Cytotaxonomy and vertebrate evolution.* Academic Press, London, N.Y., 783 pp.

Christoffersen, M.L. 1987. Phylogenetic relationships of hippolytid genera, with an assignment of new families for the Crangonoidea and Alpheoidea (Crustacea, Decapoda, Cardiea). *Cladistics,* **3:** 348-362.

Christoffersen, M.L. 1995, Cladistic taxonomy, phylogenetic systematics and evolutionary ranking. *Syst. Biol.,* **44(3):** 440-454.

Clark, E.W. and G.H. Bail. 1957. Preliminary microelectrophoretric studies on insect proteins. *Physiol. Zool.,* **29:** 206-212.

Clausen, C.P. 1942. The relation of taxonomy to biological control. *J. Econ. Ent.,* **35:** 744-748.

Clifford, H.T. 1975. *An introduction of numerical classification.* Academic Press, N.Y., 229 pp.

Cohan, F.M. and A.F. Koeppel. 2008. The origins of ecological diversity in prokaryotes. *Curr. Biol.,* **18:** 1024-1034.

Cohen, D.M. and R.F. Cressey. 1969. Natural history collection—past, present and future. *Proc. Biol. Soc. Wash.,* **8:** 559-762.

Coker, R.E. 1939. The problem of cyclomorphosis in *Daphnia. Quart. Rev. Biol.,* **14:** 137-148.

Cole, A.J. 1969. *Numerical taxonomy.* Proceedings of the colloqui in numerical taxonomy held in the University of St. Andrews, Sept. 1968. Academic Press, N.Y., 324 pp.

Coless, D.H. 1985. "On Character" and related terms. *Syst. Zool.,* **34:** 229-233.

Conard Martius, H. 1949. *Abstammungslehre.* Koesel, Muenchen, 2nd ed., 425 pp.

Consden, R., A.H. Gordon and A.J.O. Martin. 1944. Qualitative analysis of proteins: a partition chromatographic method using paper. *Biochem. J.,* **38:** 224-232.

Constance, L. 1953. The role of plant ecology in biosystematics. *Ecology,* **34:** 642-649.

Corliss, J.O. 1959. An illustrated key to the higher groups of the ciliate Protozoa, with definition of terms. *J. Protozool.*, **6:** 265-285.

Corliss, J.O. 1963. Establishment of an international type-slide collection for the ciliate Protozoa. *J. Protozool.*, **9:** 307-324.

Corliss, J.O. 1972. Common sense and courtesy in nomenclatural taxonomy. *Syst. Zool.*, **21(1):** 117-122.

Coyne, J.A., H.A. Orr and J.D. Futuyma. 1988. Do we need a new species concept? *Syst. Zool.*, **37(2):** 190-200.

Cracraft, J. 1983. Species concepts and speciation analysis. Pages 159-187 in current ornithology volume (ed) R.F. Johnston plenum, New York.

Cracraft, J. 1987. Species concepts and the Ontology of evolution. *Biol. Philos.*, **2:** 329-346.

Crenshaw, J.W. 1965. Serum protein in an interspecies hybrid swarm of turtles of the genus *Pseudemys*. *Evolution*, **19:** 1-15.

Cresson, E.T. Jr. 1934. Terminology of the types. *Ent. Afews*, **45:** 122-126.

Crick, F.C.H. 1953. On protein synthesis. In Biological replication of macromolecules. Symposium No. 12, Soc. *Exptl. Bio.*, N.Y., pp. 138-183.

Croveilo, T.J. 1970. Analysis of character variation in ecology and systematics. *Ann. Rev. Eccl. Syst.*, **1:** 55-98.

Crowson, R.A. 1970. *Classification and biology.* New York: Atherton, 350 pp.

Cullen, J.M. 1959. Behaviour as a help in taxonomy. Syst. Assoc. Publ., **3:** 131-140.

Cumley, R.W. 1940. Comparison of serological and taxonomical relationships of *Drosophila* species. N.Y., *Ent. Soc*, **48:** 265-274.

Cummins, K.W., L.D. Wilier. N.A. Smith and R.M. Fox. 1965. *Experimental entomology,* van Nostrand. Reinhold Co.

Cuvier, G. 1829. Le renge animal distribue apres son organization. Paris, Crochard et cie., p. 16.

Dall, W.H. 1877. Nomenclature in zoology and botany. *Proc. Am.* Assoc. *Adv. Sci,* **1877:** 7-56.

Dallwitz, M.J. 1974. A flexible computer program for generating identification keys. *Syst. Zool.*, **23(1):** 50-57.

Dallwitz, M.J. 1978. User's guide to KEY—A computer program for generating identification keys. *CSIRO Div. Entmol. Rep.*, No. 4, 16 pp.

Dariks, H.V. 1983. Systematics in support of Entomology. *Ann. Rev. Ent.*, **33:** 271-296.

Darlington, P.J. Jr. 1971. Modern taxonomy, reality and usefulness. *Sysl Zool.*, **20:** 341-365.

Darwin, C. 1859. *The origin of species by means of natural selection, or the preservation of the favoured races in the struggle for life.* John Murray, London, 502 pp.

Das, Dipannita. 2011 (October, 16). A losing battle for survival: pollution, over fishing and invasive species to blame. *Times of India,* page 23.

Davies, R.G. 1971. *Computer programming in quantitative biology.* Academic Press Inc., London. 493 pp.

Davis, B.J. and J. Ornstein. 1959. A new high resolution electrophoresis method. Delivered at the Society for the study of blood at the New York Academy of Medicine.

Davis, J.I. and Nixon, K.C. 1992. Populations, genetic variation and the delimitation of phylogenstic species. Syst *Biol.*, **41**: 421-435.

Davis, P.H. and V.H. Heywood. 1963. *Principles of angiosperm. taxonomy.* Edinburg; Oliver and Boyd, 558 pp.

Dayrat, B. 2005. *Towards integrative Taxonomy.* Biol. J. Linn. Soc., **85**: 407-415.

de Beer, G.R. 1940. Embryology and taxonomy, In *New Systematics* (ed.) J. Huxley. Oxford Univ. Fress, London, pp. 365-393.

de Candolle, A.P. 1313. Theoril eiemenvaire de la botanique. *Chiz Detervilie.* Paris, **8**: 27 pp.; 1844 (Ed. 3). xii, 468 pp. Roret, Paris.

de Long, R. 1983. The biological species concept and the aims of taxonomy. *J. Res. Lepid.*, **21** (1982): 226-237.

Dementev, G.P. and N.A. Giadko (eds). 1965. *Birds of the Soviet Union,* Vol. 1, Eng. Trans., Jerusalem.

Denmark, H.A., H.V. Weems and C. Taylor. 1958 Taxonomic codification of biological entities. *Science*, **128**: 990-992.

de Oueiroz, K. 2005a. Ernst Mayr and the concept of species. Proc. Natl. Acad. Sci., USA. 102: 6600-6607.

---- 2005b. Different species problems and their resoluton. *BioEssays*. **27**: 1263-69.

de Queiroz, K. 2006. The phylocode and the distinction between taxonomy and nomenclature. *Syst. Biol.*, **55**: 160-62.

---- 2007. Species concepts and species delineation. *Syst. Biol.*, **58**: 879-886.

de Queiroz, K. and M.J. Donoghue. 1908. Phylogenetic systematics and the species problem. *Cladistics*, **4**: 317 338.

de Queiroz, K. and M.J. Donoghue 1990a Phylogenetic systematics or Nelson's version of cladistics? *Cladistics*, **6**: 61-75.

de Queiroz, K. and M.J. Donoghue. 1980b. Phylogenetic systematics and species revisited, *Ibid.*, **6**: 83-90.

de Queiroz, K. and J. Gauthier. 1932. Phylogenetic taxonomy. *Ann. Rev. Ecol. Syst.*, **23**: 440-480.

de Queiroz. K. and J. Gauthier. 1994. Towards a phylogenetic system of biological nomenclature. *Trends. Ecol. Evol.*, **9**: 27-31.

Desalle, R. 2006. Species discovery vs. species identification in DNA barcoding efforts: Response to Rubinoff. *Gom.Biol..* **20**(5): 1545. .

Deutsch, H.F. and M.B. Goodloe. 1941. An electrophoretic survey of various animal plasmas. *J. Bid. Chern.*, **161**: 1-20.

Dobzhansky, V. 1937a. *Genetics arid origin of species.* Columbia Univ. Press. N.Y., xvi + 364.

Dobzhansky, V. 1937b. What is a species? *Scientm*, **2**: 280.

Dobzhansky, V. 1951. Genetics and the origin of species (3rd Ed). Columbia Univ. Press, New York. 364 pp.

Dobzhansky, V. 1961. Taxonomy, molecular biology and the peck order. *Evolution*, **15**: 263-264.

Dobzhansky, V. and O. Pavlovsky. 1971. Experimentally created incipient species of *Drosophila. Nature,* **230**: 289-292.

Donegan, T.M. 2008. New species and subspecies descriptions do not and should not always require a dead type specimen. Zootaxa. 1761: 3240; 2009. Type specimens, samples of live individuals and the Galapagos Pink Land Iguana. Zootaxa. **2201**: 12-20.

Douville, H. 1881. Regies proposees par le Comte' de la Nomenclature Palaentogique. Cong. Geol. Internet., Rend. Deux. Sen. Boulogne. 594-595 pp.

Downe, A.E.R. 1963. Comparative serology of four species of *Aedes. Science,* **139**: 1286-1287.

Downey, J.C. 1953. Host plant relations as data for butterfly classification. Sysi *Zool.,* **11**: 150-159.

Dreinsbach, R.R. 1952. Preparing and pnotographing slides of insect genitalia. *Syst. Zool.,* **1**: 134-136.

Dubois, A. 2005. Proposed rules for the incorporation of nomina for higher-ranked zoological taxa in the International Code of zoological Nomenclature. *Zocsystema,* **27**: 365-426.

Dubois, A. 2009. Endangered species and endangered knowledge. Zootaxa. **2201**: 26-29.

Dubois, A. 2011. The International Code of Zoological Nomenclature must be drastically improved before it is too late. Bionomina, **2**: 1-104.

Dubois, A., Crochet, P.A., Dickinson, E.C. and Nemesio, A. 2013. Nomenclatural and taxonomic problems related to the electronic publication of new nomina and nomenclatural acts in zoology, with brief comments on optical discs and on the situation in botany. Zootaxa. **3735(1):** 1-94.

Dubois, A. and Nemesio, A. 2007. Does nomenclatural availability of nomina of new species or subspecies require the description of vouchers in collection. Zootaxa. **1409:** 1-22.

Duchateau, W.E. and M. Florkin. 1858. A survey of aminoacidemias with special reference to the high concentration of amino acids in insect haemolymph. *Arsch. int. Physiol.,* **66**: 573 591.

Duellman, W.E. 1962. Directions for preserving amphibians and reptiles. *Univ. Kansas Mus. Nat. Hist. Misc. Publ.,* **30**: 37 40.

Dunbar, C.O. 1950. The species concept: further dⁱscussion. *Evolution.* 4; 175-176.

Dupraw, E.J. 1965. Non-Linnaean taxonomy. *Nature,* **202**. 849-852.

Ebach, F. and C. Holdrege. 2005. DNA barcoding is no substitute for taxonomy. *Nature,* **434**: 697.

Echeile, A.A. 1990a. In defence of the phylogenetic species concern and the ontological status of hybridogenotic taxa. *Herpetologica,* **46**: 109-113.

Echelle, A.A. 1990b. Nomenclature and non-Mendeiian (clonal) vertebrates. Syst. *Zool.,* **39**: 70-78.

Edwards, J.G. 1955. A new approach to infraspecific categories. *Syst. Zool.,* **3**: 1-20.

Edwards, J.G. 1956. What should we mean by subspecies? *Turtox News,* **34:** 200-202, 230-232.

Edwards, M.A. & A.T. Hopwood (eds.). 1966. Nomenclator Zoologicus, 1946-1955. Published for the proprietors by the Zoological Society of London, Vol. VI. London.

Edwards, M.A., A.T. Hopwood & H.G. Vevers (eds). 1975. Nomenclator Zoologicus, 1956-1965. Published for the proprietors by the Zoological Society of London, Vol VII. London.

Edwards, M.A., A.T. Hopwood & M.A. Tobias (eds). 1993. Nomenclator Zoologicus, 1966-1977. Published for the proprietors by the Zoological Society of London, Vol. VIII. London.

Ehrlich, P.R. 1961. Has the biological species concept outlived its usefulness? *Syst. Zool.,* **10:** 167-176.

Ehrlich, P.R. 1964. Some axioms of taxonomy. *Syst. Zool.,* **13:** 109-123.

Ehrlich, P.R. and P. Raven. 1969. Differentiation of populations. *Science,* **165:** 1228-1232.

Elton, G. 1947. *Animal ecology.* Sidgwick & Jackson Ltd., London., 209 pp.

Emerson, A.E. 1941. Taxonomy and ecology. *Ecology,* **22:** 213.

Endler, J.A. 1977. *Geographic variation, speciation and clines.* Monographs in population biology. No. 10. Princeton Univ. Press, ix + 246 pp.

Ereshefsky, M. (ed.). 1991. The units of evolution: Essays on the nature of species. MIT Press, Cambridge, Massachusetts.

Essig, E.O. 1942. The significance of taxonomy in general field of economic entomology. *Econ. Ent.,* **35:** 739-743.

Evans, H.E. 1947. *Studies on the comparative ethology of digger wasps of the genus Bombix.* Cornell Univ. Press, Ithaca, N.Y., 248 pp.

Evans, H.E. 1966. The comparative ethology and evolution of digger wasps as suggested by larval characters (Hymenoptera: Sphecoidea). *Ent. News,* **75:** 225-237.

Farris, J.S. 1974. Formal definitions of paraphyly. *Syst Zool.,* **23:** 548-554.

Farris, J.S. 1977. On the phenetic approach to vertebrate classification. In *Major patterns in vertebrate evolution.* Hecht, M.K.. D.C. Goody and B.M. Hecht (eds.), Plenum Press, N.Y., pp. 823-849.

Ferguson, A. 1979. *Biochemical systematics and evolution.* Blackie & sons Ltd., Bishoporiggs, Glasgow, U.K., 208 pp.

Fernald, H.T. 1939-40. On type nomenclature. *Ann. Ent. Soc. Am.,* **32:** 689-702.

Ferris, G.F. 1928. The principles of systematic entomology. Stanford Univ. Publ., *Biol. Sci.,* **5:** 103-269.

Fischer, F.C. 1966. List of principal kinds of types. *Ent. Ber.,* **26:** 1-5.

Fitch, W.M. and E. Margolias. 1967. Construction of phylogenetic trees. *Nature,* **155:** 279-284.

Flann, C. 2011. BioCode: third time lucky? *Zootaxa,* **2874:** 38-40.

Florkin, M. 1964. *Protides of the Biological Fluids* (H. Peeters). Elsevier, Amsterdam.

Florkin, M. and C. Jeuniaux. 1964. In *Physiology of insects* (ed.), M. Rockstein. 10-152 pp.

Florkin, M. and H.S. Mason (eds.). 1960-65. *Comparative biochemistry—A comprehensive treatise.* 7 Vols. Academic Press, N.Y.

Ford, E.B. 1957. Polymorphism in plants, animals and man. Nature, **180:** 1315-1319.

Forget, P.M., J. Lebbe, H. Puig, R. Vignes and M.H. Hideux. 1986.Microcomputer-aided identification: An application to trees from French Guiana. *Bot. J. Linn. Soc.,* **93:** 205-223.

Forsberg, F.R. 1972. The value of systematics in the environmental crisis. *Taxon,* **21:** 631-634.

Fox, A.S. 1956. Application of paper chromatography to taxonomic studies. *Science,* **123:** 143.

Frison, T.H. 1942. The application of economic entomology in the field of insect taxonomy. *J. Econ. Ent,* **35:** 749-752.

Frizzel, D.L. 1933. Terminology of types. *Am. Midi.,* **14:** 637-668.

Funk, V.A. and K.S. Richardson. 2002. Systematic data in biodiversity studies: use it or lose it. *Syst. Biol.,* **51**(2): 303-316.

Gaston, K.J. and Mound, L.H. 1993. Taxonomy, hypothesis testing, and the biodiversity of crisis. *Proc. Roy. Soc. Lond.,* **251:** 139-142.

Germigni, V. 1979. An ultrastructural approach to planarian taxonomy. *Syst. Zool.,* **28**(3): 345-355.

Ghiradella, H., R.E. Silberglied., D. Aneshansley, H.E. Hinton, and T. Eisner. 1972. Ultraviolet reflection of a male butterfly interference colour caused by thin layer elaboration of wing scales *(Eurema lisa—* Lepidoptera, Pieridae). *Science,* **178:** 1214-1217.

Ghiselin, M.T. 1966. An application of theory of definition.s to systematic principles. *Syst. Zool.,* **15:** 127-130.

Ghiselin, M.T. 1974. A radical solution to the species problem. *Syst. Zool.,* **23:** 526-544.

Ghiselin, M.T. 1987. Species concepts, individuality and objectivity. Biology and Philosophy, **2:** 127-143.

Gier, L.J. 1965. *Principles of taxonomy.* Liberty, Missouri, Published by the author.

Gilmore, J.S.L. 1951. The development of taxonomic theory since 1851. *Nature,* **168:** 400-402.

Gilmore, J.S.L. 1961. Taxonomy. In *Contemporary Botanical Thoughts.* A.M. McLead and L.S. Cobley (eds.). Oliver and Boyd, Edinburgh.

Gilmore, J.S.L. and J.W. Gregor. 1939. Demes: a suggested new terminology. *Nature,* London, **144:** 333-334.

Gingsbrug, I. 1938. Arithmatic definition of the species, sub-species and race concept, with a proposal for a modified nomenclature. *Zoologica,* **23:** 253-286.

Gisin, H. 1964. Syntheische Theorie der Systematik. *Z. Zool., Syst. Evol. Forsch.,* **2:** 1-17.

Gochfeld, M. 1940. *The material basis of evolution.* Yale Univ. Press, New Haven, xi + 436 pp. Gochfeld, M. 1974. Terms for highly similar species. *Syst. Zool.,* **23:** 156.

Godfray, H.C.J. 2002. Challenges for taxonomy. *Nature,* **417:** 17-19.

Goldschmidt, R. 1934. Lymantria. *Bibliog. Genet.*, **11:** 1-185.

Gomori, G. 1952. Microscopic histochemistry. Univ. Chicago Press.

Goldon, I. 1955. Importance of larval characters in classification. *Nature*, **176:** 911-912.

Gould, S. 1954. Permanent numbers to supplement the binominal system. *Am. Sci.*, **42:** 269-274.

Goyal, P., Royal, S., Moniz, E., Gang, W., Cafiete, M.A., and Mazrouel, S.M.A. 2015. Changing the climate, together. The Times of India, July 31, p. 24.

Graham, C.H., S. Ferrier, F. Huettman, C. Moritz, and A.T. Peterson. 2004. New deveopments in museum-based informatics and applications in biodiversity analysis. *Trends in Ecol. Evol.*, **19:** 497-503.

Grant, V. 1971. *Plant speciation.* Columbia Univ. Press, N.Y., 435 pp.

Grant, W.F. 1960. The categories of classical and experimental taxonomy and the species concept. *Rev. Canad. Biol.*, **19:** 241-262.

Green, P.E. and F.J. Carmone. 1970. Multidimensional scaling and related techniques in marketing analysis. Allyn and Bacon, Boston, 203 pp.

Gregg, J.R. 1954. The language of taxonomy: an application of symbiotic logic to the study of classification systems. Columbia Univ. Press, N.Y., 70 pp.

Greuter, W., G. Garrity, D.L. Hawksworth, R. John, P.M. Kirk, S., Knapp, J., McNeill, E. Michel, D.J. Patterson, R. Pyle and B.J. Tindall, 2011. Draft Biocode (2011): Principles and Rules Regulating the Naming of Organisms. Taxon, **60:** 201-212.

Greuter, W., D.L. Hawksworth, J. McNeill, M.A. Mayo, A. Minelli, P.H.A. Sneath, B.J. Tindall, P. Trehane, P. Tubbs and IUBS/IUMS International Committee for Bionomenclature. 1996. Draft Biocode: The Prospective International Rules for the Scientific Names of Organisms. *Taxon*, **45:** 349-372.

Grewal, J.S. 1982. Cytotaxonomy and evolution of fruit flies (Diptera, Tephritidae). Ph.D. Thesis, Punjab Agricultural University, Ludhiana, India, pp. 166 + 42 plates.

Griffiths, G.C.D. 1976. The future of Linnaean nomenclature. *Syst. Zool*, **25(2):** 168-173.

Griffiths, G. 1946. The ills of taxonomy. *Proc. Trans. Texas Acad Sci.*, **30:** 69-73.

Guerra-Garcia, J.M., F. Espinosa and J.C. Garcia-Gomez. 2008. Trends in Taxonomy today: an overview about the main topics in Taxomony. *Zool. Baetica*, **19:** 15-49.

Gunther, K. 1962. Sustematik und Stammegschichte der Tiere. *Fortschritte der zoologie.*, **14:** 268-547.

Gurney, A.B. 1962. What was taxonomy in 1970? *Syst. Zool.*, **11:** 92-93.

Gyllenberg, H.G. 1963. A general method for deriving determinative schemes for random collections of microbial isolates. *Ann. Acad. Sci. Fenn. Ser. A, V. Biologica*, **69:** 5-22.

Hackett, L.W. 1937. *Malaria in Europe.* Oxford Univ. Press, London, xvi + 366 pp.

Hagmeir, E.M. 1958. Inapplicability of the subspecies concept to North American marten. *Syst. Zool.*, **7(1):** 1-7.

Hall, A.V. 1967. Methods for demonstrating resemblance in taxonomy and ecology. *Nature*, **214:** 830-831.

Hall, A.V. 1968. Methods for aiding identification of critical groups in taxonomy. *Nature*, **218:** 203.

Hall, A.V. 1970. A computer-based system for forming identification keys. *Taxon*, **19:** 12-18.

Handler, P. (ed.) 1964. Biochemistry symposium: Biochemical evolution. *Fed. Proc*, **23:** 1229-1266.

Hardy, D.E. 1982. The role of taxonomy and systematics in integrated pest management programs. *Prot. Ecol.*, **96:** 231-238.

Harlin, M. 2005. Definitions and Phlygenetic nomenclature. *Proc. Calif. Acad. Scs.*, **56:** 216-224.

Harlow, R.D., R.H. Lumb and R. Wood. 1969. Insect lipids: carbon number distribution of triglycerides in five species. *Comp. Biochem. Physiol.*, **30:** 761-769.

Hatch, M.H. 1941. The logical basis of the species concept. *Am. Nat.*, **75:** 193-212.

Hausdorf, B. 2011. Progress toward a general species concept. 2011. *Evol.*, **65(4):** 923-931.

Hausdorf, B. and C. Hennig. 2010. Species delimitation using dominant and codominant multilocus markers. *Syst.Biol.*, **59:** 491-503.

Hawkes, J.C. (ed). 1968. *Symposium on chemotaxonomy and serotaxonomy.* Academic Press, London. & N.Y., 299 pp.

Hawksworth, D.L. 2011 Introducing the Draft Biocode (2011). *Taxon.* **60:** 199-200.

Herbert P.D.N, and T.R. Gregory. 2005. The promise of DNA barcoding for taxonomy. *Syst.Biol.*, **54:** 852-859.

Hebert, P.D.N., A. Cywinska, S.L. Ball and J.R. de Waard. 2003. Biological idenfications through DNA barcodes. *Proc. R. Soc. London* (B), **270:** 313-322.

Hedberg, O. 1958. The taxonomic treatment of vicarious taxa. *Upps. Univ Arsskr.*, **6:** 186-195.

Hegberg, D. 1977. *Systematics.* Systematics Development Inc. P.O. Box 52, Pasadena CA 91102, xii + 748.

Heller, J.C. 1964. The early history of binominal nomenclature. *Huntia*, **1:** 33-70.

Hennig, W. 1950. Grudnzuege einer theorie der phylogenetischen systematik, Deut zentral-verlage., Berlin, 370 pp.

Hennig, W. 1966. *Phylogenetic systematics.* Univ. Illinois Press, III, 263 pp. Herrara, J. and M. Ethcheverry. 1960. El concepto de especie (meaning of the term species). *Rev. Uni. Catol.* Chile, Santiago, 44-45: 157-163.

Herrera, A.L. 1899. See opinion 72, Internat. Commission on Zool, nomenclature. *Smithosn. Inst. Misc. Coll.*, **73:** 19, 1922.

Heslop-Harrison, J. 1962. Purposes and procedures in the taxonomic treatment of higher organisms. *Symp. Soc. Gsn. Microbiol.*, No. **12:** 14-36.

Heslop-Harrison, J. 1963. Species concepts: Theoretical and practical aspects. In *Chemical plant taxonomy*, (ed. T. Swan). Academic Press Inc., N.Y., pp. 17-40.

Heywood, V.H. (ed.). 1968. *Modern methods in plant taxonomy*. New York, Academic Press, 312 pp.

Heywood, V.H. (ed.). 1973. *Taxonomy and Ecology*. Systematic Association special Vol. 5. Academic Press, London & New York. 370 pp.

Hoffmann, C.H., H.K. Townes, H.H. Swift and R.I. Sailor. 1949. Field studies on the effects of aeroplane application of DDT on forest invertebrates. *Ecological Monographs*. **19**: 1-46.

Hogben, L. 1940. Problems of the origin of species. In *New Systematics* (ed. J.S, Huxley). Oxford Univ., Press, London. Holland, W.J. 1929. Forum on types. *Trans. 4th Internal Cong. Ent.*, N.Y., pp. 688-694.

Hopwood, A.T. 1950. Animal classification from Greeks to Linnaeus. In T.A Sprague *et al.; Linn. Soc. London*.

Horn, W. 1929. The future of insect taxonomy. Trans *4th Int Congr. Ent.*, Ithaca. 1928, **2**: 34-51.

Horovitz, I. and A. Meyer. 1995. Systematic of new world monkeys (Platyrrhim, Primates) based on 16S mDNA sequences: A comparative analysis of different working methods in cladistic analysis. *Mol. Phyl. Evol.*, **4(4)**: 448-456.

Hoyer, H.B., B.J. McCarthy and E.T. Bolton. 1964. A molecular approach in the systematics of higher organisms. *Science*, **144**: 959-967.

Hubbell, T.H. 1954. The naming of geographically variant populations. *Syst. Zool.*, **3**: 113-121.

Hubby, J.L. 1963. Protein differences in *Drosophila melanogaster*. I. *Genetics*, **51**: 671-679.

Hubby, J.L. 1964. Protein differences in *Drosophila*. II. Comparative species genetics and evolutionary problems. *Genetics*, **52**: 203-215.

Hubby, J.L. and R.C. Lewontin. 1966. A molecular approach to the study of genie heterozygosity in natural populations. I. The number of alleles at different loci in *Drosophila pseudoobscura*. *Genetics*, **54**: 577-594.

Hubby, J.L. and L.H. Throckmorton. 1960. Evolution and pteridine metabolism in the genus *Drosophila*. *Proc. Nat. Acad. Sci. U.S.A.*, **46**: 65-78.

Hughea-Schrader, S. 1958. The DNA content of the nucleus as a tool in the cytotaxonomic study of insects. *Proc. 10th Internat. Congr. Ent.*, Montreal.,1956, **2**: 935-944.

Hull, D.L. 1964. Consistency and monophyly. *Syst. Zool.*, **25(2)**: 174-191.

Hull, D.L. 1965. The effect of essentialism on taxonomy. *Brit. J. Phil Sci.*, **15**: 314-326,

Hull, D.L. 1976. Are species really individuals? *Syst. Zool.*, **25(2)**: 174-191.

Hull, D.L. 1980. Individuality and selection. *Ann. Rev Ecol. & Syst.*, **11**: 311-332.

Hunter, R.L. and C.L. Market. 1957. Histochemical demonstration of enzymes separated by zone electrophoresis in starch gels. *Science*, **125**: 1294-1295.

Huxley, J.S. 1939. Clines: an auxiliary method in taxonomy. *Bijdr. Dierk.*, **27**: 491-520

Huxley, J.S. (ed). 1940. *The New Systematics.* Oxford Univ. Press. London. 538 pp.

Huxley, J.S. 1942. Evolution. The modern synthesis. Harper & Brothers. New York.

ICZN. 2014. Zoological Nomenclature and Electronic Publication- a reply to Dubois, *et. al.*, (2013). Zootaxa, **3779(1)**: 3-5.

Inger, R.E. 1958. Comments on the definition of genera. *Evolution,* **12**: 370-384.

Inger, R.E. 1961. Problems in the application of the subspecies concept in vertebrate taxonomy. In *Vertebrate speciation* (ed. W.F. Blair) Univ. Texas Press. Austin, 262-285.

Inglis, W.G. 1S70. The purpose and judgement of biological classification. *Syst. Zool.* **19**: 240-250.

International Commission on Zoological nomenclature. 2008. Proposed amendment of the International Code of Zoological Nomenclature to expand and refine methods of publication. *Bull. Zool. Nomencl.,* **65(4)**: 265-275. 2008 *Zootaxa.* 1908: 57-67: also in others like African Invertebrates in 2008, *Zool. J. Linn. Soc.* 2009, and *J. Crust. Biol.,* 2008.

Irwin, M.R. 1947 Immunogenetics. *Adv. in Genetics* **1**: 133-159.

Isaac, N.J.B., J. Mallet and G.M. Mace. 2004 Taxonomic inflation: its influence on macroecology and conservation *Trends Ecol. Eyol.,* **19**: 464-469.

IUBS. 2010. International Union for biological Sciences: Biocode on website-http:/www.iubs.org/prg/biocode.dp.html.

IUSE8. 2011. Resolution voted on by the symposium on "Modernizing the nomenclatural codes to meet future needs of scientific communities (BioCode)" — Available from http://www.bionomenclature.net/documents/IOSEB-vii-resolution.pdf (March 2011).

Jaeger, E.C. 1950. *A source boo of biological names and terms (2nd ed.).* Springfield, llinois, C.C. Thomas.

Jameson, D.L. 1955. Evolutionary trends in the courtship and mating behaviour of *Silientia. Syst. Zool.* **4**: 105-119.

Jardine, C.J., N. Jardine and R. Sibson. 1967. The structure and construction of taxonomic hierarchies. *Math. Biosc.* 1: 173-179.

Jardine, N. 1969. A logical basis for biological classification. *Syst. Zool.,* 18: 37-52.

Jardine, N. and R. Sibson. 1971. *Mathematical taxonomy.* John Wiley & Sons Ltd., 286 pp.

Jeffrey, C. 1977. *Biological nomenclature.* Indian Ed., Oxford & IBH Pub. Co. New Delhi, 72 pp.

Jensen, R.J. 1993. Report. The 25th International Numerical Taxonomy Conference. *Syst. Biol.,* 42: 599-601.

Johnsgard, P.A. 1965. *Handbook of waterfowl behaviour.* Cornell Univ. Press, Ithaca, N.Y.

Johnson, F.M., H.E. Schaffer, J.E. Gillsdpy and E.S. Rockwood. 1968-69. Isozyme genotype-environment relationship in natural populations of the harvester ant, *Pogonomyrmex barbatus* from Texas. *Biochem. Genetics,* **3(5):** 429-450.

Johnson, L.A.S. 1970. Rainbow's end: The quest for an optimal taxonomy. *Syst. Zool.,* **19:** 203-239.

Johnson, M.L. 1974. Mammals. In *Biochemical and Immunological taxonomy of Animals* (ed. by C.A. Wright). Academic Press, pp. 1-87.

Johnson, N.K. and C. Cicero. 2004. New mitochondrial data affirming the importance of Pleistocene speciation in North American birds. *Evol.,* **58(5):**1122-1130.

Kaplan, N.O., M.M. Ciotti, M. Hamolsky and R.E. Bieber. 1960. Molecular heterogenecity and evolution of enzymes. *Science.* **131:** 392-398.

Kapoor. V.C. 1973. The importance of zootaxonomy in economic biology. *Everyday Science,* **10(4):** 25-31.

Kapoor, V.C. 1982. *Insect Classification: A History.* Vijaya Publ., Ludhiana, 84 pp.

Kapoor, V.C. and M.L. Agarwal. 1980. Recent trends in animal taxonomy. *Everyday Science,* **18:** 21-27.

Keck, D.D. 1957. Trends in systematic botany. In *Survey of biological progress* (ed. B. Glass), **3:** 47-107.

Keen, A.M. and S.W. Mulier. 1956. *Procedure in taxonomy.* Stanford Univ. Press, California.

Keifer, H.H. 1944. Importance of applied entomological taxonomy. *Pan Pacific Ent.,* **20:** 1-6.

Keirans, J.E., Harry Hoogstraal and C.M. Clifford. 1979. Observations on the subgenus *Argas* (Ixodoidea: Argasidae, *Argas*). *J. Med. Entomol.,* **15(3):** 246-252.

Kelly, R.P., I.N. Sarkar, D.J. Eernisse and R. Desalle. 2007. DNA barcoding using chitons (genus *Mopalla*). *Mol. Ecol. notes.* **7:** 177-83.

Kessel, E.L. 1955a. The mating activities of balloon flies. *Syst. Zool.,* **4:** 96-104.

Kessel, E.L. (ed). 1955b. *A century of progress in the natural sciences.* San Francisco. California Acad. Sciences, pp. 75-87.

Kiauta, B. 1967. Distribution of the chromosome numbers in Trichoptera in the light of phylogenetic evidence. *Genen en Phaemen,* **12:** 110-113.

Kinsey, A.C. 1930. The gall wasp genus *Cynips*. A study in the origin of species. *Ind. Univ. Studies,* 84-86: 577 pp.

Kiriakoff, S.G. 1956. On the subspecies concept in taxonomy. *Lep. News,* **10:** 207-208.

Kiriakoff, S.G. 1962. On the Neo-Adansonian School. *Syst. Zool.,* **11:** 180-185.

Kiriakoff, S.G. 1965. Some remarks on Sokal and Sneath's principle of numerical taxonomy. *Syst. Zool.,* **14:** 61-64.

Kirk, R.L., A.R. Mian and F.G. Beyer. 1954. The use of paper partition chromatography for taxonomic studies of land snails. *Biochem J.,* **57(3):** 440-442.

Klass, K.D., O. Zompro, N.P. Kristensen, and J. Adis 2002. Mantosphasmatodea: a new order with extant members in Afro-tropics. *Science,* **296:** 1456.

Kleinschmidt, O. 1962. Die Formenkreislehre und das weltwerden des lebens. Gebauer-Schweitzer, Halls-S., 188 pp.

Knapp, S., A. Polaszek and M. Watson. 2007. Spreading the word. *Nature,* **446:** 261-62.

Knudsen, J.W. 1966. *Biological techniques.* Harper & Row, Inc., New York, 525 pp.

Kohn, A.J. and G.H. Orians. 1962. Ecological data in the classification of closely related species. *Syst. Zool.,* **11:** 119-127.

Kohne, D.E. 1968. Taxonomic application of DNA hybridization techniques. In *Chemo- and Sero-taxonomy* (ed. Hawkes), 117-130 pp.

Krauss, O. and W.D.L. Ride. 1995. International Code of Zoological Nomenclature: discussion draft of the proposed fourth edition. *Syst. Ent.,* **20:** 375-377.

Kraus, R. 1897. Wien, Klin. *Wochenschr.,* **10:** 736.

Krzysztof, S., Harry Hoogstraal, C.M. Clifford and H.Y. Wassef. 1979. Observations on the subgenus *Argas* (Ixodoidea: Argasidae: *Argas).* 17. *Argas (A.) polonicus* n. sp. parasitizing domestic pigeons in Krahow, Poland. *J. Parasitol.,* **65(1):** 170-181.

Lahni, F. 1964. Viruses and molecular taxonomy; In *Plant Virology.* Univ. of Florida Press, Gainesville, 527 pp.

Lambert, D.M., B. Michaux, and C.S. White. 1987. Are species self defining? *Syst. Zool.,* **36:** 196-205.

Lanham, V. 1965. Uninomenclature. *Syst. Zool.,* **14:** 144.

Lanier, G.N. 1972. Biosystematics of the genus *Ips* (Coleoptera: Scolytidae) in North America. *The Canadian Ent.,* **104:** 361-388.

Laubenfels, M.W. de. 1953. Trivial names. *Syst. Zool.,* **2:** 42-45.

Laurin, M. and P.D. Candino. 2007. Second meeting of the International Society for Phylogenetic Nomenclature: a report. *Zool.Scripta,* **36:** 109-117.

Lebbe. J. 1984. Manuel d' utilisation de logicel XPER. Microapplication, Paris. pp. 1-150.

Leone, C.A. (ed). 1964. *Taxonomic biochemistry and serology.* The Ronald Press Co.. N.Y., 728 pp.

Lerman, S.C. 1970. *Les bases de la classification auiomatique* (Publ. Inst. Programmation Fac. Sci. Paris): Gauthier-Villars. 136 pp.

Le Quesne, W.J. 1969. A method for selection of characters in numerical taxonomy. Sysf. *Zool.,* **18:** 201-205.

Levi, C. 1956. Etude des Halisara de Roscoff. Embrylogie et systematique des Demospongs. *Arch. Zool. Exptl. Gen..* **93:** 1-181.

Levi, C. 1957. Ontogeny and systematics. *Syst. Zool.,* **6:** 174-183.

Levi, H.W. 1966. The care of alcoholic collections of small invertebrates. *Syst. Zool,* **15:** 183-188.

Lewis, H. 1957. Genetics and cytology in relation to taxonomy. *Taxon.* **6:** 42-46.

Lewis, H. 1969. Comparative cytology in systematics. pp. 523-533. In *Systematic Biology,* U.S. Nat. Acad. Sci. Publ. No. 1692.

Lewontin, R.L. and J.L. Hubby. 1966. A molecular approach to the study o* genie heterozygosity in natural populations. II. Amount of variation and degree of heterozygosity in natural populations of *D. pseudoobscura. Genetics,* **54:** 595-609.

Liddell, H.G. and R. Scot. 1864. *A Greek English Lexion,* 5th ed. Clarendon Press, Oxford.

Linnaeus, C. 1758. Systema naturae per regma tria naturae, secundum classes, ordines, genera, species cum characteribus, differntis, synonymis, locis. Edito decima, reformata, Tom. I. Laurentii Salvii. Holmiae, 824 pp.

Linsley, E.G. 1944. The naming of infraspecific categories. *Ent. News,* **55:** 225-232.

Linsley, E.G. and R.L. Usinger. 1959. Linnaeus and the development of the Code of Zoological Nomenclature. *Syst. Zool,* **8(1):** 39-47.

Linsley, E.G. 1961. Taxonomy. In *Encyclopedia of the biological sciences,* (ed. P. Gray). Reinhold Pub. Corp., New York.

Long, Samuel S. 1996. Tautonyms in Biology. *Word Ways,* **29(3):** 146-150; **29(4):** 253-257.

Lotsy, J.P. 1925. Species of Linnean. *Genetica,* **7:** 487-506.

Love, A. 1964. The biological species concept and its evolutionary structure. *Taxon,* **13:** 33-45.

Macfayden, A. 1955. A comparison of methods for extracting soil arthropods. Soil *Zool,* **1955:** 315-332.

Maeki, K. 1958. A use of chromosome numbers in the study of taxonomy of the Lepidoptera. *Lep. News,* **11:** 8-9.

Maerz, A. and M.R. Paul. 1950. *A dictionary of colour,* 2nd ed. McGraw Hill Book Co., N.Y., 208 pp.

Malaise, R. 1948. Keeping specimens free from mould and pests in any climate. *Proc. 8th. Intrn. Congr. Ent,* Stockholm, pp. 926-927.

Mallet, J. 1995. A species definition for the Modern Synthesis. *Trends Ecol. Evol.,* 10:294-99.

Mallet, J. 2005. Hybridization as an invasion of the genome. *Trends Ecol. Evol.* **20:** 229-37.

Mallet, J. 2008. Hybridization, ecological races and nature of species: empirical evidence for the case of speciation. *Phil.R.Soc.Lond. B* **363:** 2971-2986.

Mallet, J. and K. Willmott. 2003. Taxonomy: renaissance or Tower of Babel? *Trends Ecol.Evol.,* **18:** 57-58.

Manfredi Romanini, M.G. 1973. The DNA nuclear content and the evolution of vertebrates. In *Cytotaxonomy and vertebrate evolution* (ed. A.B. Chiarelli and E. Campana). Academic Press, pp. 39-81.

Manwell, C. 1970. *Molecular biology and the origin of species.* Sidgwicks Jackson, U.S.A., 394 pp.

Manwell, C. and CM.A. Baker. 1963. A sibling species of sea cucumber discovered by starch-gel electrophoresis. *Comp. Biochem. Physiol.,* **10:** 39-53.

Manwell, C, CM.A. Baker and W. Chliders. 1967, The genetics of haemoglobin in hybrids. I.A molecular basis for hybrid vigour. *Comp. Biochem. Physiol..* **10:** 103-120.

Marrnur, J., S. Falkow and S. Mandel. 1963. New approaches to bacterial taxonomy. *Ann. Rev Microbiol.* **17**: 329-372.

Martin, P.G. and D.L. Hyman. 1965. A quantitative method for comparing the karyotypes of related-species. *Evolution.* **19**: 157.

Martin, S. and F. Cotner. 1934. Serological studies of proteins with special reference to their phyiogenetic significance. *Ann Ent Soc. Am.* **27**: 372-383.

Maslin, T.P. 1952. Methodological criteria of phyletic relationships. *Syst. Zool.,* **1**: 49-70.

Mason, H.L. 1950. Taxonomy, systematic botany and biosystematics. *Madrono,* **10**: 193-208.

Mauly, P. & M.A. Tobias (eds), 1996. Nomenclator Zooiogicus. 1978-1994. Published for the proprietors by the Zoological Society of London, Vol. IX. London.

Mayr, E. 1931. Birds collected during Whitney South Sea expedition. 12. Notes on *Halcyon chloris* and some of its subspecies. *Am. Mus. Novitates,* **469**: 1-10.

Mayr, E. 1940. Speciation phenomena in birds. *Amer. Nat,* 74: 249-278.

Mayr, E. 1942. *Systematics and the origin of species from the viewpoint of a zoologist* Columbia Univ. Press, N.Y., 334 pp.

Mayr, E. 1949. The species concept: semantics versus sematics. *Evolution,* **3**: 371-373.

Mayr, E. 1953. Concepts of classification in higher organisms and microorganisms. *Ann. N.Y. Acad. Sci.,* **56**. 391-397.

Mayr, E. 1954. Notes on nomenclature and classification. *Syst. Zool.,* **3**: 86-89.

Mayr, E. 1957. Species concept and definitions in species problem. *Am. Assoc. Adv. Sci. Publ.* No. **50**: 39.

Mayr, E. 1958. Behaviour and systematics, In *Behaviour and Evolution* (eds. A. Roe and G.G. Simpson). Yale Univ. Press, New Haven, 341-362.

Mayr, E. 1963. *Animal species and evolution.* The Belknap Press. Harvard Univ. Press, Cambridge, Mass., 797 pp.

Mayr, E. 1964a. From molecules to organic diversity. *Fed. Proc.,* **23**: 1231-1235.

Mayr, E. 1964b. The new systematics. In *Taxonomic biochemistry and serology,* C.A. Leone (ed.). The Ronald Press. Co. N.Y., pp. 13-32.

Mayr, E. 1965a. Classification and phylogeny. *Am. Zool.,* **5**: 165-174.

Mayr, E. 1965b. Classification, identification and sequence of genera and species. *L' Oiseau,* 35 special No., pp. 90-95. Mayr, E. 1966. The proper spelling of taxonomy. *Syst. Zool.,* **15**: 88.

Mayr, E. 1968a. The role of systematics in biology. *Science,* **159**: 595-599. Mayr, E. 1968b. Theory of biological classification. *Nature.* **220**: 545-548.

Mayr, E. 1969. *Principles of systematic zoology.* McGraw-Hill, N.Y., 428 pp.

Mayr, E. 1970. *Populations, species and evolution.* Cambridge, Mass. Harvard Univ. Press, 453 pp.

Mayr, E. 1971. Methods and strategies in taxonomic research. *Syst. Zool.,* **20**: 426-433.

Mayr, E. 1976. is species a class or an individual? *Syst. Zool.*, **23**: 192.

Mayr, E. 1978. Origin and history of some terms in systematics and evolutionary biology *Syst. Zool.*, **27**: 112-118.

Mayr, E. and Amadon. A. 1951. A classification of recent birds. *Am. Mus. Novit.*, No. **1496**: 42 pp.

Mayr E. and R. Goodwin. 1956 Biological materials. Pt. I. Preserved materials and museum collections. Wash. D C. Publ. 399, NAS/NRC, *Bio. Council Div. Bio. Agric.*

Mayr, E., E.G. Linsley and R.L. Usinger. 1953. *Methods and principles of systematic zoology* McGraw-Hill, N.Y., 328 pp. McCafferty, W.P. and L. Chandler. 1974. Denotations of some comparative systematics terminology. *Syst. Zool.*, **23**: 139-140.

McCarthy, B.J. and E.T. Bolton. 1963 Approach to the measurement of genetic relatedness among organisms. *Proc. Nail. Acad. Sci.*, U.S.A., **50**(1): 156-164.

McMillan, C. 1954. Parallelisms between ecology and taxonomy. *Ecology*, **35**: 92-94.

McNeill, J. 1996. General Introduction to Draft BioCode. In: International Committee on Bionomenclature **(1996),** pp. 7-18.

Meglitsch, P.A. 1954. On the nature of species. *Syst. Zool.*, **3**: 49-65.

Meikle, R.D. 1957. What is subspecies? *Taxon*, **6**: 102-105. Menon, M.G.R. 1965. *Systematics of Indian insects.* In *Entomology in India.* Ent Soc. India, pp. 70-87.

Menon, M.G.R. 1981. Physiological approach in insect taxonomy. In *Insect Physiology and Anatomy* (eds.) N.C. Pant and S. Ghai. I.C.A.R., New Delhi, pp. 30-41 (1st edn. 1973).

Merril, E.D. 1943. Some economic aspects of taxonomy. *Torreya*, **43**: 50-64.

Meryman, H.T. 1960. The preparation of biological museum specimens by freeze-drying. *Curator*, **3**: 5-19.

Metcalf, Z.P. 1954. The construction of keys. *Syst. Zool.*, **3**: 38-45.

Meyer, A. 1926. Taxonomic terminology. Logik der Morphologie, pp. 126-137, Delm.

Meyer, C.P. and G. Paulav. 2005. DNA barcoding: error rates based on comprehensive sampling. Plos *Biol.*, **3**(12): 422.

Michener, C.D. 1953. Life-history studies in insect systematics. *Syst. Zool.*, **2**: 112-1.18.

Michener, C.D. 1957. A quantitative approach to a problem in classification.*Evolution*, **11**: 130-162.

Michener, C.D. 1963. Some future developments in taxonomy. *Syst Zool.*, **12**: 151-174.

Michener, C.D. 1964. The possible use of uninominal nomenclature to increase the stability of names in biology. *Syst. Zool.*, **13**: 182-190.

Michener, C.D. 1956. Paper chromatography in insect taxonomy. *Ann. Ent. Soc. Amer.*, **49**: 576-581.

Micks, D.W. 1954. Paper chromatography as a tool for mosquito taxonomy. The *Culex pipiens* complex. *Nature*, **174**: 217-221.

Micks, D.W. and A.A. Benedict. 1953. Infrared spectrophotometry as a means for identification of mosquitoes. *Proc. Soc. Exp. Biol. Med.,* **84:** 12-14.

Micks, D.W., A. Rehmet and J. Jennings. 1966. Biochemical differentiation of morphologically indistinguishable strains of *Aedes aegypti* (Diptera: Culicidae). *Ann. Ent. Soc. Am.,* **59(2):** 239-244.

Miller, R.R. 1956. Plastic bags for carrying and shipping live fish, *Copeia,* **43(2):** 118-119.

Miller, R.R. 1957. Utilization of X-rays as a tool in systematic zoology. *Syst. Zool.,* 6: 29-40.

Minelli, A. 1993. Biological Systematics: the state of the art. Chapman & Hall, London, xvi, 387 pp.

Minelli, A. 1995. The changing paradigms of biological systematics: new challenges to the principles and practice of biological nomenclature. *Bull. Zool. Nomencl,* **52(4):** 303-309.

Minelli, A. 2008. Zoological vs. botanical nomenclature: a forgotten 'BioCode' experiment from the times of the Strickland's Code. *Zootaxa,* **1950:** 21-38.

Minkoff, E.C. 1956. Effects on classification of slight alterations in numerical taxonomy. *Syst. Zool.,* **14:** 196-213.

Mishler, B.D. and M.J. Donoghue. 1982. Species concepts: A case for pluralism. *Syst. Zool.,* **31:** 491-503.

Minst, K.J. 1958. Origin of Linnaean signs z and z. *Anz. Schadt,* Berlin, **31:** 185-188.

Mittal, O.P., V. Sawhney and V.C. Kapoor. 1974. On the status of *Labidura bengalensis* (Dohrn). *Oriental Ins.,* **8(4):** 541-543.

Monagham, M.T., R. Wild, M. Elliot, T. Fugisawa, M. Balke, D.J.G. Inward, D.C. Lees, R. Ranaivosolo, P. Eggletion, T.G. Barra Clough and others. 2009. Accelerated species inventory to Madagascar using coalescent-based models of species delimitation. *Syst.Biol.,* **58:** 298-311.

Moritz, C. and C. Cicero. 2004. DNA barcoding: Promise and Pitfalls; http://biology:plosjournals.org/perlserv/?request-pdf&file=10.1371.

Morse, L.E. 1970. A computer-based system for forming identification keys. *Taxon,* **28:** 35-38.

Morse, L.E. 1971. Specimens identification and key construction with time sharing computers. *Taxon,* **20:** 269-282.

Morse, L.E. 1974. Computer programmes for specimen identification, key construction and description printing using taxonomic data matrices. *Publ. Mus. Mich. Stat. Univ. Biol. Ser.,* **5:** 1-128.

Morse, L.E. 1975. Recent advances in the theory and practice of biological specimen identification, pages 11-52. In Biological identification with computers R.J. Pankhurst, (ed.). The Systematics Association Special Vol. No. 7. Academic Press, London & New York.

Moss, W.W. 1967. Some new analytic and graphic approaches to numerical taxonomy with an example from the Dermanyssidae (Acari). *Syst. Zool.,* **16:** 177-207.

Moss, W.W. and J.A. Hendrickson Jr. 1973. Numerical Taxonomy. *Ann. Rev. Ent,* **18**: 227-258.

Moylan, E., H. Harold, P. Harris, C. Foote, A. Arme, A. Menelli, M. Kowalczuk and C. Black. 2014. Online-only publishers are here to stay and will continue to work closely with ICZN. Zootaxa, **3779** (1): 6.

Moynihan, M. 1959. A revision of the family Laridae (Aves.). *Amer. Mus. Novitates,* No. **19**: 42 pp.

Muesebeck, C.F.W. 1942. Fundamental taxonomic problems in quarantine and nursery inspection. *J. Econ. Ent,* **35**: 753-758.

Munroe, E. 1960. An assessment of the contribution of experimental taxonomy to the classification of insects. *Rev. Canad. Biol.,* **19**: 293-319.

Munroe, E. 1964. Problems and trends in systematics. *Canadian Ent.,* **96**: 368-377.

Muttkowski, R.A. 1910. Catalogue of the Odonata of North America. *Bull. Public. Mus,* City of Milwankee, **1**: 1-207.

Myers, G.S. 1952. The nature of systematic biology and of a species description. *Syst. Zool.,* **1**: 106-111.

Myers, G.S. 1956a. Manual of tropical herpetologica collecting, Cirrc. No. 4 (Mimeo). Nat. Hist. Mus. Stanford Univ.

Myers, G.S. 1956b. *Brief directions for preserving and shipping specimens of fishes, amphibians and reptiles,* (Mimeo). Nat. Hist. Mus. Stanford Univ., Circ. no. **5**: 1-3.

Neave, S.A. (ed.) 1939. Nomenclator Zoologicus. A list of the names of genera and subgenera in zoology from the tenth edition of Linnaeus, 1758, to the end of 1935. Published for the proprietors by the Zoological Society of London, Vol. I-IV (A-C; D-L; M-P; Q-Z). London.

Neave, S.A. (ed). 1950. Nomenclator Zoologicus, 1936-1945. Published for the proprietors by the Zoological Society of London, Vol. V. London.

Nekrutenko, Y.P., 1964. On the method of quantitative analysis of the wing-pattern of Pieridae (Lepidoptera) (In Ukranian with English & Russian Summaries). *Nauk Ukr. RSR, Kiev,* **3**: 405-407.

Nelson, G.J. 1971. Paraphyly and Polyphyly: Redefinitions. *Syst. Zool,* **20**: 471-472.

Nelson, G.J. 1979. Cladistic analysis and synthesis. Principles and definitions, with a historical note on Adanson's families des Plantes (1763-1764). *Syst. Zool.,* **28(1)**: 1-21.

Nelson, G.J. 1989. Cladistics and evolutionary models. *Cladistics,* **5**: 275-289.

Nelson, G.J. and Platnick, N. 1981. Systematics and biogeography: Cladistics & vicariance. Columbia Univ. Press, New York, pp. 567.

Nemesio, A. 2009. On the live holotype of the Galapagos Pink Iguana, Conolophus marthae Gentile and Smell, 2009 (Squamata: Iguanidae): is it an acceptable exception. Zootaxa. **2201**: 21-25.

Newell, I.M. 1955. An autosegregator for use in collecting soil-inhabiting arthropods. *Trans. Am. Micr. Soc,* **74:** 389-392.

Newell, I.M. 1970. Construction and use of tabular keys. *Pacific. Ins.,* **12(1):** 25-37.

Newell, I.M. 1972. Tabular keys—Further notes on their construction & use. *Trans. Conn. Acad. Arts Sci.,* **44:** 259-267.

Newell, I.M. 1976. Construction of tabular keys: Addendum. *Syst. Zool.,* **25(2):** 243-250.

Newell, N.D. 1947. Infraspecific categories in invertebrate palaeontology. *Evolution.* **1:** 163-171. Nichols, D. 1958. Taxonomy versus stratigraphy. *J. Wash. Acad. Sci.,* **48:** 113-114.

Nichols, D. (ed.). 1962. Taxonomy and Geography. *Syst. Assoc. Publ.* No. **41,** 158 pp.

Nilsson, H. 1953. *Synthetische Artbiidung.* C.W.K. Glear up, Lund, 2 Vols., 1303 pp.

Nixon, K.C. and Wheeler, Q.D. 1990. An amplification of the phylogenetic species concept. *Cladistics,* **6:** 211-223.

Noor, M.A.F. 2002. Is biological species concept showing its age? *Trends Ecol.Evol.,* **17:** 153-54.

Norris, K.P. 1959. Infrared spectroscopy and its application to microbiology. *J. Hyg.,* London, **59:** 326-345.

Norton, B.G. (ed). 1986. The preservation of species: the value of biological diversity **Princeton N.J. Princeton University Press.**

Nuttal, G.H.F. 1901. On the formation of specific antibodies in the blood following upon treatment with the sera of different animals together with their use in legal. *J. Hyg.* Lond., **1:** 357-487.

Nuttal. G.H.F. 1904. *Blood immunity and blood relationship.* Cambridge Univ. Press. London.

Nybakken. O.E. 1959. *Greek and Latin in terminology.* Ames., Iowa, Iowa State College Press; 321 pp.: 1985: 322 pp.

O'Hara, J.R. 1993. Systematic generalization, historical fate and the species problem. *Syst. Biol.,* **42(3):** 231-246.

Ohno, S. in Atkins, N.B. et al. 1965. Comparative DNA content of 19 species of placental mammals, reptiles and birds. *Choromosoma,* **17:** 1-10.

Oosterbroek, P. 1987. More appropriate definitions of paraphyly and polyphyly. with a comment on the Farris 1974 model. *Syst. Zool.,* **36(2):** 103-108.

O'Rourke, F.J. 1974. *Fish—in biochemical immunological taxonomy of animals* (ed. C.A Wright). Academic Press, London, pp. 243-303.

Oldroyd, H. 1966. The future of taxonomic entomology. *Syst. Zool.,* **15:** 253.

Oldroyd, H. 1970. *Collecting, preserving and studying insects.* New York, Macmillan Co., 336 pp.

Oman, P.W. and A.D. Cushmann. 1948. Collection and preservation of insects. *U.S. Dept. of Agric. Misc. Pub.,* **601:** 1-42.

Orians, G.H. and A.J. Kohn. 1962. Ecological data in the classification of closely related species. *Syst. Zool.,* **11:** 119-127.

Orr, H.A. 2001. Some doubts about (yet another) view of species. *J.Evol. Biol.,* **14:** 870-71.

Orton, G.L. 1953. The role of ontogeny in systematics and evolution. *Evolution.* **9:** 75-83.

Ouchterlony, O. 1968. *Handbook of immunodiffusion and immunochemistry.* Ann. Arbor Science Publ. Inc., Ann. Arbor, Michigan, 215 pp.

Panchen, A.L. 1992. Classification, evolution and the nature of biology Cambridge Univ. Press, London, 403 pp.

Pankhurst, R.J. 1970. Key generation by computer. *Nature,* **227:** 1269-1270.

Pankhurst, R.J. 1970. A computer programme for generating diagnostic keys. *Comp. Jour.,* **13:** 145-151.

Pankhurst, R.J. (ed). 1975. Biological identification with computers. Academic Press for the Syst. Asson. Spec, 7, pp. 1-333.

Pankhurst, R.J. 1978. Biological identification. Edward Arnold Ltd.. London, 104 pp.

Pankhurst, R.J. 1984. Online identification programme, version 4. British Museum (Natural History), London.

Pankhurst, R.J. 1991. Practical taxonomic computing. Cambridge Univ. Press, England, 202 pp.

Pankhurst, R.J. and R.R. Aitchinson. 1975. An online identification programme. Pages 181-192. In Biological Identification with Computers (ed. R.J. Pankhurst), Academic Press, London & New York.

Papp, C.S. 1968. *Scientific illustration: theory and practice.* W.C. Brown, Dubuque. Iowa, xiv + 318 pp.

Park, O., W.C. Allee and V.E. Shelford. 1939. *A laboratory introduction to animal ecology and taxonomy with keys, etc.* Univ. Chicago Press, Chicago, x + 272 pp.

Parr, A.E. 1958. Systematics and museums. *Curator,* **1(2):** 13-16.

Pasteur, G. 1976. The proper spelling of taxonomy. *Syst. Zool.,* **25(2):** 192-193.

Paterson, H.E.H. 1985. The recognition concept of species, pp. 21-29 in species and speciation (ed. E.S. Vrba). *Transvaal Mus. Monog.* 4.

Transvaal Museum, Pretoria. Patterson, D.J., D. Ramsen, W.A. Marino and Cathy Norton. 2006. Taxonomic Indexing—Extending the role of taxonomy. *Syst. Biol.,* **55:** 367-373.

Patterson, J.T. and W.S. Stone. 1952. *Evolution in the genus Drosophila.* Macmillan, New York, 610 pp.

Pauley, L.K. 1955. Serological relationships among orthopteroid insects, determined by their whole blood. *J.N.Y. Ent. Soc,* **63:** 83-93.

Payne, R.W. 1975. Genkey: A programme for constructing diagnostic keys. Pages 65-72. In Biological identification with computers (ed. R.J. Pankhurst). Academic Press, London & New York.

Pemberton, C.E. 1941. Contributions of the entomologists to Hawaii's Welfare. *Hawaian Planters Rec,* **45:** 107-111.

Petrunkevitich, A. 1926. The value of instinct as a taxonomic character in spiders. *Brial. Bull.,* **50:** 427-432.

Petrunkevitch, A. 1952. Principles of classification as illustrated by studies of Arachnida. *Syst Zool.*, **1:** 1-19.

Pettersen, B. 1966. Problems and methods in modern systematics. *Norsk. Ent. Tidsskr.*, **13:** 245-254.

Piechocki, R. 1966. Techniques for the collection of macroscopic specimens. Akad. Verlags. Geest & Portig K.-G. Leipzig (in German), 339 pp.

Polaszek, A. 2006. Zoo bank: ICZN's open-access web-based register of allnew animal names and original descriptions. *Bull. Zool. Nomencl.*, **63** (2): 88-99.

Polaszek, A. et al. 2005. A universal register for animal names. *Nature*, 437-477.

Prosser, C.L. 1955. Physiological variation in animals. *Biol. Rev.*, **30:** 229-262.

Quicke, D.L.J. 1993. Principles and techniques of contemporary taxonomy. Chapman and Hall, London, pp. 311.

Randall, H.M., D.W. Smith, A.C. Colm and W.J. Nungester. 1951. Correlation of biological properties of strains of *Mycobacterium* with infra-red spectrums. I. Reproducibility of extracts of *M. tuberculosis* as determined by infra-red spectroscopy. *Ann. R. Tuberc*, **63:** 373-380.

Raubenheimer, R. and T.M. Crowe. 1987. The recognition species concept: Isvit really an alternative? *S. Afr. J. Sci.*, **83:** 530-534.

Raw, A. 1955. A. flotation extraction process for soil micro-arthropods. *Soil Zool.*, 341-346. In D.K. McKevan (ed.), Soil Zoology. Butterworth & Co. Ltd., London.

Raymond, S. and L. Weintraub. 1959. Acrylamide gel as a supporting medium for zone electrophoresis. *Science*, **130:** 711.

Redfield, A.C. 1936. The distribution of physiological and chemical peculiarities in the 'natural' groups of organisms. *Am. Nat*, **70:** 110-122.

Remane, A. 1952. Die grundzuege des naturelichen systems der vergleichendess anatomie und der physiogonetik. Acad. Verl., Keipzig., 400 pp.

Rensch, B. 1929. Das prinzip geographischer Rasenkreisse und das problem der artbildung. Berlin, 206 pp.

Rensch, B. 1960. *Evolution above species level.* Columbia Univ. Press, N.Y., 419 pp.

Ride, W.D.L., C.W. Sabrosky, G. Bernardi and R.V. Melville (eds.). 1985.International Code of Zoological Nomenclature (3rd Ed.). Int. Trust for Zool. Nomenclature. Nat. Hist. Mus., London, 338 pp.

Ride, W.D.L., H.G. Cogger, C. Dupius, O. Kraus, A. Minelli, F.C. Thompson and P.K. Tubbs (eds.). 1999. International Code of Zooloigical Nomenclature (4th ed.). Int. Trust Zool. Nomen., Nat. Hist. Mus., London, pp. 306.

Ridgway, R. 1912. *Colour standards and colour nomenclature.* A. Hoen Co., Wash. D.C. iv + 43 pp.

Rieppel, O. 2006. The Phylocode: a crucial discussion of its theoretical foundamtion. *Cladistics*, **22:** 186-197.

Rivas, L.R. 1960. The fishes of the genus *Romacentrus* in Florida and the western Bahamas. *Quart. J. Florida Acad. Sci.*, **23**: 130-162.

Rivas, L.R. 1962. *Cichlasoma pasionis*, new species of cichlid fish of the Thorichthys group, from the Rio de al Passion, Guatemala. *Ibid.*, **25**: 147-156.

Rivas, L.R. 1964. A reinterpretation of the concepts "sympatric" and "allopatric" with syntopic and allotopic. *Syst. Zool.*, **13**: 42-43.

Robinson, W.H. 1975. Type specimen vs. voucher specimen. *Syst. Zool.*, **24(1)**: 110-111.

Robson, G.C. and C.W. Richards. 1936. *The variation of animals in nature.* Longmans Green, London.

Rodman, J.E. and J.H. Cody. 2003. The taxonomic impediment overcome: NSF's Partnerships for enhancing expertise in taxonomy (PEET) as a model. *Syst. Biol.*, **52(3)**: 428-435.

Rodrigues, D.P. 1986. On the term character. *Syst. Zool.*, **35(1)**: 140-141.

Rogers, D.J. 1963. Taximetrics—new name, old concept. *Brittonia*, **15**: 285- 290.

Rogers, D.J., H.S. Fleming and G. Estabrook. 1976. Use of computers in studies of taxonomy and evolution. *Evol. Biol.*, **1**: 169-196.

Rogers, D.J. and T.T. Tanimoto. 1960. A computer program for classifying plants. *Science*, **132**: 1115-1118.

Rohlf, F.J. 1965. Character correlation in numerical taxonomy. *Proc. 12th Int. Congr. Ent*, London, pp. 109-110.

Rohlf, F.J. I970. Adaptive hierarchical clustering schemes. *Syst. Zool.*, **19**: 58-82.

Rohmberger, J.A. (ed). 1978. *Biosystematics in agriculture.* Allanheld, Osman & Co., Distr. by John Wiley & Sons, N.Y. x + 340 pp.

Rollings, R.C. 1954. Plant taxonomy today. *Syst. Zool.*, **2**: 180-190.

Rollings, R.C. 1965. On the bases of biological classification. *Taxon*, **14**: 1-16.

Ronald, S., K. Martens and R.F. Schram. 2011. The Phylocode: naming of biodiversity at a crossroads: avail, at www.sciencedirect.com.

Rosenbers, G., E.T. Krell and R. Ryle. 2012. Call to register new species in Zoo bank. Nature. **489**: 178.

Rosen, D. 1986. The role of taxonomy in effective biological control programmes. *Agric. Ecosyst. Environ.*, **15**: 121-129.

Rosenkerg, M.S. 2001. The systematic and taxonomy of fiddler crabs; a phylogeny of the genus *Uca.J.Crust.Biol.*. **21(3)**: 839-869.

Ross, E.S. 1956. What is species describing? *Syst. Zool.*. **2**: 180-190.

Ross, H.H. 1941. How to collect and preserve insects. Illinois Nat. Hist. Survey, Circ, 39, pp. 1-48.

Ross, H.H. 1974. Biological systematics. Addison-Wesley Publ. House, Inc., 345 pp.

Rossetto, C.J. 1966. Significance of behavioural characteristics in species identification. Ciena Cult., S. Paulo, **18**: 323-325 (English summary).

Rothanaler, W. 1954. Terminoiogie des subdivisions de I' espece. VIII. Congr. Int. Bot. Rapp. and Comm. Sect. **2-6**: 67-74.

Rubinoff, D. 2006. Utility of mitochondrial DNA barcodes in species conservation. *Cons.Biol,* **20** (5): 1026-1033.

Rubinoff, D., S. Cameron and K. Will. 2006. A genome perspective on the shortcomings of mitochondrial DNA for "barcoding" identification. *J.Hered.,* 97(6): 581-594.

Rudolphi, C.A. 1801. Beobachtungen uber die Eingeweirdwurner. *Erchir zool. u. Zootomie.,* **2:** 1-65.

Rundle, H.D., F. Breden, C. Griswold, A.Q. Mooers, A. Vosil, and J. Whitton. 2001. Hybridization without guilt: gene flow and the biological speciesconcept. *J.Evol.Biol.,* **14:**868-69.

Ruse, M. 1969. Definitions of species in biology. *Brit. J. Phil. Sci.,* **20:** 197-219.

Sabrosky, C.W. 1950. Taxonomy and Ecology. *Ecology,* **31:** 151-152.

Sabrosky, C.W. 1965a. The interrelations of biological control and taxonomy. *J. Econ. Ent,* **48:** 710-714.

Sabrosky, C.W. 1965b. The objectives of a museum entomology department *Pac. Ins.,* 7: 14-20.

Sabrosky, C.W. 1967. Zoological nomenclature in the jet age. *Proc. Helminth. Soc. Wash.,* **34(2):** 236-240.

Sabrosky, C.W. 1970. Quo vadis, taxomomy. *Bull. Ent. Soc. Am.,* **16:** 3-7.

Sabrosky, C.W. 1975. The language of scientific names in zoology. *Syst. Zool,* **24(1):** 109-110.

Sachtleben, H. 1947. Importance of taxonomy in plant protection. Fetscher. 8° ten aenburtst Olto. Appel, Berlin, pp. 6-8.

Sailor, R.I. 1969. A taxonomist's viewpoint of environmental research and habitat manipulation. Proc. *Tall Timbers conference on ecological animal control by habitat management.* No. 1. 1969. Publ. by Tall Timbers Res. Station, Tallahasse, Florida. Salt, G. 1941. The effects of hosts upon their insect parasites. *Biol. Rev.,* **16:** 239-264.

Sarkar, I.N., P.J. Planet and R. DeSalle, 2003. CAOS software for use in character based DNA barcoding. *Mol.Ecol.Res.,* 8: 1256-59.

Sande, M. Van and D. Karcher. 1960. Species differentiation of insects by hemolymph electrophoresis. *Science,* **131:** 1103-1104.

Savory, T. 1962. Naming the living world. John Wiley & Sons, N.Y., 128 pp.

Savory. T. 1970. *Animal taxonomy.* Heinmann Educational Books Ltd., London, 101 pp.

Saxena, K.N., B.R. Seshachar and J.R. Gandhi. 1965. Taxonomic value of biochemical characteristics of animals with reference to aminoacids. *Syst. Zool,* **14:** 33-46.

Schenk, E.T., J.H. McMasters, A. Myrakeen and S.W. Muller. 1956. *Procedure in taxonomy,* 3rd. ed. Stanford Univ. Press, Stanford, 119 pp.

Schildkrant, Carl L, J. Marmur, and Paul Doty. 1961. The formation of hybrid DNA molecules and their use in studies of DNA homologies. *J. Molecular Biol.,* **3(5):** 595-617.

Schindewolf, O.H. 1950. *Grundfragen der palaentologie.* Schweizerbart. Stuttgart, 506 pp.

Schlee, S.D. 1975a. Review of numerical taxonomy (The principles and practices of numerical classification eds. P.H.A. Sneath and R.R. Sokal.) *Syst. Zool.,* **24:** 266-268.

Schlee, S.D. 1975b. An analysis of numerical phenetics. *Entomon. Scan.,* **6:** 1-9.

Schmidt, R.S. 1955a. The evolution of nest-building behaviour in *Apicotermes* (Isoptera). *Evolution,* **9:** 157-181.

Schmidt. R.S. 1955b. Termite *(Apicotermes)* nests—Important ethological material. *Behaviour,* **8:** 344-356.

Schmidt, W.L. 1953. Applied systematics. In *Conference on importance and needs of systematics. Washington Nat. Acad. Sci. Nat. Res. Counc,* pp. 4-12.

Scott Ram, N.R. 1990. Transformed Cladistics Taxonomy and Evolution. Cambridge Univ. Press, England, 238 pp.

Scudder, G.G.E. 1974. Species concepts and speciation. *Can. J. Zool.,* **52:** 1121-1134.

Selander, R.K. 1969. The ecological aspects of the systematics of animals. *Systematic Biology. U.S. Nat. Acad. Sci., Publ. No.* **1692:** 213-247.

Senn, H.A. 1960. The species concept and taxonomy – a summary. *Rev. Canad. Biol.,* **19:** 320-325.

Sereno, P.C. 2005. The logical basis of phyllogenetic taxonomy. *Syst. Biol.,* **54(4):** 595-619.

Seshachar, B.R. (ed.). 1966. Proceedings of the symposium on "Newer trends in taxonomy". *Bull. Nat. Inst. Sci.,* New Delhi, No. 34, 409 pp.

Seshachar, B.R., K.N. Saxena and Kamal Bhaila. 1967. Usefulness of certain patterns of proteins and their derivatives as taxonomic criteria. *Ibid.,* pp. 387-409.

Settler, R. 1963. Phenetic contra Phyletic. *Syst. Zool.,* **12:** 94-95.

Sharov, A.G. 1965. Evolution and taxonomy. *Zool. Syst. Evol.,* **3:** 349-358.

Sherrock and Felsenstein. 1976. Phylogeny of the sea snakes (Hydrophilidae) using monothetic character set matching method. Univ. Illinois, Urbana. USA.

Sibley, C.G. 1954. Hybridization in the red-eyed towhees of Mexico. *Evolution,* **8:** 252-290.

Sibley, C.G. 1960. Electrophoretic patterns of avian egg-white proteins as taxonomic characters. *Condor.,* **102:** 215-284.

Sibley, C.G. 1962. The comparative morphology of protein molecules as data for classification. *Syst. Zool.,* **11:** 108-118.

Sibley, C.G. 1964. In *Taxonomic Biochemistry and Serology* (ed. C.A, Leone), pp. 435-450.

Sibley, C.G. 1967. Discovery New Haven. *Conn.,* **3:** 5-30.

Sibley, C.G., K.W. Corbin, J.E. Ahlquist and A. Ferguson. 1974. Mammals. In *Biochemical and Immunological Taxonomy* (ed. C.A. Wright), pp. 89-176.

Silvestri, F. 1929. The relation of taxonomy to other branches of entomology. *4th Internat. Congr. Ent,* **2:** 52-54.

Simonetti, J.A. 1997. Biodiversity and taxonomy of Chilean taxonomists.*Biodiversity and Conservation,* **6:** 633-637.

Simpson, G.G. 1945. The principles of classification and a classification of mammals. *Bull. Am. Mus. Nat. Hist,* **85:** 1-350.

Simpson, G.G. 1952. The species concept. *Evolution,* **8:** 285-298.

Simpson, G.G. 1959. Note on biometry and systematics. *Bull. Am. Mus. Nat. Hist,* **117:** 51-54.

Simpson, G.G. 1961. *Principles of animal taxonomy.* Columbia. Univ. Press, N.Y., 247 pp.

Simpson, G.G. 1964. Organisms and molecules in evolution. *Science,* **146:** 1535-1538.

Smirnov, E.S. 1969. *Taxonomic analysis.* Moscow State Univ., Moscow, 187 pp.

Smith, H.M. 1952. Definitions of species. *Turtox News,* **30:** 110-112, 180-182.

Smith, H.M. 1955. The perspective of species. *Turtox News,* **33(4):** 74-77.

Smith, H.M. 1965. Phenetic taxonomy: An example of essentialism in science. *Syst. Zool.,* **14:** 148-149.

Smith, H.M. and G. Perez-Higarda. 1986. Nomenclatural name-forms. *Syst. Zool.,* **35(3):** 421-422.

Smith, H.M. and K.B. Smith. 1972. Chresonymy ex synonymy. *Syst. Zool.,* **21:** 445.

Smith, R.C. and R.H. Painter. 1967. *Guide to the literature of the zoological sciences,* 7th ed. Burgess Publ. Co. Minneapolis, 238 pp.

Smith, R.I. 1952. Cooperation between systematics and experimental biologists. *Science,* **116:** 152-153.

Smith, S.G. 1958. Animal cytology and cytotaxonomy. *Proc. Genet. Soc. Canada,* **3:** 57-64.

Smith, W.J. 1966. Communication and relationships in the genus *Tyrannus.* Pub. Nuttal Ornith. Club, Cambridge, No. 6, 250 pp.

Smithies, O. 1955. Zone electrophoresis in starch gels; group variations in the serum proteins of normal human adults. *Biochem. J.,* **61:** 629-641.

Sneath, P.H.A. 1957. The application of computers to taxonomy. *Gen. Microbiol.,* **17:** 201-226.

Sneath, P.H.A. 1961. Recent development in theoretical and quantitative taxonomy. *Syst. Zool.,* **10:** 118-139.

Sneath, P.H.A. 1962. The construction of taxonomy groups. In G.C. Ainsworth's *12th symposium of the society for general microbiology,* pp. 289-332. Cambridge Univ. Press, Cambridge.

Sneath, P.H.A. 1995. Thirty years of numerical taxonomy. *Syst. Biol.,* **44(3):** 281-298.

Sneath, P.H.A. 1971. Numerical taxonomy: criticisms and critiques. *Biol. J. Linn. Soc,* **3:** 147-157.

Sneath, P.H.A. and R.R. Sokal. 1962. Numerical taxonomy. *Nature,* **193:** 855-860.

Sokal, R.R. 1965. Statistical methods in systematics. *Biol. Rev. Cambridge. Phil. Soc,* **40:** 209-211.

Sokal, R.R. 1966 Numerical taxonomy. *Sci. Am.,* **215:** 106-116.

Sokal, R.R. 1969. Principles of taxonomy (review). *Science,* **156:** 135.

Sokal, R.R. 1971. Animal taxonomy: Theory and practice. *Quart. Rev. Biol.*, **44:** 209-211.

Sokal, R.R. 1973. The species problem reconsidered. *Syst. Zool.*, **22(1):** 360-374.

Sokal, R.R. 1975. Mayr on cladism and his critics. *Syst Zool.*, **24(2):** 257-262.

Sokal, R.R. and J.H. Camin. 1975. The two taxonomies. *Syst. Zool.*, **14:** 176-195.

Sokal, R.R., J.H. Camin and F.J. Rohlf. 1965. Numerical taxonomy: some points of view. *Syst. Zool.*, **14:** 137-143.

Sokal, R.R. and T.J. Crovello. 1970. The biological species concept: a critical evaluation. *Am. Natur.*, **104:** 127-153. Sokal, R.R. and F.J. Rohlf. 1969. Biometry—*The principle and practice of statistics in biological research.* W.H. Freeman, S. Fransisco. xii + 776 pp.

Sokal, R.R. and F.J. Rohlf. 1975. *Introduction to Biostatistics.* W.H. Freeman & Co., S. Francisco, 368 pp. Sokal, R.R. and P.H.A. Sneath. 1963. Principles of numerical taxonomy. W.H. Freeman & Co., Lond., 359 pp.

Sokal, R.R. and P.H.A. Sneath. 1966. Efficiency in taxonomy. *Taxon,* **15:** 1-21.

Solbrig, O.T. 1966. *Evolution and Systematics.* The Macmilian Co., New York, 122 pp.

Solow, A.R., L.A. Mound and K.J. Gaston. 1995. Estimating the rate of synonymy. *Syst. Biol.*, **44(1):** 93-96.

Staniland, L.N. 1953. *The principles of line illustration.* Burke Publ. Co., London, 224 pp.

Stanley, J.T. 2006. The bacterial species dilemma and the genomic phylogenetic species concept. *Phil.Trans.R.Soc.Lond.*, **B 361:** 1899-1909.

Stebbins, G.L. 1947. Types of polyploids: their classification and significance. *Adv. in Genetics,* **1:** 403-430.

Stebbins, G.L. 1966. Chromosomal variation and evolution. *Science,* **152:** 1463.

Stephen, W.P. 1958. Haemolymph proteins and their use in taxonomic studies. *Proc. 10th Internal Congr. Ent. Montreal,* 1956, **1:** 395-400.

Stephen, W.P. 1961. Phylogenetic significance of blood proteins among some orthopteroid insects, *Syst. Zool.*, **10:** 1-9.

Stephen, W.P. 1974. Insects, pp. 303-349. In *Biochemical and Immunological Taxonomy of Animals* (ed. C.A. Wright), 490 pp.

Stevens, P.F. 1994. The development of biological systematics. Antonie-Lauxent de Jussicu. Nature, and the natural system. Columbia Univ. Press, New York, xxiii + 616 pp.

Stevenson, H.J.R. and O.E.A. Bolduan. 1952. Infra-red spectrophotometry as a means for identification of bacteria. *Science,* **116:** 111-113.

Steyskal, G.C. 1949. An indexing system for taxonomists. *Coleopt. Bull.*, **3:** 65-71.

Steyskal, G.C. 1965. Trend curves of the rate of species descriptions in zoology. *Science,* **149:** 880-882.

Steyskal, G.C. 1967. Another view of the future of taxonomy. *Syst. Zool.*, **16:** 265-268.

Steyskal, G.C. 1968. The number and kind of characters needed for significant taxonomy. *Syst. Zool.,* **17:** 474-477.

Steyskal, G.C. 1972. The meaning of the term sibling species. *Syst. Zool.,* **21:** 446.

Stone, N.D., R.N. Coulson, R.E. Frisbie and D.K. Loh. 1986. Expert systems in entomology: Three approaches to problem solving. *Bull. Entomol. Soc. Am.,* **32:** 161-166.

Storey, M. and N.J. Wilmowsky. 1955. Curatorial practices in zoological research collection 1. Preliminary report on containers and closures for storing specimens preserved in liquid. *Circ. Nat. Hist. Mus. Stanford Univ.* No. **3,** pp. 1-22.

Strickland, H.E. 1842. Rules of zoological nomenclature. Report of the 12th meeting of British Association held at Manchester in 1842. *Brit. Assoc. Adv. Sci. Rept,* **1842:** 7-18.

Sylvester Bradley, P.C. 1954. The superspecies. *Syst. Zool.,* **3:** 145-146. Sylvester Bradley, P.C. 1956. The species concept in Palaeontology. *Syst. Zool. Publ.,* London, No. 2, p. 145.

Tchistovitch, T. 1899. I' Ettudes sur immunization centre le serum anguieles. *Ann. inst. Pasteur.,* **13:** 406-425.

Telechow, A., P. Hammerstein and J.H. Werren 2005. The effect of Wolbachia versus genetic incompabilities on reinforcement and speciation. *Evol.,* **59:** 1607-19.

Templetion, A.R. 1989. The meaning of species and speciation. Pp. 3-27 in D. Otte and J.A. Endler (eds). Speciation and its consequences. **Sinaeur, Sunderland, MA.**

Thielcke, G. 1964. Lautausserum gen Vogel in iher Bedeutung fur die Taxonomie. *J. Ornith.,* **105:** 78-84.

Thompson, W.R. 1937. *Science and common sense.* Longmans Green & Co., London, 224 pp.

Thorpe, W.H., 1940. Ecology and the future of systematics. In *The New Systematics* (ed. by J.S. Huxley), London, Oxford Univ. Press, pp. 341-364.

Throckmorton, L.H. 1962. The problem of phylogeny in the genus *Drosophila.* In *Study in Genetics* (ed. M.R. Wheeler). Univ. Texas Publ., No. **6205:** 207-343.

Throckmorton, L.H. 1965. Similarity versus relationship in *Drosophila.* *Syst. Zool.,* **14:** 221-236.

Throckmorton, L.H. 1968a. Biochemistry and taxonomy. *Ann. Rev. Ent,* 13: 99-144.

Throckmorton, L.H. 1968b. Concordance and discordance of taxonomie characters in *Drosophila* classification. *Syst. Zool.,* **17:** 355-387.

Throckmorton, L.H. 1969. Round table discussion: Molecular Systematics—A view of the future. In *Systematic Biology, U.S. Nat. Acad. Sci. Publ.* No. **1692,** pp. 382-384.

Tilden, J.W. 1961. Certain comments on the subspecies problem. *Syst. Zool.,* **10:** 17-23.

Tinbergen, N. 1959. Comparative studies of the behaviour of gulls (Laridae). A progress report. *Behaviour,* **15:** 1-70.

Tiselius, A. 1937. Electrophoresis. Tram Faraday Soc, **33:** 524-531.

Turner, B.L. 1966. Chemosystematics: present and future. In *Newer Trends in taxonomy* (ed. B.S. Seshachar). Bull. Nat. Inst. Sci. No. 34, pp. 189-211.

Uprety, D.R. and V.C. Kapoor. 1979. *Collection, preservation and identification of insects.* Tribhuvan Univ. Press, Kathmandu, Nepal, pp. 94.

Valentine, D.H. and A. Love. 1958. Taxonomie and biosystematic categories. *Brittonia,* **10:** 153-179.

van der Kloot, W.G. and C.M. Williams. 1953a. Cocoon construction by the *Cecropia* silkworm. I. The role of the external environment. *Behaviour,* **5:**141-157.

van der Kloot, W.G. and C.M. Williams. 1953b. Cocoon construction by the *Cecropia* silkworm. II. The role of the internal environment. *Ibid.,* **5:** 157-174.

van Emden, F.I. 1957. The taxonomie significance of the characters of immature insects. *Ann. Rev. Ent.,* **2:** 91-106.

, M. and D. Karcher. 1960. Species differentiation of insects by haemolymph electrophoresis. *Science,* **131:** 1103-1104.

van Son, G. 1955. A proposal for restriction of the use of the term subspecies. *Lep. News,* **9: 1**-3.

van Tyne, J. 1952. Principles and practices in *Collecting and taxonomie work. Auk,* **69:** 205-342.

van Valen, L. 1976. Ecological species, multispecies and oaks. *Taxon,* **25:** 233-239.

Voss, E.G. 1952. The history of keys and phylogenetic trees. In *Systematic Biology. J. Sci. Labs.* Denison Univ., **43:** 1-25.

Wagele, J.W. 2005. Foundations of Phylogenetic Systematics. Verlag Dr. Friedrich Pfeil. Munchen, Germany.

Wagner, R.P. 1944. Nutritional differences in the *muleri* group. Univ. Texas. Publ. No. 4920, pp. 39-41.

Wagner, W.H. Jr. 1969. The construction of a classification. *Syst. Biol. U.S. Nat. Acad. Sci. Publ.* No. 1692, pp. 67-103.

Wagstaffe, R. and J.H. Fidler. 1955. *The preservation of natural history specimens. Vol. 1. Invertebrates.* H.F. Witherby and G. Witherby, London, 205 pp.

Wainstein, B.A. 1960. On the criteria of taxonomic categories. *Zool. Zhurnal.,* **39:** 1774-1778.

Walbank, B.E. and D.F. Waterhouse. 1970. The defensive secretions of *Polyzosteria* and related cockroaches. *J. Insect. Physiol.,* **16:** 2081-2096.

Walker, T.J. 1964. Cryptic species among sound producing ensiferan Orthoptera (Gryllidae and Tettigoniidae). *Quart. Rev. Biol.,* **39:** 345-355.

Warburton, F.E. 1967. The purposes of classification. *Syst. Zool.,* **16(3):** 241-245.

Warren, B.C.S. 1958. On the recognition of the species. *Proc. 10th Internal Congr. Ent,* Montreal, **1956,** pp. 111-123.

Waterston, J. 1929. Discussion on types. Trans. *4th Internal. Congr. Ent,* Ithaca, N.Y., pp. 695-699.

Watson, L. and P. Milne. 1972. A flexible system for automatic generation of special-purpose dichotomous keys, and its application to Australian grass genera. *Aust. J. Bot,* **20:** 331-352.

Weizenbaum, J. 1972. On the impact of the computer on society. *Science,* **176:** 609-614.

Wetmore, A. 1960. A classification for birds of the world. *Smithson. misc. coll.,* **139:** 1-37.

Wheeler. Q. D. 2005. Losing the plot.: DNA "barcoding" and taxonomy. *Cladistics,* **21:** 405-07.

Wheeler, Q.D. (Ed.). 2008. The New Taxonomy. *CRC Press,* New York, pp. 1-237.

Wheeler, Q.D. and F.T. Krell. 2007. Codes must be updated so that names are known to all. *Nature,* **447:** 142.

Wheeler, Q.D. and K.C. Nixon. 1990. Another way of looking at the species problem: A reply to de Queiroz and Donoghue. *Cladistics,* **6:** 77-81.

Wheeler. Q.D., P.H. Raven and E.O. Wilson 2004. Taxonomy: Impediment or *Experiment?* Science, **303:** 285.

Wheeler, Q.D. and A.C. Valdecasas. 2005. Ten challenges to transform taxonomy. Graellsia, **61:** 151-160.

White, M.J.D. 1957. Cytogenetics and systematic entomology. *Ann. Rev. Ent,* **2:** 71-90.

White, M.J.D. 1973a. *Animal cytofogy and evolution,* 3rd. ed. Cambridge Univ. Press, Cambridge, England, 961 pp.

White, M.J.D. 1973b. Chromosomal rearrangements in mammalian population, polymorphism and speciation. In *Cytotaxonomy and vertebrate evolution* (ed. Chiarelli and Campana) pp. 95-128.

Wiley, E.O. 1978. The evolutionary species concept reconsidered. *Syst. Zool.,* **27:** 17-26.

Williams, C.A. and M.W. Chase. 1967. *Methods in lmmunology and Immunochemistry.* Vol. ll. 1968. Vol, ll. Academic Press. N.Y.

Williams. P.M., H.M. de Klerk and T.M. Crowe 1999. Interpreting boundariesamong Afro tropical birds: spatial patterns in richness gradients andspecies replacement. *J. Biogeography.* **26:** 419-474.

Williams, W.T. and G.N. Lance 1965 Logic, of computer based intrinsic classification. *Nature,* **207:** 159-161.

Willis, E.O. 1981. is a species interbreeding or an internally similar part of a phylogenetic tree? *Syst. Zool,* **30:** 84-85.

Wilmoth, J.H. 1967. *Biology of Invertebrates.* Prentice-Hall, Inc., Englewood Cliffs, New Jersey. Wilson. E.O 1971. The plight of taxonomy. *Ecology.* **52:** 741.

Wilson, A.C. and N.O. Kaplan. 1964. Enzyme structure and its relation to taxonomy. In. Taxonomic biochemistry and serology by Leone, C.A (ed.), pp. 321-346.

Wilson, E.O. 1985. Time to revive systematica. *Science,* **230:**1227.

Wilson, E.O. (ed). 1988. Biodiversity. Washington: National Academy of Sciences/Smithsonian Institution.

Wilson, E.O. 1992. The diversity of life. Combridge: Belknap Press.

Wilson, E.O. 2003. The encyclopedia of life. Trends *Ecol.Evol.*, **18**: 77-80.

Wilson, E.O. and W.L. Brown Jr. 1953. The subspecies concept and its taxonomic application. *Syst Zool.*, **2**: 97-111.

Wilson, J.B. and T.R. Partridge. 1986. Interactive plant identification. *Taxon.* **35**:1-12.

Witworth, T.L., R.D. Dawson, H. Malagon and E. Baudry. 2007. DNA barcoding can not reliably identify species of the blow fly genus. *Protocalliphora* (Dipt.: Calliphoridae). Pro. *Biol.Sci*, **274**: 1731-1739.

Wolfe, S. 1969 Strandedness of chromosomes. *Int. Rev. Cytol.*, **25**: 279.

Woltereck, H. 1931. Verebung wed erbaenderung. In Driesch und oltereck.Das Lebonsproblim. pp 225-310.

Womble, W.H. 1951. Differential systematica *Science.* **114**: 815-322.

Woodford, F. Peter (ed). 1968. *Scientific writing for graduate students.* The Rockefeller Univ. Press N.Y., pp. 192.

Woods, R.S. 1944. The Naturalists Lexicon. *A list of classical Greek and Latin words used or suitable for use in biological nomenclature.* Pasadena. California. Abbey Garden Press. (Addende to the Naturalists Lexicon. 1974.)

Woolley. J.B and N.D. Stone. 1987 Application of artificial intelligence to systematica: Systex—A prototype expert system for species identification. *Syst. Zool.*, **36(3).** 248-267

Woods, R.S. 1966. *An English classical dictionary for the use of taxonomists. Pomona.* College, xiv, 331 pp.

Wright, C.A. 1953. The application of paper chromatography to the taxonomic study of the molluscan genus *Lymnaes. J. Linn. Soc* Lend., **44**: 222-237

Wright, C.A. 1966. Experimental taxonomy: A review of some techniques and their applications. *Int. Rev. Gen. Exptl. Zool.* **2**: 1-42.

Wright, C.A. (ed.) 1974. *Biochemical and Immuncological Taxonomy of Animals.* Academic Press, London & N.Y.. 490 pp.

Wu, C.I. 2001. The genic view of the process of speciation. *J Evol.Biol.*, **14**: 851-65.

Wu, C.I. and C.T. Ting. 2004. Genes and speciation *Nat.Rev.Genet.*, **5**: 114-22.

Wyatt, G.R. 1961. The biochemistry of insect haemolymph. *Ann. Rev. Ent.*, **6**: 75-102.

Young, F.N. 1953. The approach to taxonomic problems. *Prco. Indiana Acad. Sci*, **62**: 172-175.

Zhang, Z. 2014. Electronic publication in zoological nomenclature and taxonomy: problems, reponses and solution. Zootaxa, **3779**(1): 1-2.

Zimmermann, W. 1948. *Groundtragen der evolution.* Klostermann, Frankfort a Main, 221 pp.

Zweig, G. and J.W. Crenshaw Jr. 1957. Differentiation of species by paper electrophoresis of set serum proteins of pseudemys turtles *Science,* **126**: 1065-1066

Glossary

Abode—Home

Absorption—The taking of energy from radiation by the medium through which the radiation passes.

Adaptation—A change in structure, function or form that produces better adjustment of an animal to its environment.

Adaptive radiation—Evolutionary divergence of members of one phyletic line into a series of rather different niches or adaptive zones.

Affinity—Likeness, especially in relationship.

Agamic—Species or generation which does not reproduce sexually.

Agamospecies—An asexual species whose members are of common origin.

Agar gel—A gel formed of agar-agar which is a dried mucilaginous substance obtained from certain oriental seaweeds.

Algae—A large group of simple organisms, mostly aquatic; contain chlorophyll and/or other photosynthetic pigments and have simply organised reproductive organs, e.g., spirogyra, etc.

Algorithm—The repetitive calculations used in finding the greatest common divisor of two numbers.

Allele—A Greek word meaning "belonging to one another" and is used to refer to the individual members of a gene pair.

Allochronic species—Species not occurring at the same time level; not contemporary.

Allometry—The study and measurement of the relative growth of a part of an organism in combination with the whole.

Allopatric—Populations or species which are mutually exclusive, but usually inhabit adjacent geographical regions.

Allopatric hybridisation—The crossing of individuals belonging to two allopatric populations along a well defined contact zone.

Allopatric speciation—Formation of species during geographical isolation.

Allotype—A paratype opposite in sex to holotype.

Ally—Any organism akin to another by structure, etc.

Alpha taxonomy—A taxonomic level concerned with the characterisation and naming of species.

Altitude—Height, measured as distance along the extended earth's radius above a given datum, such as average sea level.

Amino acid—A group of nitrogenous organic compounds that serve as unit of structure of the proteins and are essential to metabolism.

Ancestor—An early form or type from which a later organism is developed.

Antibody—A protein usually found in serum whose presence can be demonstrated by its specific reactivity with an antigen.

Antigen—Any substance, but usually a protein, foreign to the host, which stimulates the production of antibodies.

Antigenic—Having the properties of an antigen. Antiserum—A serum containing antibodies.

Apomictic—Parthenogenetic populations.

Archive – A depository for works; to place a work in an archive with the intent that it be permanently preserved there.

Ascorbic acid—Vitamin C; a white crystalline substance, stable when dry but very easily oxidised in solution, especially in neutral or alkaline solution.

Asexual reproduction—In which there is no fusion of the nuclei of different gametes.

Aspirator—Any instrument or apparatus that utilises vacuum to draw up small insects.

Attractants—Various chemical substances used to lure the insects.

Available name—A scientific name published in accordance with the requirements specified in Articles 8-20 of the Code.

Bait—Any substance used to lure a particular species.

Beetle—Insects belonging to the order Coleoptera; insects in which fore-wings are hard, opaque and called elytra.

Behaviour—The entire complex of observable, recordable or measurable activities of a living animal.

Beta taxonomy—A taxonomic level concerned with the arrangement of species into a natural system of lower and higher taxa.

Bibliography—The study of the editions, dates, authorship, etc. of books and other writings.

Binomen—The designation of a scientific name of a species consisting of generic and specific parts.

Binominal nomenclature—The Linnaean system of nomenclature of naming scientific names of species having both generic and specific components, and adopted by the International Congress of Zoology in International Code of Zoological Nomenclature.

Biochemical—Having to do with chemical processes in living organisms.

Biochemistry—The branch of chemistry that deals with plants and animals and their life processes.

Biodiversity—Totality of different kinds of species on earth.

Biological classification—The arrangement of animals into taxa on the basis of inferences concerning their genetic relationships.

Biological control—Control of pests and weeds by other living organisms.

Biological race—Strains of a species which are alike morphologically but differ in some physiological way, such as a parasite or saprophy.e with particular host requirements or a free living organism with a food or habitat preference.

Biological Resources—Those components of biodiversity of direct, indirect or of potential use to humanity.

Biological species concept—A concept stressing reproductive isolation and the possession of a genetic programme effecting such isolation (for true species).

Biome—A major portion of the living environment of a particular region characterised by its distinctive vegetation and mainland by local climatic conditions.

Bionomics—The branch of biology that deals with the adaptation of living things to their environment.

Biosphere—It is life supporting zone of earth, where atmosphere (air), hydrosphere (water) and lithosphere (land) meet, interact, and make life possible.

Biosphere reserves—These are multipurpose protected areas to conserve diversity and integrity of plants, animals and microorganisms; to promote ecological conservations; and to educate, train and create awareness about environmental aspects and ecofriendly living.

Biosynthesis—The method of synthesis of complex molecules within the living organism.

Biotechnology—The application of scientific and engineering principles to the processing of material by biological agents.

Biotope—A small area with uniform environment occupied by unified community of organism.

Biotype—All the individuals of identical genotype.

Brackish water—Water having less salt than sea water but undrinkable.

Breeding season—A particular period of year in which a particular animal species breeds.

Carbohydrates—A group of organic compounds composed of carbon, hydrogen and oxygen only, e.g., sugar, starch and cellulose.

Catastrophe—A sudden disaster, a sudden and violent upheaval in some part of the surface of the earth.

Category—in taxonomy, designates rank or level in a hierarchic classification; or a class, the members of which are all taxa assigned a given rank.

Chaetotaxy—The arrangement of bristles or chaetae.

Character correlation—Degree of association between two or more variable characters.

Chromatography—Method of chemical analysis developed from the fact that if a liquid mixture is allowed to trickle through a column of adsorbing material (e.g., chalk) the components of the mixture may be adsorbed in separate layers in the column.

Chromomere— One of the chromatin granules of which a chromosome is formed and which may correspond to a gene.

Chronological—Pertaining to arrangement of data in order of time of appearance.

Clade— A delimitable monophyletic unit formed by cladogenesis (a mode of evolution which consists in the splitting of an evolutionary line, such as a species, genus, family and other higher categories).

Cladism—When the organisms are ordered and ranged entirely on the basis of "recency of common descent".

Cladistic— Based on the principle of cladism.

Cladogram—A branching from a common ancestor; a branching diagram showing the development of a clade.

Class—A division of the animal kingdom lower than a sub-kingdom and higher than an order.

Classification—The systematic arrangement of animals in series showing their relation or agreement in structure, life habits, or other characters, i.e., delimitation, ordering and ranking of taxa (see also biological classification).

Cline—A series of form changes: gradient of biotypes: character gradient.

Cluster—Group of related or similar species into species groups or higher taxa.

Code—The International Code of Zoological Nomenclature.

Cohort—An indefinite taxonomic group used in different ways by different authorities, such as a group in rank above a superorder, a group between class and order, or group of related families.

Commission—The International Commission on Zoological Nomenclature.

Competition—The struggle between organisms for the necessities of life.

Complex—Used for a number of related taxonomic units, especially those in which taxonomy is difficult or confusing.

Computer—Any mechanic, electric or electronic apparatus or assembly which performs calculations, simple or including integration and differentiation, or simultaneous equations.

Condensation—Chemical change in which two or more molecules react with the elimination of water.

Congeneric—A species agreeing in all characters of generic value with others compared with it.

Conspecific—Individuals of the same species.

Contemporaneous species—Species occurring at the same time period.

Continental—Inhabiting large and extensive land masses.

Convergence—Development of similarities between animals of different groups resulting from adaptation to similar habitats.

Cosmopolitan—Belonging to whole world.

Coxa—The basal segment of the leg by which it is articulated with body.

Cryostat microtome—Microtome having low-temperature thermostat, to cut section direct from the tissue without embedding it in wax or jelly.

Cryptic species—Those which are genetically and sexually distinct but cannot be differentiated on the basis of observed characteristics.

Curating—Care and superintendence of museum containing animal collection.

Cytogenetics —It is the comparative study of chromosomal mechanisms and behaviour in populations and taxa and their effect on inheritance and evolution.

Cytology—The science dealing with structure, functions and the life "history of cells.

Darwinism—An evolutionary theory propounded by Darwin.

DDT—Dichloro-Dimethyl-Trichloro-Ethane; a common insecticide used for killing pests, especially insects.

Deme—An assemblage of taxonomically closely related individuals; a local population of species; the community of potentially interbreeding individuals at a given locality.

Dendrogram—A diagram with branches indicating the relationships of items in a classification.

Descendant—Deriving from an earlier form.

Desiccation—Drying; removal of moisture.

Differential diagnosis—A statement of important characters which are used for clearcut differentiation of a given taxon from other specifically mentioned equivalent taxa.

Dimorphism—Occurrence of two morphs or phena (morphological types) in a single population.

Dinosaur—Any extinct Mesozoic reptile of the order Dinosauria in length from two to eight feet.

Dip net—A net used to collect aquatic organisms.

Diversity—Multiformity; unlikeness; of various kinds.

DNA—Deoxyribonucleic acid.

Double citation—Citation of the name of both the original author of a species-group name and the person who transferred the taxon to which it applies to another genus, the latter author's name is placed after the parentheses that enclose the name of the original author; in new combinations.

Drooping—Leaning downward.

Ecology—The scientific study of the interrelations between living organisms and their environment.

Ecosystem—Any area of nature which includes living organisms and non-living substances interacting to produce an exchange of materials between the living and non-living parts, e.g., a pond, lake, or forest; it comprises four components—abiotic substances, producers, consumers, and decomposers.

Egg-shell—Hard outermost layer of the egg of some organisms.

Electron microscope—In this a parallel beam of electrons is used. The object, which must be in the form of a very thin film of the material, allows the electron beam to pass through it. The much magnified image is finally received on a fluorescent screen and recorded by using a camera.

Electrophoresis—Cataphoresis; the migration of the electrically charged solute particles present in a colloidal solution towards the oppositely charged electrode, when two electrodes are placed in the solution and connected externally to a source of Electro Motive Force (the source of electric energy required to produce an electric current).

Elytron—In Coleoptera or beetles, the hardened, chitinised fore wing which forms horny sheath to protect hind wing when the latter is not in use.

Endangered—To bring into danger or peril.

Endemic—Restricted to a particular area; used to describe a species or organism that is confined to a particular geographical region, e.g. on island or river basin.

Endochorion—The inner layer of chorion in insects.

Enzyme—Any of various organic substances that are produced in living cells and causes chemical changes in other substances by catalytic action.

Epiphenomenon—Phenomenon which holds that metal activities have no casual role and are merely the accomplishment of physical activity.

Epistemological—Scientific study of knowledge, its acquisition and its communication.

Epithet—A descriptive name or title.

Equal weighting—A pheneticist's method in which all taxonomic characters are treated as equally important.

Essentialistic species concept—The concept of Plato and Aristotle and later of Linnaeus in which the observed diversity of universe reflects the existence of a limited number of "universals" or types, and the variation is not taken care of.

Ethology—The science dealing with the comparative study of animal behaviour.

Evolution—It is the change in the genetic composition of a population the starting point of which is the formation of individuals with different genotypes.

Evolutionary—Based on evolution—The process of biological and organic change in organisms by which descendants come to differ from their ancestors.

Evolutionary divergence—The degree of divergence, at the intra- and interspecific levels, of two or more populations, which presumably have evolved from a common ancestor.

Extant—Still existing.

Extinct—Having no living descendant; no longer in existence.

Family—A taxonomic division consisting of one or more genera agreeing in one or a set of characters and so closely related that they are apparently descended from one stem.

Family-group—Consisting of superfamily, family, subfamily, tribe and subtribe.

Faunistic—Pertaining to fauna of a region; or animal life characteristics of a particular region or environment.

Fecundity—The number of young produced by a species or individual.

Fertilisation—The union of two sexually differentiated gametes to form a zygote.

Fertility—The reproductive potential of an individual or population as measured by the ability to produce viable offspring.

Fixation—Chemical treatment to check biochemical changes in cells at a particular stage; for histochemicai studies.

Flotation method—A process used to separate micro-organisms from soil by causing them to float.

Fluorescence—Emission of light, generally visible from material irradiated (generally from higher frequency source) or from impact of electrons as in phosphorus.

Food chain—The transfer of food energy from the source in plants through a series of organisms, beginning with a herbivore, which successively depend on each other for food.

Fossil—Any hardened remains or traces of plant or animal life of some previous geological period, preserved in rock formation in the earth's crust.

Fruit fly—An insect belonging to the family Tephritidae of the order Diptera.

Galaxy—A duster or groups of billions of stars.

Gamma taxonomy—A taxonomic level which deals with various biological aspects of taxa, ranging from the study of intraspecific populations to the studies of specification and of evolutionary rates and trends.

Gene—A particular sequence or cistron of nucleotides along a molecule of DNA (or on occasion RNA; certain viruses) which represents a functional unit of inheritance.

Gene bank—A comprehensive database containing publicly available DNA sequences for named animals, established in 2005 with National Institute of Health, USA.

Gene flow—The spread of genes from one breeding population to others owing to the dispersal of gametes or zygotes; it may give rise to changes of allele frequency and is thus a factor of evolution.

Gene pool—The total genetic information encoded in the sum total of the genes in a breeding population existing at a given time

Genetic diversity—Variation of genes within species.

Genetic erosion—Decline or total loss of local varieties, breeds or species of plants and animals.

Genitalia—All the genital structures collectively.

Genome—The total chromosomal contents (genetic factors) of the nucleus of a gamete.

Genotype—Genetic constitution of an individual or taxon; its use is now forbidden by the Code.

Genus—A taxonomic category including one or more species, presumably of one common phylogenetic origin, and separated from other related genera by some distinct characteristics.

Genus-group—Consisting of genus, subgenus.

Geocline—A gradual or continuous change in a character over a considerable area as a result of its adjustment to changing geographical conditions.

Geographical isolation—Prevention of gene exchange between a population and others by geographical barriers.

Germplasm—Genetic material.

Global warming—The slow increase in the earth's surface air temperature caused by man-made CO_2 in the atmosphere trapping excessive amounts of solar radiation which would otherwise be reflected back into space (also called greenhouse effect).

Grade—A unit of biological improvement from an evolutionary point of view comprising a group of individuals similar in their level or organisation.

Habitat—The sum total of environmental conditions of a specific place that is occupied by an organism, by a population, or a community; or the place or type of site where an organism or population lives naturally.

Habitat diversity—Varieties of habitats, biotic communities and the ecological differences.

Haemoglobin—A respiratory pigment occurring in the erythrocytes of all Carinata and in blood plasma of certain invertebrates.

Haemolymph—The watery fluid containing leucocytes believed to represent blood, found in the haemocoelic body-cavity of certain invertebrates.

Hapantotype—One or more preparations of directly relaxed individuals showing different life cycle stages forming a name-bearing type in an extant species of Protistans.

Haustellate—Those insects having a haustellum (Proboscis) or sucker, a portion of the mouth.

Hemimetabolous—Insects having an incomplete metamorphosis, i.e., no pupal stage in the life-history.

Hereditary—Inherited; handed down from parent to offspring.

Heredity—Inheritance from progenitors; organic resemblance based on descent; organic relation between successive generations.

Heterochromatic—Containing heterochromatin (chromatin rich in nucleic acid and involved in nuclear and cytoplasmic nucleic metabolism).

Heterogenous—Composed of dissimilar or nonuniform constituents.

Hierarchy—The system of ranks in an animal classification indicating the categorical levels of various taxa.

Higher category—A taxonomic category from subgenus and above.

Histochemistry—The chemistry of living tissue.

Holometabolous—Insects having complete metamorphosis, with a pupal stage in their life histories.

Holophyletic—A monophyletic group including all the progeny of the original ancestor.

Holotype—A type; a specimen on which the original description of the species is based, name bearer.

Homeostasis—In higher animals, the maintenance of an internal constancy independence of the environment.

Homology—A characteristic feature in two or more taxa winch can be traced back to the same feature in their common ancestor.

Homonym—A name identical in orthography with another and based on a different type.

Homoplasy—Correspondence between organs or structures in different organisms acquired as a result of evolutionary convergence or of parallel evolution.

Homozygous—A condition of having inherited a given genetical factor from both parents, and therefore of producing gametes of only one kind as regards that factor.

Honey—The concentrated secretion deposited as a food store by certain species of the bee genus Apis; results from enzyme action upon nectar.

Hormone—Any type of chemical messenger synthesised in discreet glands and execute their action at other sites in the same individual; responsible for the timing and the regulation of growth and development and homeostatic maintenance of the fully developed organisms.

Hybrid—Any offspring of a cross between two genetically unlike individuals.

Hybridisation—Any cress-mating of two genetically different individuals which leads to hybrid progeny.

Ichnotaxon—A taxon based on fossilised work of an animal.

Igneous rock—Rock masses generally accepted as being formed by the solidification of magma injected into the earth's crust, or extruded on Its surface.

Immunised—To render invulnerable to a toxin; usually by injecting the toxins in small quantities at short intervals, without appearance of servere symptoms.

Immunology—The branch of medicine dealing with immunity to diseases.

Incipient species—A population which has not attained all the attributes of reproductive isolation from parent population of a species.

Incisor teeth—The front teeth of mammals; have single root, adapted for cutting and are the only teeth borne by the premaxillae in the upper jaw.

Index— A list of record surrogates arranged in order of some attribute.

Inductive—Leading to inferences by method of logical induction.

Infrared rays—Invisible heat radiation, radiant heat; infra-red radiation has the power of penetrating fog or haze which would scatter ordinary visible light, and thus photographs taken on a plate made sensitive to infra-red radiation may often disclose detail invisible on an ordinary plate or to the naked eye.

Infraspecific—Within one species.

Infrasubspecific—Within one subspecies, e.g., variety or form (not accepted in the Code).

Inheritance—The acquisition of characteristics by transmission of germplasm from ancestor to descendant.

Insular—Living or inhabiting an island.

Insulin—A protein hormone which is produced by the islets of Langerhans of the pancreas.

Interbreeding—Individuals capable of actual or potential gene exchange by hybridisation.

Intersterility—Cross-sterility confined to specific genetically determined group of individuals.

Intraspecific—Within the species.

Introgression—The incorporation of genes of one species into the gene pool of another species by hybridisation and back-crossing.

Invasive species—Outside (exotic) species brought in a country at the expense of local ones.

In vitro—A biological process occurring experimentally in isolation from an organism.

Irreducible—That can not be reduced.

Isolating mechanism—Prevention from mating between breeding group owing to spatial, topographical, physiological, genetic behaviour or other barriers.

Isophene—A contour line delimiting area corresponding to a given frequency of variant form.

Karyomorphology—The branch of genetics dealing with structure and form of chromosomes.

Karyotype—Group of individuals with the same chromosome number and similar linear arrangement or genes in homologous chromosome.

Key—Tabulation of diagnostic characters of taxa in dichotomous couplets facilitating rapid identification.

Lac—A resinous substance, an excretion product of certain coccid insects (e.g., Tachardia, etc.) in certain forest trees; used in the manufacture of shellac.

Larva—After hatching from the egg and before reaching the pupa are the growing stages undergoing complete metamorphosis.

Lectotype—One of the syntypes which, subsequent to the publication of the original description, is selected and designated through publication to serve as a "type".

Lethal gene—A gene which, when expressed, is fatal to its carrier.

Linnaean hierarchy—When in the categorical ranks of taxa each category except the lowest includes one or more subordinate categories.

Lineage—Direct descent from an ancestor; ancestry; family; stock.

Lipid—A class of diverse chemical compounds that include fats, waxes, phospholipids, carotenoids and sterols.

Locus—The location of gene on genetic map.

Maggot—An acephalous, apodous, eruciform larva such as that of certain Diptera; footless larva of Diptera.

Mandibulate—Having biting jaws.

Matrix—A set of numbers or terms arranged in rows and columns between parentheses or double lines.

Mega-diversity—Refers to richness in diversity of plant and animal life.

Metamerism—Repetitions of parts along the long axis of an animal.

Metaphase—The second stage in the cell division process.

Microbiology—The branch of biology that deals with microorganisms.

Microcomputer—A tiny computer containing a microprocessor, often used as the control unit for some instrument, tool, etc., or as a personal computer; microprocessor itself.

Microspecies—One of the number of true breeding, morphologically slightly different lines within a complex of largely interbreeding forms.

Mimicry—Adoption by one species of the colour, habits, sounds or structure of another species.

Mineral—A naturally occurring substance of more or less definite chemical composition and physical properties.

Missing link—A hypothetical form of animal believed to have existed in the evolutionary process intermediate between two closely related groups of animals.

Mitosis—The series of changes through which the nucleus passes during ordinary cell division and by which each of the daughter cells is provided with a set of chromosomes similar to that possessed by the parent cells.

Molecular biology—A branch of modern biology in which biological phenomena and processes are studied not only from the phenomenological point of view but by physical, chemical and biochemical investigation at the molecular level.

Monoculture—Growing of same crop on same piece of land year after year.

Monophyletic—Of individuals derived in the course of evolution from a single interbreeding population or phyletic stock (as opposed) to polyphyletic.

Monophyly—The formation of a taxon through one or more lineages from one immediately ancestral taxon of the same lower rank.

Morph—Any of the individual variants that accounts for polymorphism.

Morph bank—It is an open web repository of images serving the biological research community; established in 1998 and housed at the School of Computational Sciences at the Florida State University, USA.

Morphology—The study of the structure and form of an organism.

Mosaic evolution—The evolution involving unequal rates for different structures, organs, etc., of the phenotype.

Mould—A popular name for any of numerous small fungi.

Moulting—Or ecdysis; the act of casting off the outer layers of the integument.

Museum—A building, room, etc., for keeping animal specimens or their parts, and other objects of natural history, etc.

Mutation—Any detectable and heritable change in the genetic material not caused by genetic segregation or genetic recombination, which is transmitted to daughter cells and even to succeeding generations giving rise to mutant individuals provided it does not act as dominant lethal factor.

Mutation theory—Postulates a sudden variation in some inheritable characteristics of an individual animal as distinguished from variation resulting from generations of gradual changes.

Natural group—A group of organisms having a common ancestor.

Natural Park—An area dedicated by statue for all time to conserve wild life.

Natural selection—A theory of the mechanism of evolution which states the survival of the best-adapted forms, with the inheritance of fitness determining characteristics which arise as random variation

Neotype—A specimen is designated when the original types are known to have been destroyed or were suppressed by the Commission.

Niche—The position or status of an organism within its community or ecosystem.

Nomenclatural Act—Any published action creating a new nomen or modifying the nomenclatural status of a nomen (eg. a subsequent onomatophore designation or a first reviser action).

Nominalistic species concept—which governs that only the individuals exist in nature while species are man-made.

Nominate—Containing the type of the name of the higher taxon to which it is subordinate.

Non-dimensional specks—Species lacking dimensions of space and time.

Nucleic acids—Those found within the nucleus of a cell.

Nucleotide—Basic structure unit of DNA composed of nitrogenous base, pentose sugar and phosphate group.

Nymph—Young stage of insects, differing from the adults in having incompletely developed wings and the genitalia.

Objective synonym—Each of two or more scientific names (synonyms) based on same type.

Official Register—An abbreviated title for the Official Register of Zoological Nomenclature (Article 78.2.4), maintained by the Commission to record information about works, names and nomenclatural acts (see Zoobank).

Omnispective—Capable of beholding everything.

Onomatophore—Simpson's name in place of type.

Ontogeny—The developmental history of an organism from its egg to adult.

Operator—It is a stretch of bases along a DNA molecule located at the proximal end of an operon (a group of continuous structural genes showing coordinate expression and closely associated controlling sites).

Optical disc—Laser-readable data storage medium. Compact disc read-only memory (CD-ROM) and digital video disc read only memory (DVD-ROM) are optical disc formats that could be used to produce available works after 1985 and before 2013 (Article 8.4.2).

Order—A taxonomic group of related organisms ranking between family and class.

Palaeo—A fossil species.

Palaeontology—The science dealing with the life of past geological periods.

Panmictic—Of a randomly interbreeding population of individuals which mate at random, i.e.. each individual is equally likely to mate with any individual of the opposite sex.

Paralectotype—A confusing term used in the Code to designate "each specimen of a former syntype series renaming after the designation of Lectotype.

Parallelism—The independent acquisition of similar characters in related evolutionary lines.

Paranym—Different names of taxa but similar in orthography due to derivation from words of same origin and spelling.

Parapatric—Of allopatric populations whose ranges are in contact and between which gene exchange is geographically possible even without sympatry.

Parasite—An organism which lives in or on another organism and derives subsistence from it without rendering it any service in return.

Paraspecies—In palaeontology, isolated parts of animals or fragments when, named equally as species or other taxa.

Paratype—A specimen other than the holotype available with the author at the time of describing new species.

Parthenogenesis—The production of an embryo from a female gamete without the participation of a male gamete, with or without eventual development into an adult.

Pathogen—Any disease-producing micro-organism.

Peptides—A combination of amino acids formed by the linkage of amino group of some of the amino acids with the carboxyl group of others.

Pesticide—Any chemical used for killing insect pests, etc.

Phenetic—The classification based on appearances of organisms rather than on evolution from a common ancestor.

Pheneticist—One who studies the phenotypic characters.

Phenon (pl. Phena)—A sample of phenotypically similar specimens.

Phenotype—The totality of characteristics of an individual due to interaction between genotype and environment.

Pheromone— A chemical substance secreted by animal, which Influences the behaviour of other animals, generally of the same species.

Philapatry—The tendency of an organism to stay in, or return to, its home area,

Phylogenetic—Evolutionary development of any animal species.

Phylogenetic tree—A graphic representation showing the descent relations of different organisms.

Phylogeny—The evolutionary history of an organism or taxonomic group.

Plankton net—A net used to collect planktons.

Plasmon—The deoxyribonucleic acid contents in the cytoplasmic organelles.

Plenary powers—Special powers granted to the Commission.

Poaching—Unlawful capturing or killing of animals.

Pollutant—One which creates pollution.

Pollution—Term applied to environmental state or manifestation which is harmful or unpleasant to life, resulting from man's failure to achieve or maintain control over the chemical, physical or biological consequences or side effects of his scientific, industrial or social habits.

Polymeric molecule—A molecule that can be represented by more than two whorls.

Polymorphism—The simultaneous occurrence of several discontinuous phenotypes or genes in a population; the condition of having several forms in the adult.

Polypeptide—A polymeric, covalently bonded arrangement of amino acids joined together by peptide linkages.

Polyploidy—The state of having the chromosomes in somatic cells or tissue more than twice the haploid number.

Polytene chromosome—Giant chromosome found in most of the dipterous insects during development.

Polytypic—A taxon containing two or more taxa in the immediately subordinate category, e.g., genus with several species and species with several subspecies.

Population—A group of interacting individuals of the same species.

Precipitin reaction—The formation of visible precipitate at the interface when an antigen is brought in contact with the corresponding antiserum.

Predator—An organism which directly attacks, kills and eats one of the other species (prey).

Predictive value—Usefulness of a classification in making predictions on newly employed characters or newly discovered taxa.

Prospecting—Area which shows sufficient promise of mineral wealth to warrant exploration.

Protein—Any of class of nitrogeneous substances consisting of a complex union of amino acids and containing carbon, hydrogen, nitrogen, oxygen and frequently sulphur.

Protein taxonomy—An approach to classify organisms on the basis of the differences in the structure of protein.

Psychic—Beyond natural or known physical processes.

Publication, Electronic—A publication issued and distributed by means of electronic signals.

Pupa—The resting stage of insects with complete metamorphosis, between larva and adult.

Quarantine—Isolation or restrictions placed on the movements of individuals associated with a case of a communicable disease; place or period of detention of ships coming from suspected infected ports; or of animals or plants on importation.

Race—Equated with subspecies in taxonomy; in general a category of variant individuals within a species and differing slightly in characteristics from the typical members of the species.

Radioactive—Giving off, or capable of giving off, radiant energy in the form of particles or rays as alpha, beta and gamma rays, by the disintegration of atomic nuclei.

Recapitulation—The theory that ontogeny recapitulates phytogeny.

Register—To enter into the Official Register Information above a work, name, author, nomenclatural act, or other item tracked for purpose of zoological nomenclature.

Registration number—A unique identifying number or alpha numeric code assigned in the Official Register to a particular item.

Repressor—Any of the specific allosteric protein molecules (product of regulatory genes) which bind to the operator of operon and prevent RNA polymerase from proceeding into the operon to transcribe messenger RNA from DNA (negative control).

Reproductive isolation—Prevention or restriction of gene flow by genetically controlled differences in the reproductive behaviour and fertility relationships of the individuals which are members of different populations.

Reticulate evolution—Evolution depending on repeated intercrossing between a number of lines, and thus, once, both convergent and divergent.

Retrieve—To recover (information) from data stored in a computer.

RNA—Ribonucleic acid.

Salamander—Any of a group of tailed amphibians (order Caudata), related with frogs and toads, with a soft, moist skin.

Sanctuary—An area where killing and capturing of any species of wild life is prohibited.

Scala naturae—Ladder of nature or "Great Chain of Being" is a Latin concept of Greek Aristotle—means natural scale that was linear in evolution from simple (insects & others) to comblex (humans) organisms, complex ones being closer to God; species could not ever change.

Sedimentary rocks—All those rocks which result from the wastage of pre-existing rocks; include the fragmentary rocks deposited as sheets of sediment on the floors of seas, lakes and rivers and on land.

Semispecies—A taxonomsc group intermediate between a species and subspecies especially as a result of geographical isolation.

Serum—Watery fluid which separates from blood on coagulation.

Sexual dimorphism—The presence of marked differences (in shape, size, structure, colour, etc.) between male and female individuals of the same species.

Shellac—The purified product of lac.

Sibling species—Reproductively isolated populations that are morphologically similar or identical and frequently sympatic.

Silk—The fabric made from the natural product of the silk-worm, a moth (insect), of which there are many varieties, both wild and cultivated.

Sonograph—A graph of sound analysis of an animal.

Special adaptation—The occurrence of genetic changes in a population or species as the result of natural selection so that it adjusts to new or altered environmental conditions.

Speciation—The evolutionary formation of species which in essence, occurs by the production of barriers to gene exchange (reproductive isolation) between genetically diverging populations.

Speciationist—one who studies speciation.

Species diversity—Varieties of species.

Species-group—Includes species and subspecies.

Spectrophotometry—Instrument to investigate chemical nature of a substance by the examination of its spectrum using infrared rays.

Sperm—male gamete.

Sphagnum—A genus of mosses (peat or bog moss) of family Sphagnaceae.

Sponge—Any organism belonging to the animal phylum Porifera.

Successional cline—Cline that succeeds parental cline during ecological succession process.

Stimulus—An agent which provokes active response or reaction in a living organism.

State vector (character state vector)—The process of transferring control (in a programme) to an intermediate vector.

Suction—The production of a vacuum or partial vacuum in a container so that the external atmospheric pressure forces the surrounding organisms into it (aspirator).

Subjective synonym—Each of the two or more synonyms based on different types but of the same taxon.

Sustainable development—Development that meets the needs of present without compromising the ability of future generations to meet their own demands.

Sweeping—To cover a wide area with insect net by moving it to and fro.

Sympatric—Two or more populations occupying the same geographical area or of a population existing in a breeding condition within the cruising range of individuals of another population.

Synchronic—Contemporary; existing at the same time.

Synonym—Each of the two or more different names for the same taxon.

Taxon (pl, taxa)—A taxonomic group which is sufficiently distinct and can be given a name and ranked in a definite category.

Temporal—Limited by time; lasting only for a time.

Tissue—An aggregate of similar cells forming a definite and continuous fabric, and usually having a comparable function.

Trabecular layer—A row of cells bridging a cavity; a band or plate of fibrous tissue forming part of the internal supporting framework of an organ.

Transient—A distinct individual mode in a developmental series, such as a line of descent corresponding in the time-character concept to a species in a taxonomic character.

Tribe—A taxonomic division between genus and subfamily.

Tundra—A frozen Arctic plain, with lichens, mosses, and dwarfed vegetation.

Turtle—A reptile belonging to order Chelonia of Reptilia, distinguished by the two bony shells enclosing the body.

Ultraviolet light—Electromagnetic radiations in the wavelength range 4×10^{-5}cm to 5×10^{-7} cm approximately, covering the gap between X-rays and visible light rays.

Universe—The vast space surrounding us; includes everything that exists-the most distant stars, planets, satellites, as well as own earth and all objects on it.

Urodeles—Members of the order Urodela of Amphibia.

Variation—The occurrence of heritable or non-heritable differences among the individuals of one population or among populations.

Variety—Heterogenous group of a taxonomic taxon having varied phenotype; it is now not recognised by the Commission.

Vassopressin—A hormone, produced by the posterior lobe of the pituitory which is anti-diuretic and also causes contraction of muscles, especially of the arterioles.

Vector—An agent (usually an insect) which transmits a viral, bacterial, or fungal disease from one host to another.

Vernacular name—Native name of a country or place for a scientific name and there may be many such names for the same taxon and so causes confusion.

Weighting—A way for finding out phyletic information content of a character for sound classification.

Yeast—Micro-organisms producing zymase which induces the alcoholic fermentation of carbohydrates.

Zoobank—It is an initiative of International Commission of Zoological Nomenclature for open-access registry for animals and taxonomic act in zoology; the online version of the Official Register of Zoological Nomenclature.

Zoogeography—The science dealing with the geographical distribution of animals; specifically, the study of the relationship between specific animal form and the region in which they live.

Zooplankton—Floating and drifting animal life.

Index

A

Aberrans, 244
Abode, 51
Absorption, 96
Abstracts, 216, 219, 221, 222, 230
  – biological, 219
  – dissertation, 219
  – entomology, 221
  – helminthology, 221
Acanthocephala, 217
Acari, 157
Acarina, 81
Acarines, 200, 201
Acrididae, 88
Acrocentrics, 88
Acrylamide gel, 94
ACTH, 95
Adams, R.P., 273
Adanson, M., 273
Adaptation special, 98
Adaptive radiation, 307
Adhesive, 172, 173
Adis, J., 288
Aerial, 78
Aerodromes, 8
Aeroplane, 10, 286
Affinities, 90, 97, 98, 105
Afrodacus, 244
Agamospecies, 132, 246
Agapow, P.M., 273
Agar, 94, 277

AGA solution, 156
Agamic (Agamo) species, 307
Agassiz, 218
Age differences, 119
Agerartina trapezoideum, 7
Agrindex, 223
AGRINDEX, 222
AGRIS, 222, 223
Agrochemicals, 35
Agronomie, 1
Ahlquist, 300
Aitchinson, 183, 296
Albrech, P., 273
Alcohol, 154–158, 162, 166, 168, 173, 200
Aleurodidae, 78
Alexander, R.D., 273
Algae, 10, 32, 239
Algorithm, 183
Allah, H.H., 273
Alleles, 286
Allochronic, 307
Allometry, 226
Allomorphic, 130
Allopatric, 131, 140
Allotopic, 131
Allotopy, 131
Allotype, 260, 261, 265
Ally, 307
All Species Foundation, 148
Alpha taxonomy, 4
Alston, R.E., 273

Altitude, 37, 166, 266
Amacrogeographically, 137
Amadon, 95, 273, 292
Amino acid, 90, 92, 95, 129
Amlol parua, 29
Ammonia, 165, 174
Amoeba, 41
Amphibia, 147, 218
Amphibians, 42, 46, 48, 85, 86, 144, 145, 281, 294
Anatomy, 14, 41, 207
Ancestor, 106–108, 110–113, 128, 241
Anderson, R.M., 273
Anderson, S., 273
Andreev, S.V., 274
Angel Hair, 43
Animal husbandry, 3
Annelida, 158, 217
Anodonta, 256
Anopheles, 8, 78, 79, 247, 275
Ant, 29, 288
Anthidium, 81
Anthony, H.E., 274
Antibodies, 92, 295
Antigens, 92
Anti-mosquito, 30, 31
Anti-Taxonomy, 86
Anti-venom vaccines, 31
Anzu wyliei, 108
Apes, 108
Aphanic species, 130
Aphids, 6, 147, 200
Apicotermes, 81, 300
Apis mellifera, 20
Apomictic, 121, 132
Apotype, 261, 263
Appendices, 242
Aquatic, 10, 36, 54, 78, 155, 158, 166
Arachnida, 217, 297
Araneae (spiders), 157
Archaeocyatha, 217
Archaeopteryx, 107
Archive, 308
Archytas incertus, 7
Argas, 72, 73, 288, 289

Arginine, 95
Aristotelian type-concept, 260
Aristotle, 24, 63, 64, 66, 78, 119
Arme, A., 294
Armyworm, 7
Arnett, R.H., 274
Artenkreis, 140, 141
Arthropleurida, 217
Arthropoda,
Articules, 66
Ascorbic acid, 90, 91
Asexual reproduction, 308
Ashlock, P.D., 274
Asiatic Lion, 35
Aspirator bulb type, 161
Astraptes fulgerator, 84
Astronomie, 1
Atkins, D.E., 274
Atkins, N.B., 295
Atlas, 208
Atmosphere, 24, 28, 37–40, 44, 56
ATREE, 27
Attractants, 151
Attributes, 7, 97, 101, 106, 193, 203, 226
Australopithecus afracnsis, 108
Autotype, 261, 263
Available name, 225, 251, 253, 257
Aves, 111, 218, 257, 294
Avise, J.C., 274

B

Bacon, 278, 284
Bacteria, 30, 31, 44, 79, 107, 232, 240, 264, 302
Bacteriology, 252
Bactrocera oleae, 77, 78
Bactrocera zonata, 86, 87
Bag, 153, 159, 160, 201
Bait, 35, 153
Baker C.M.A., 290
Balke M., 293
Ballast, 10
Ball, C.R., 274
Ball, G.H., 275
Ball, I.R., 275
Ball, S.L., 285

Ballooning, 43
Balloons, 10
Balsa wood, 172
Barber, H.S., 275
Barcode, 84, 85
Barr, A.R., 275
Basford, N.L, 275
Basionym, 248
Basu Chaudhury, R.C., 275
Bates, M., 275
Bather, F.A., 275
Bats, 28, 154
Baudry, E., 306
Baum, D., 275
Beandry, J.R., 275
Beating, 154, 275
Becak M.L., 275
Becak, W., 275
Beckner, M., 275
Bee, 11, 81, 247
Beer, J.R. de, 275
Beetles, 6, 29, 79, 135, 144, 153, 164, 165, 173, 198, 204
Behaviour approach, 79
Behavioural characteristic, 68, 79, 298
Benazzi, M., 275
Benedict, A.A., 293
Bengal Tiger, 35
Benton, M.J., 275
Benzene, 165
Bergan, T., 275
Bergstrom, J., 275
Berlese funnel, 153, 164
Bernardi G., 297
Bessey, C.E., 275
Beta taxonomy, 4
Beyer, F.G., 288
Bibliography, 207, 208, 211, 212, 215, 218
Bieber, R.E., 288
Bier, M., 275
Bigelow, R.S., 275
Binary nomenclature, 236, 244
Bininda-Edmonds, O.R.P., 273
Binomen, 244
Binomial, 63, 64, 148, 236, 244
Binominalism, 247

Binominal nomenclature, 64, 65, 227, 236, 241, 244, 247, 285
Biochemical approach, 89
– characters, 90
– patterns, 91
– taxonomy, 90
– variation, 91
Biochemistry, 15, 41, 68, 129, 179, 283, 285, 289, 291, 300, 303, 305, 306
Biocode, 240, 284, 285, 287
Biodiversity, 9, 13, 14, 16, 17, 19–27, 29, 31–37, 40, 42, 44, 46, 48–55, 58, 59, 84, 85, 101, 106, 143, 148–150, 217, 229–233, 235, 240, 242, 273, 283, 284, 298, 300, 305
– agricultural, 53
– boon, 16
– Conservation Information, 52
Biodiversity Hot Spots, 35
Biodiversity of Life, 14
Biogeographically, 26
Bioinformatics, 1, 233
Biological Control, 7, 21, 278, 298, 299
Biological Diversity, 14, 16, 53
Biological resources, 21, 51, 53
Biome, 24
BioNET International, 22
Bionomics, 69
Bioprospecting, 31
Bioresearch Index, 219, 230
Biorights, 27
Bio-safety, 23
Biosis, 216, 230
Biosphere reserves, 48, 49
Biosynthesis, 129
Biosystematics, 68, 278, 291
Biota, 23, 54, 146
Biotechnology, 14, 50, 231
Biotypes, 122, 138
Biparental, 124, 126
Bird, 9, 30, 31, 36, 41, 44, 48, 54, 79, 84, 107, 108, 151, 204
Bird Droppings-Facial, 31
Bishy, F.A., 276
Bitty, 30

Black, C., 294
Blackith, R.E., 276
Blackwelder, R.E., 276
Blaker, A.A., 276
Blatta orientalis, 20
Bleaching, 32, 45, 166
Blomback, B., 276
Blood, 43, 63, 64, 92, 93, 98, 157,
 280, 295, 296, 302
Bloodless, 66
Blotting paper, 162–164, 168
Blum, M.S., 276
Bock, W.J., 276
Bogert, C.M., 276
Bolduan, O.E.A., 302
Bolton, E.T., 276
Bombs, 10
Books, 226
Borgmeier, T., 276
Bossert, W., 276
Boyden, A.A., 276
Brachiopoda, 217
Brackish water, 152
Branching tree, 66
Brand, J.M., 276
Breden, F., 299
Breed, local,
Breeding season, 6, 78
Bristol board, 172
Britain Wildlife Vets International
 (WVI), 35
British Association Code, 237
Britton-Davidian, J., 276
Brown, F.M., 276
Brown, R.W., 212, 277
Brown, W.J., 277
Brown, W.L., 277
Brues, C.T., 277
Bryozoa, 217
Bryson, V., 277
Buffalo, 13
Buffer, 49, 94
Bug, 29
Bulinus, 95
Bulletin of Zoological
 Nomenclature, 224, 225, 239,
 268, 269

Burma, B.H., 277
Burt, W.H., 277
Bush crickets, 80
Butler, J.E., 277
Butlin, R., 277
Butterflies, 28, 75, 92, 135, 144, 178,
 191, 195, 209
Buzzati-Traverso, 93, 277

C

Cable, R.M., 277
Cactus, 79
Cafiete, M.A., 284
Cain, A.J., 277
Calanopsoidea, 97
Cambrian, 42
Cambrian explosion, 42
Camel, 13, 156, 157
Camin, J.H., 302
Campanna, E., 278
Campbell, D.H., 277
Camp, W.H., 277
Candino, P.D., 277, 278
Canvas, 159
Carbohydrates, 96
Carbon Credits, 55, 56
Carbon Emission Reduction
 (CER), 56
Carbon Footprints, 56
Carboniferous, 42
Carboniferous period, 42
Carbon tetrachloride, 165
Carding, 156, 173
Carmone, F.J., 284
Carnivoures, 108
Carnufex carolensis, 107
Carpomyia, 87, 185
Cat, 36, 43, 83
Catalogue of Life Consortium, 148
Cataloguing, 14, 84, 150, 167, 176
Catastrophes, 143, 145
CATE, 17
Category, 3, 35, 49, 113, 114, 117,
 136, 140, 141, 179, 181, 238, 277
 – collective, 140
 – lowest, 113

Caterpillars, 6, 81
Cattle, 13
Cazier, M.A., 278
Cell biology, 14
Cell-lines, 82
Centre for Environment
  Education, Centre for Science
  and Environment, 27
Century of Dictionary, 220
Ceratitis capitata, 77, 78
Cercariae, 78
Chadwick, C.E., 278
Chaetognatha, 217
Chaetotaxy, 71
Chalk, 93
Chamberlin, W.J., 278
Chandler, L., 292
Chapul, 30
Characters, corelation, 97
Chase, M.W., 305
Chatterjee, J.B., 275
Checklist, 149, 208, 209
Cheetah, 107, 144
Chen, P.S., 278
Chesapeake Bay, 40
Chiarelli, A.B., 278
Chiggers, 130
Childers, W., 290
Chilisaurus diegosauranzi, 107
Chimpanzee, 83
Chirotype, 261
Chloral hydrate, 157
Chlorofluoro-carbons (CFCs), 37
Chloroform, 165
Cholera, 9, 44
Chresonymy, 255, 301
Christoffersen, M.L., 278
Chromatography, 93, 277, 283,
  288, 292, 306
  – Column, 93
  – Paper, 93
Chromomere, 141
Chromosome fusion, 86
Chromosome painting, 83
Chronocline, 140
Chronological, 141, 211
Cicada, 31

Cicero, C., 288, 293
Ciotti, M.M. 288
Civilisation, 117
Clades, 108, 109, 241, 275, 277
Cladism, 302
Cladistic, 3, 108, 128, 230, 286
  – classification, 106, 108
  – phylogeny, 106
  – taxonomy, 3, 278
Cladogram, 108
Clark, E.W., 278
Classification, biological
  components, 104
  – evolutionary, 104, 109
  – future of, 110
  – natural, 104, 105
  – omnispective, 104, 110
  – phenetic, 104, 105
  – phylogenetic, 104, 105
Clausen, C.P., 278
Clean Development Mechanism,
  56
Clean Energy Ministerial, 38
Clifford, C.M., 288
Clifford, H.T., 278
Climate change, 26, 32, 35, 36,
  38–40, 46, 55, 56
Cline, 140
Clough, T.G. Barra, 293
Cluster, 124, 125, 129
Coastal, 26, 32, 39, 45, 46, 54, 103,
  144
Cockroach, 13, 30
Code, origin, 236
Cody, J.H., 298
Coelenterata, 217
Cohan, F.M., 278
Cohen, D.M., 278
Cohort, 115
Coker, R.E., 278
Cole, A.J., 278
Coleoptera, 63, 72, 81, 88, 156, 170,
  217, 275, 277, 289
Collecting ways, 151, 153, 167
Collection, data, 166
  – taxonomic, 143
Collembola, 154

Colm, A.C., 297
Commerce, 11
Commission, 34, 149, 150, 223–225, 231, 237, 238, 242, 243, 247, 252, 259, 267–270, 285, 287
Comparative anatomy, 14, 63, 66, 207
Competition, 78, 79, 97, 132
Competitors, 6
Complex, 8, 9, 11, 24, 66, 78, 79, 82, 83, 89, 93–95, 121, 123, 136, 178, 204, 267, 275, 292
Composite variable, 100
Computer, 98, 105, 182, 223, 230, 231, 233, 279, 285, 293, 296, 298, 305
Computerised, Key 180, 184,
– programme, 183
Computing methods, 101
Conard Martius, H., 278
Condensation, 162
Congeneric, 258
Congress, 114, 135, 237, 238, 242, 260, 268, 270
Conodonta, 217
Consden, R., 278
Conservation, 14, 16, 19–23, 26, 27, 34, 36, 38, 39, 46, 47, 50–53, 84, 101, 106, 144, 146, 242, 254, 287, 299, 300
Conservationists, 3, 24, 26, 43, 44, 47, 84, 149, 230, 266
Conservation Law Foundation (CLF), 38
Conspecific, 119, 263, 264
Constance, L., 278
Convention of Biodiversity, 22, 23, 51, 53, 148
Convention on Biodiversity (or Biodiversity Convention), 51
Convergence, 100, 109
Cook, E.F., 275
Coral bleaching, 32
Coral reef, 21, 32, 33, 40, 45
Corals, 32, 35, 44, 45
Cork, 163, 172, 197, 198, 199
Corliss, J.O., 279
Corytophanidae, 107

Cotton, 160, 164, 168, 198–201
Cotype, 260
Couplet, 180, 181, 184, 186–188
– primary, 184, 186, 187
– secondary, 184, 186, 187, 188
– tertiary, 186, 187, 188
Cow, 83
Coxa, 170
Coyne, J.A., 279
Cracraft, J., 279
Cramer, 277
Crandali, K.A., 273
Crayfish, 41
Crenshaw, J.W., 279
Cressey, R.F., 278
Cresson, E.T., 279
Cretaceous, 42, 48
Cricket, 29, 30, 273
Crick, F.C.H., 279
Critica Botanica, 236
Crochet, P.A., 281
Crocodilian ancestor, 107
Crofton weed, 7
Crop, 6, 28, 54, 55, 222
Cross-fertilisation, 132
Crovello., 302
Crowbar, 29
Crowe, T.M., 297
Crowson, R.A., 279
Crustaceans, 158, 200, 201
Cryostat microtomes, 96
Cryptic species, 84–86, 121, 131
Ctenophora, 217
Cullen, J.M., 279
Cumley, R.W., 279
Cummins, K.W., 279
Curating, 147, 167
Curator, 146, 167, 175, 177, 178
Cushmann., 295
Cuvier, G., 279
Cyanide bottle, 163, 165
Cyanins, 94
Cyber Taxonomy, 232
Cycliophora, 20
Cyprinus carpio, 11
Cytogenetics, 275, 305
Cytological Approach, 81
Cytology, 86, 179, 289, 301

Cytoplasmic inclusions, 75
Cywinska, A., 285

D

D. californica, 204
Dall's Code, 237
Dall, W.H., 279
Dallwitz, M.J., 279
Dariks, H.V, 279
Darlington, P.J., 279
Darwin, C., 279
Darwinism, 67, 109
Davies, R.G., 279
Davis, B.J., 280
Davis, J.I., 280
Davis, P.H., 280
Dawson, R.D., 306
Dayrat, B., 280
DDT, 10, 286
de Beer, G.R., 280
de Candolle, A.P., 280
Declarations, 224, 269
Deep-sea, 44
Defence, 10, 90, 281
Deforestation, 26, 51
De Long, R., 280
Deme, 139
Dementev, G.P., 280
Dendrogram, 108
Denmark, H.A., 280
Deoxyribonucleic acid (DNA), 81
de Queiroz, K., 280
Dermanyssus, 97
Dermaptera, 154, 170, 206
Desalle, R., 280
Descendants, 111, 112, 121, 124
Descent, 67, 109, 117, 118, 124, 129, 241
Description, 16, 20, 43, 65, 76, 103, 149, 150, 180, 202, 203, 208, 210, 211, 213, 227, 236, 253, 259–267, 281, 293, 294
 – language for taxonomy, 230
 – taxonomic, 230
Desert, 26, 107
Desiccation, 162
Detoxification, 28

Deutsch, H.F., 280
Developmental, 97, 100, 143, 204
de *Waard, J.R.*, 285
Dhrangadhra, 35
Dhruva, 29
Dianthidium, 81
Dickinson, E.C., 281
Diclofenac, 49
Dictionary, 213, 220, 290, 306
Dicymennea abelis, 204
Differential diagnosis, 202
 – systematics, 100
Digital Taxonomy, 232
Dimorphism, 119
Dinosaur, 26, 46, 48, 107, 108, 146
Dioecious, 132
Dipannita, Das, 279
Diplura, 154
Diplurans, 201
Dip net, 153, 158
Diptera, 63, 78, 86–88, 157, 170, 172, 206–208, 217, 250, 251, 284, 293
Dipterous flies, 86
Directions, 281
Directories, 220
Direct pinning, 169, 171
Distribution, 20, 21, 25, 90, 92, 100, 103, 131, 132, 147, 151, 203, 213, 226, 285, 297
Divergence, 68, 89, 109, 114, 141
Divergent evolution, 109
Diversity, 2, 3, 12, 13, 16, 19, 20, 21, 24–26, 53–55, 91, 103, 104, 106, 119, 129, 143, 147–149, 166, 167, 177, 178, 278, 291, 295, 306
 – barcodes biologial, 14, 16, 19, 21, 24, 53, 54, 129, 295,
 – genetic, 13, 25, 55
 – global, 147, 148
 – habitat, 25
 – organic, 20, 291
 – phylogenetic, 25
 – species, 25
DNA, 71, 81–85, 90, 101, 124, 129, 149, 276, 280, 281, 285, 286, 288–290, 292, 293, 295, 299, 305, 306

DNA barcoding, 84, 85, 280, 281, 285, 288, 292, 293, 299, 306
– barcoding, 83, 84, 85
– hybridisation, 81, 82
– matching, 82
– Mitochondrial, 84
– sequences, 84
– tags, 84
– Taxonomy, 83
DNA-based methods, 85
Donkey, 120
Donoghue, M.J., 280, 293
Doty, P., 299
Double-citation, 248
Double mounting, 169
Douville Code, 237
Douville, H., 281
Downe, A.E.R., 281
Downey, J.C., 281
Dredging, 151, 155
Dredonoughtus sehrani, 108
Dreinsbach, R.R., 281
Droegeneier, K.K., 274
Drooping, 170
Drosophila, 79, 86, 111, 130, 279, 281, 286, 296, 303
– *aldrichi*, 79
– *mulleri*, 79
Dubois, A., 281
Duchateau, W.E., 281
Duellman, W.E., 281
Dunbar, C.O., 281
Dupraw, E.J., 281
Dyeing, 166

E

Earth, 12, 20, 21, 24–26, 28, 34–37, 42, 44, 45, 50, 51, 53, 107, 132, 143, 145, 146, 148, 154, 240
Earthworms, 158
Earwig, 86
Ebach, F., 281
Echelle, A.A., 281
Echinodermata, 217
Echiura, 217
Ecocline, 140
Ecological, approach balance, 78

Ecology, 5, 14, 15, 27, 63, 68, 69, 89, 141, 179, 233, 278, 279, 282, 285, 286, 292, 296, 299, 303, 305
Ecosystem, 10, 22, 24, 26, 28, 32, 34, 41, 45, 47, 48, 51, 54
Ecotourism, 31
Ecotypes, 122
Ectoparasites, 154, 155, 157, 275
EDIT, 17
Editorial Committee, 238, 270
Edwards, J.G., 281, 282
Edwards, M.A., 282
Eernisse, 288
Egg, 29, 50, 63, 77, 78, 81, 95, 300
Egg-shell, 78
Eggleton, P.B., 293
Egyptian vulture, 44
Ehrlich, P.R., 282
Electrode, 94
Electronic Publication, 239, 287
Electron microscopes, 96
– scanning, 72, 78
– transmission, 78
Electrophoresis, 92, 94, 95, 275, 280, 286, 290, 297, 299, 301, 304, 306
Elephants, 108, 144
Elk, 235
Elliot, M., 293
El Nino, 33, 40
Elton, G., 282
Elytron, 170
Embioptera, 155
Embryological approach, 76
Embryological pattern, 68
Embryology, 63, 78
Emendations, 150
Emerson, A.E., 282
Empirical, 67, 290
Emu, 108
Endangered, 9, 16, 33, 35, 36, 42, 44, 46–51, 54, 239, 266, 267, 281
Endemic, 48, 143
Endemism, 21, 226
Endler, J.A., 282
Endochorion, 77, 78
Endochorionic, 78
End-populations, 122

Energy Sector Management
  Assistance Programmed
  (ESMAP), 38
Enhydrus setboldi, 44
Entomologica Photosynthesis, ????
Entomological pins, 169
Entoprocta, 217
Enzymes Oil-eating, 31
Ephemeroptera, 155
Epiphenomenon, 136
Epistemological, 129
Epithet, 245
Equal weighting, 97, 98
Equality, 256
Ereshefsky, M., 282
Eriophyidae, 81
ERMS, 149
Espinosa, 284
Essentialism, 119, 286, 301
Essig, E.O., 282
Estabrook, G., 298
Ethcheverry, M., 285
Ether, 165
Ethologists, 15
Ethology, 69, 79, 282
Ethyl acetate, 164
Ethyl chloride, 165
European carp, 11
Eutrophication, 28
Evans, H.E., 282
Evaporation, 173
Evolution, 2, 3, 14, 26, 66, 67, 84,
  98, 100, 105, 107–110, 112, 117,
  120, 122, 144, 226, 241, 275–279,
  282–285, 288, 290, 291, 296, 298,
  300–302, 305, 306
  – biology, 292
  – divergence, 109
  – history, 3, 91, 106, 232
  – species concept, 119, 123, 127,
    128, 305
  – taxonomy, 105, 109
  – theory, 65, 67
Evolutionary, 3–5, 20, 41, 65–68,
  71, 81–83, 88, 91, 100, 103, 105,
  106, 108–110, 114, 117, 123, 125,
  127–130, 133, 138, 141, 204, 232,

238, 248, 278, 286, 290, 292, 294,
  305
Exotic species, 21, 48
Experimental biologists, 68, 301
Expert Center for Taxonomic
  Identification, 230
Extant, 12, 263, 266, 288
Extinct, 3, 9, 12, 36, 41–44, 48, 50,
  54, 66, 101, 107, 108, 110, 144,
  146, 147, 167, 232, 240
Extinction, 16, 21, 28, 35–37, 42, 44,
  45, 55, 121, 144, 145, 226
Extras, author's, 214

F

Fabricius, 135, 236, 253, 258
Facies, 136
Falkow, S., 291
Family, 33, 81, 114, 115, 131, 179,
  181, 196, 241, 243–245, 247, 251,
  257, 259, 294
Family-group, 243
Fauna and Flora International, 23
Faunistic, 138
Fecundity, 118, 133
Feldman, S.L., 274
Ferguson, A., 282
Fernald, H.T., 282
Fernbank Museum of Natural, 30
Ferrier, S., 284
Ferris, G.F., 282
Fertilisation, 123–125, 132
Fertility, 11, 27, 118
Fertilizer, 28
FESTO, 33
Fibre, 33
  – polyester, 33
Fidler, J.H., 304
Fin, 144
Finches, 41
Finger prints, 84
Fireant, 90, 276
Fireflies, 81, 82, 275
Fischer, F.C., 282
Fish, 10, 11, 13, 26, 29, 36, 40, 43,
  44, 46–48, 54, 93, 103, 107, 158,
  293, 298

– stickle back, 41
Fisheries, 145
Fitch, W.M., 282
Fixation, 96
Flagellates, 204
Flann, C., 282
Flashes, 81, 82
Flatworms, 75
Flavonoides, 90
Fleas, 153
Fleming, H.S., 298
Flies, 6, 30, 86, 87, 153, 207, 284, 288
Floristic wealth, 13
Florkin, M., 282, 283
Flotation method, 153, 163
Fluorescence, 83
Fluorescent, label, 83
– patterns, 94
FMCG, 30
Folk Taxonomy, 235
Food and Agricultural Organization, 29
Food chains, 10, 26, 42
Food security, 20, 46
Foote, C., 206, 294
Forcipula indica, 250, 251
– pugnax, 251
Ford, E.B., 283
Forest, 10, 26, 30, 39, 43, 47, 53, 54, 103, 144, 156, 286
Forestry, 6, 7, 54
Forficula riparia, 258, 259
Forget, P.M., 283
Form seasonal, 4
Formalin, 156, 158
Formenkreis, 140
Forsberg, F.R., 283
Fossil, 9, 38, 40, 49, 66, 71, 83, 91, 103, 104, 106, 107, 123, 133, 266, 275
Fox, R.M., 279
Fox, A.S., 283
Fragments, 133
Freezing, 28, 247
Frequencies, 100, 140
Frison, T.H., 283
Frizzel, D.L., 283

Fruit flies, 6, 87, 207, 284
Fuels, 9, 31, 38, 40
Fugisawa, T., 293
Fungi, 16, 23, 31, 154, 156, 162
Fungus, 145, 172
– Chyrtid, 145
Funk, V.A., 283
Funnel, 153–158, 162, 164
Futuyma, J.D., 279

G

Galaxy, 25
Gall, 7, 288
Gamma taxonomy, 4
Gandhi, J.R., 299
Gang, W., 284
Garcia-Gomez, J.C., 284
Garcia-Molina, Klein, 274
Garrity, G., 284
Garvey, J.S., 277
Gaston, K.J., 283
Gastroenteritis, 44
Gastrotricha, 217
Gauthier, J., 280
Gelatin capsules, 198
Gene, 50, 83–85, 89, 100, 114, 118, 121, 123–125, 136, 149, 299
Genealogical concordance, 126
Gene Bank, 149
Gene flow, 125, 299
Gene pool, 121
General Assembly, 242, 268, 270
Genetic, 12, 13, 25, 41, 50, 51, 55, 68, 82–84, 88, 89, 91, 100, 109, 117, 119–121, 126, 140, 147, 204, 262, 280, 292, 303
– divergence, 89
– diversity, 13, 25, 55
– erosion, 55
– similarity, 68, 109
– variation, 25, 280
Genetical unit, 120
Genetics, 13–15, 20, 68, 69, 125, 126, 179, 262, 286, 290
Genitalia, 71, 168, 281
Genoholotype, 262
Genome, 82, 83, 290, 299

Genosyntype, 262
Genotype, 128, 262, 287
Genritype, 262
Genus, 44, 48, 64, 72, 81, 82, 86, 93, 95, 97, 111, 114, 115, 141, 176, 196, 202, 206, 208, 210, 220, 227, 228, 236, 241, 243–248, 251, 256–259, 262–265, 275, 279, 282, 286, 288, 289, 296, 298, 301, 303, 306
Geocline, 140
Geographical distribution, 147
Geographical isolation,
Geological, 138, 210
Geosphere, 24
Geotype, 263
Germigni, V., 283
Germplasm, 50
Ghiselin, M.T., 283
Gier, L.J., 283
Gills, 64
Gillsdpy, J.E., 287
Gilly, C.L., 277
Gilmore, J.S.L., 283
Gingsbrug, I., 283
Gisin, H., 283
Gittleman, J.L., 273
Glacial acetic acid, 156
Gladkov, N.A.,
Global Biodiversity Information Facility, 52, 148, 231
Global Taxonomic Initiative, 22
Global warming, 25, 40, 44, 45, 145
Glossary, 213
Glycerine, 157, 158, 162, 168, 200
Gnathostomulida, 217
Goat, 13
Gochfeld, M., 283
Godfray, H.C.J., 283
Goldon, I., 284
Goldschmidt, R., 284
Gomori, G., 284
Goodloe, M.B., 280
Goodwin, R., 292
Gordon, A.H., 278
Gould, S., 284
Goyal, P., 284
Gradation, 140

Grades, 109
Gradient, 140
Graham, C.H., 284
Grant, V., 284
Grant, W.F., 284
Grasshopper, 15, 29, 88, 132, 165
Grasslands, 26, 28
Great Barrier Reef, 32, 60
Great Indian Bustards, 35
Greek, 1, 63, 97, 212, 227, 248, 253, 255, 290, 295, 306
Green Domestic Product, 36
Greenhouse gases, 25, 37, 40
Green, P.E., 284
Gregg, J.R., 284
Gregor, J.W., 283
Greuter, W., 284
Grewal, J.S., 284
Griffiths, G., 284
Griffiths, G.C.D., 284
Griswold, C., 299
Gross Domestic Product, 36
Gryllenberg, H.G., 284
Gryllidae, 80, 304
Guerra-Garcia, J.M., 284
Guides, 209, 212, 220, 225
Gums, 31
Gunther, K., 284
Gurney, A.B., 284

H

Habitat, 21, 25, 42, 47–49, 78, 79, 119, 124, 125, 139, 144, 146, 256, 299
Hackett, L.W., 284
Haemoglobin, 95, 96, 290
Haemolymph, 94, 281, 304, 306
Haeuser, C., 276
Hagmeir, E.M., 285
Hall, A.V., 285
Hammerstein, P., 303
Hamolsky, M., 288
Handbook, 209, 210, 287, 296
Handler, P., 285
Hapantotype, 263
Haplotype, 263
Hardy, D.E., 207, 285

Harlin, M., 285
Harlow, R.D., 285
Harold, H., 294
Harris, P., 294
Hatch, M.H., 285
Hausdorf, B., 285
Haustellate, 63
Hawkes, J.C., 285
Hawksworth, D.L., 285
HCFC-22, 39
Heautotype, 260, 263
Hebert, P.D.N., 285
Hedberg, O., 285
Heffron, H.N. 276
Hegberg, D., 285
Heller, J.C., 285
Helminthology, 218, 221
Hemimetabolous, 94
Hemiptera, 156, 170, 207, 218
Hendrickson, J.A., 100, 294
Hennig, C., 112, 285
Hennig, W., 285
Herbivoures, 107
Hereditary, 89, 136, 262
Heredity, 81, 86
Heritage, 32, 41, 46, 48, 53, 120
Hermaphrodite, 132
Herrera, A.L., 285
Heslop-Harrison, J., 285, 286
Heterochromatic, 81
Heterogeneous, 138, 251
Heterospecific, 127
Heterotypic synonyms, 256
Heywood, V.H., 286
HFC-23, 39
Hideux, M.H., 283
Hierarchy, 3, 12, 113, 114, 118, 141, 247
 – Linnaean, 114
Higher category, 4, 12, 90, 114
Histochemical, 96, 286
Histochemistry, 284
Histological, 166
Hoffmann, C.H., 286
Hogben, L., 286
Holdrege, C., 281
Holland, W.J., 286
Holocene, 42

Holometabolous, 94
Holophyletic, 112
Holotype, 146, 176, 260, 261, 263, 265–267
Holuhraun volcanic eruption, 35
Homeostasis, 100
Homeotype, 263
Homoeotype, 260, 265
Homology, 83, 91
Homonym, 246, 249, 254, 256–258
 – Primary, 258
 – Secondary, 257, 258
Homonymy, 151, 239, 247, 252, 257, 258
 – Law, 257
Homoplasy, 98
Homoptera, 78, 208
Homo sapiens, 20, 108
Homotype, 260, 263
Homozygous, 118
Honey, 11, 20, 29
Honey bee, 11
Hoogstraal, Harry, 288, 289
Hooves, 64
Hopwood, A.T., 286
Hormones, 5, 89
Horn, W., 286
Horovitz, I., 286
Horse, 120, 275
Hoyer, H.B., 286
Hubbell, T.H., 286
Hubby, J.L., 286
Huettman, F., 284
Hughea-Schrader, S., 286
Hull, D.L., 286
Human beings, 8, 25, 31
Human Placenta Facial, 31
Humidity, 175, 205
Hunter, R.L., 286
Hurricanes, 45
Huxley, J.S., 287
Hybrid, 123, 125, 126, 133, 246, 254, 279, 290, 299
Hybridization, 81–83, 124, 126, 273, 289, 290, 299, 300
Hybridogenetic, 246
Hydro-Chlorofluorocarbons (HCFCs), 37

Hydrosphere, 24
Hydroxyapatite, 83
Hylon, 37
Hyman, D.L., 291
Hymenoptera, 156, 170, 218, 282
Hypodigm, 263
Hypotype, 263

I

IBM card, 177
Ichnotaxon, 245
Icotype, 263
Identification, 178, 179, 196
 – from combination of methods,
   179, 196
 – from direct comparison, 179,
   191
 – from keys, 179, 180
 – from literature, 179, 180
 – problems encountered, 196
 – through pictures, 179, 191
Ideotype, 263
Igneous rocks, 9
Illustrations, 12, 211, 212, 225
Images, 149
Immature stages, 76, 207
Immune serum, 92
Immunisation, 92
Immunised, 92
Immunological Taxonomy, 93,
   300, 302
Incalculable, 51
Incertae sedis, 229
Incipient species, 136, 137, 281
Incisor teeth, 64
Index 218
 – animalia, 218,
Indexes, rejected and approved
   names, 224, 269
Indian Stripped Hyena, 44
Indices, 100
Inductive, 12, 104, 277
Information, storage, 175
Infraclass, 115
Infracohort, 115
Infrafamily, 115
Infraorder, 115

Infraphylum, 115
Infrared, 96, 293, 295
Infraspecific, 295
Infrasubspecific, 245
Inger, R.E., 287
Inglis, W.G., 287
Insecticides, 5, 7
Insect net, 153
Insulators, 28
Insulin, 95
Integrated Taxonomic Information
   System, 23, 148, 231
Integrative Taxonomy, 85
Intellectual Property Rights, 37
Interbreeding, 41, 68, 114, 118,
   120–124, 127–129, 136, 138, 141,
   305
International Atomic Energy
   Agency (IAEA), 222
International Code of Zoological
   Nomenclature, 65, 150, 224, 236,
   237, 239, 242, 266, 269, 271, 274,
   281, 287, 289, 297
International Commission of
   Zoological Nomenclature
   (ICZN), 150
International Congress of
   Zoology, 114, 135, 237, 238, 242,
   260, 268, 270
International Day for the
   Preservation of the Ozone
   Layer, 38
International Trust of Zoological
   Nomenclature, 238, 270
International Union for
   Conservation of Nature, 34, 36
International Union of Biological
   Sciences (IUBS), 223, 238
Internet, 229, 233
Interspecific competition, 79
Intersterile, 122
Intersterility, 127
Intrafertility, 118
Intraspecific, 4, 84, 120
Introgression, 96
[illegible], 130,
[illegible], 286,
289, 295

In vitro, 82
Inward, D.J.G., 293
Ions, 96
Ips, 72, 289
Irreducible, 117, 129
Irwin, M.R., 287
Isaac, N.J.B., 287
Isolating mechanisms, 79, 105, 118
Isolation, 105, 122, 123, 138, 141, 205
 – reproductive, 105, 122, 123,
 – temporal, 138
Isophenes, 140
Isoptera, 155, 300
Isosyntype, 264
Isotype, 264
ITIS, 23, 148
IUCN, 34–36, 49, 52, 144
IUCN Species Survival Commission, 34

J

Jaeger, E.C., 287
Jameson, D.L., 287
Jardine, C.J., 287
Jardine, N., 287
Jassids, 198
Jeffrey, C., 287
Jellyfish, 31, 103
Jennings, J., 293
Jensen, R.J., 287
Jeopardy, 49
Jesus lizard group, 107
Jeuniaux, C., 282
John, R., 284
Johnsgard, P.A., 287
Johnson, F.M., 287
Johnson, L.A.S., 288
Johnson, M.L., 288
Johnson, N.K., 288
Journal, 107, 167, 205, 211–213, 221, 222, 226, 231–233
Jungle, 31, 145
Jurassic Park, 107
Juvenile stages, 76

K

Kant, 135
Kaplan, N.O., 288
Kapoor, V.C., 206, 207, 288
Karcher, D., 299, 304
Karyological, 86
Karyomorphology, 89
Karyotaxonomy, 86
Karyotype, 86
Keck, D.D., 288
Keen, A.M., 288
Keifer, H.H., 288
Keirans, J.E., 288
Kelly, R.P., 288
Kessel, E.L., 288
Keys, 12, 24, 72, 85, 98, 179–184, 187, 188, 191, 196, 210, 211, 230, 279, 285, 292, 293, 295, 296, 304, 305
Kiauta, B., 288
Kidney, 90, 91
Killing bottle, 159–161, 164, 165
Kinds of Names, 253
Kinorhyncha, 217
Kinsey, A.C., 288
Kiriakoff, S.G., 288
Kirk, P.M., 284
Kirk, R.L., 288
Klass, K.D., 288
Kleinschmidt, O., 289
Klerk, H.M. de, 305
Knapp, S., 289
Knudsen, J.W., 289
Koeppel, A.F., 278
Kohn, A.J., 289
Kohne, D.E., 289
Kormaed Ocean Sanctuary, 46
Kosmoceratops, 107, 116
Kowalczuk, M., 294
Kraus, R., 289
Krell, E.T. 298
Krell, F.T., 305
Kristensen, N.P., 288
Krzysztof, S., 289
Kure Atoll, 46

L

Labidura, 86, 87, 258, 259, 293
Lac, 11
Laccognathus embryi, 107
Lahni, F., 289
Lamarck, 65–67
Lamarckism, 65
Lambert, D.M., 289
Lampyridae, 81
Lance, G.N., 305
Land erosion, 28
Landslides, 28
Langurs, 144
Lanham, V., 289
Lanier, G.N., 289
Lankester, 89
Larvae, 4, 6, 32, 33, 78, 79, 84, 156,
    165, 202
Larval cases, 81
Latin Abbreviations, 227
    – terminology 227
    – words, 229
Laubenfels, M.W. de, 289
Laurin, M., 289
Lebbe, J., 183, 283
Lectotype, 176, 261, 264, 267
Lectotypifications, 150
Leech Facial, 31
Lees, D.C., 293
Legitimate name, 253
Leone, C.A., 289
Lepidoptera, 4, 63, 86, 88, 156, 159,
    170, 209, 218, 283, 290, 294
Lepiota mansueta, 29
Le Quesne, W.J., 289
Lerman, S.C., 289
Levi, C., 289
Levi, H.W., 289
Lewis, H., 289
Lewontin. R.C., 286
Library, 65, 223
Lice, 153, 201, 204
Liddell, H.G., 290
Light traps, 153, 155–157, 274
Lineage, 108, 110, 123, 128, 133,
    146, 231
Linen, 159

Linnaean Signs, 229
Linnaean-Stricklandian, 150
Linnaean system, 65
Linnaeus, C., 290
Linsley, E.G., 290
Lipids, 96, 285
Literature cited, 211
Liver, 90, 91, 165
Lizards, 107, 108
Logotype, 264
Long Now Foundation, 148
Long, Samuel S., 290
Loricifera, 20
Lotion, 37
Love, A., 290
LUCY, 108
Lumb, 285
Lungs, 64
Lysine, 95

M

Mace, G.M., 273
Macfayden, A., 290
Macrogeographically, 138
Macrohabitat, 131
Macromolecular, differentiation,
    95
Macromolecules, 129, 279
Maeki, K., 290
Maerz, A., 290
Maggots, 6
Magnesium sulphate, 163
Malagon, H., 306
Malaise, R., 290
Malaise traps, 153
Mallet, J., 290
Mallophaga, 204
Mammalia, 64, 111, 218
Mammals, 16, 31, 35, 43, 86, 92,
    107, 135, 144, 147, 153, 155, 157,
    162, 168, 178, 191, 209, 274, 277,
    288, 295, 300, 301
Mandel, S., 291
Mandibulate, 63
Manfredi Romanini, M.G., 290
Mangroves, 45
Maniophasmatodea, 20

Man of the Biosphere, 51
Mantodea, 81
Manuscript, 261
Manwell, C., 290
Maps, topographic, 138
Margolias, E., 282
Marijuana, 33
Marine, 26, 31, 32, 44–46, 48, 130, 145, 149, 266
Marino, W.A., 296
Market, C.L., 286
Marmur, J., 299
Marshall, J.C., 273
Martens, K., 298
Martin, A.J.O., 278
Martin, P.G., 291
Martin, S., 291
Maslin, T.P, 291
Mason, H.L., 291
Mason, H.S., 283
Massachusetts Institute of Technology's (MIT), 42
Mathematical methods, 12
Matrix (character state), 182
Mauly, P., 291
Mayo, M.A., 284
Mayr, E., 291, 292
Mazrouel, S.M.A., 284
mBio, 107
McCafferty, W.P., 292
McCarthy, B.J., 292
McMasters, J.H., 299
McMillan, C., 292
McNeill, J., 292
Measurements, Multiple, 100
Meat, 29, 55, 144
Mecoptera, 156, 170
Mega diverse, 26, 35
Megadiversity, 25
Meglitsch, P.A., 292
Meikle, R.D., 292
Melville, 297
Mendel, 5
Mendelian species, 246
Menelli, A., 294
Menon, M.G.R., 292
Merril, E.D., 292
Merozoa, 204

Meryman, H.T., 292
Mesozoa, 217
Messer-Schmitt, M.L., 274
Messina, P., 274
Metaphase, 81
Metatype, 260, 261, 264
Metcalf, Z.P., 208, 292
Meyer, A., 292
Meyer, C.P., 292
Mian, A.R., 288
Michel, E., 284
Michener, C.D., 292
Micks, D.W., 292, 293
Microbes, 20, 23, 24, 27, 42, 45
Microbiology, 100
Microcomputer, 283
Microgeographically, 138
Microlepidoptera, 209
Micromolecular taxonomy, 90
Microorganisms, 16, 21, 291
Microscopic slides, 168, 201
Microspecies, 246
Microtechniques, 96
Military, 10
Miller, R.R., 293
Milne, P., 305
Mimicry, 5
Minelli, A., 293
Mineral, 9, 46
  – prospecting, 8
MINITEL system, 183
Minkoff, E.C., 293
Minst, K.J., 293
Miscellarea, 217
Mishler, B.D., 293
Missing link, 107
Mist net, 152
Mites, 72, 81, 97, 130, 157, 161, 163, 219
Mitosis, 81
Mittal, O.P., 293
Modern taxonomy, 65–67, 71, 276, 279
Molecular, 14, 17, 85, 89, 90, 94, 126, 232, 240, 246, 280, 286, 289, 290
  – biology, 14, 280, 290
  – systematics, 85, 303

– taxonomy, 14, 90, 289
Molehanova, V.A., 274
Molluscs, 27, 66, 81, 158, 176
Monagham, M.T., 293
Moniz, E., 284
Monkeys, 144
Monoculture, 35
Monoecious, 132
Monograph, 149, 150, 206, 207,
    245, 282, 286
Mononominalism, 247
Monophyletic, 110–112, 141, 242
– group, 110, 111, 112, 141, 242
– lineage, 110
– species-group, 112
Monophyly, 109, 111, 112, 274, 286
Monothetic, 134, 182
Monotype, 264
Montreal protocol, 37
Monura, 111
Mooers, A.Q., 299
Moon, 8
Moritz, C., 293
Morph Bank, 149
Morphocline, 203
Morphology, 15, 20, 65, 68, 86, 89,
    97, 108, 119, 126, 141, 179, 196,
    300
Morphotype, 139, 264
Morphs, 4, 89, 139, 140
Morse, L.E., 293
Mosaic evolution, 98
Mosquitoes, 8, 30, 93, 96, 145, 247,
    275, 293
Mosses, 155, 162
Moss, W.W., 293, 294
Mould, 156, 166, 168, 175, 199, 201,
    290
Mound, L.H., 283
Mount Everest, 45
Mounting, 168, 169, 172, 173, 197
Mount Nimba, 43
Moylan, E., 294
Moynihan, M., 294
Muesebeck, C.F.W., 294
Mule, 120
Muller, S.W., 299

Multilateralism, 38
Multiple Entry Keys, 182
Multiplication of species, 5
Munroe, E., 294
Museology, 167
Museum, 22, 24, 30, 65, 98, 108,
    146, 147, 167, 177, 216, 224, 231,
    238, 273–275, 296
Museum Journal, 167
Mushroom, 29, 183
Muslin cloth, 153, 160
Mussel, 256
Mutation, 5, 84, 124, 126, 137
Mutation theory, 5
Muttkowski, R.A., 294
Mycrognathozoa, 20
Myers, G.S., 294
Myrakeen, A., 299
Myriapoda, 217

N

Name Bank, 233
NASA, 41
Nascent, 141
National Agricultural Library, 223
National Biodiversity
    Authority, 53
National Center for Biotechnology
    Information, 231
National Oceanic and
    Atmospheric Administration
    (NOAA), 43
National Park System, 48
National Science Foundation, 17,
    231, 232, 274
Natural disasters, 28
Natural enemies, 7
– group Natural gas,
Natural parks, 48
Natural populations, 68, 120, 128,
    286, 287, 289, 290
Natural selection, 5, 67, 279
Neave, S.A., 294
Nekrutenko, Y.P., 294
Nelson, G.J., 294
Nematoda, 217
Nematomorpha, 217

Nemertinea, 217
Nemesio, A., 281, 294
Neoallotype, 264
Neoparatype, 264
Neotype, 259–261, 264, 267
Nests, 81, 152, 154, 155, 157, 162, 300
New combination, 227, 248
   – systematics, 68, 273, 280, 286, 287, 303
Newell, I.M., 295
Newell, N.D., 295
New onymorph, 248
Niche, 6, 78, 125
Nichols, D., 295
Nightingale, 144
Nilsson, H., 295
Ninhydrin treatment, 93
Nitrogen, 10
Nixon, K.C., 280, 295
Nodasaur, 107
Nomenclator Zoologicus, 220, 282, 294
Nomenclatural Acts, 150, 281
Nomenclatural Rules, 150
Nomenclatural System, 151
Nomenclatural type, 259, 261
Nomen conservadum, 228, 254
   – dubium, 254
   – hybridum, 254
   – novum, 254
   – nudum, 254
   – oblitum,254
   – rejicundum, 254
   – triviale, 254
Nomina, 151, 281
Nominate, 23, 44
Non-HFC, 37
Non-Mendelian species, 246
Non-monophyletic groups, 112
Non-monophyly, 112
Noor, M.A.F., 295
Norris, K.P., 295
Norton, B.G., 24, 295
Norton, Cathy, 296
Nucleic acids, 71, 89, 91, 96
Nucleotide, 41, 90
   – sequences, 90

Numerical phenetics, 106, 300
Numericals in Compound Names, 249
Nungester, W.J., 297
Nut, 30
Nuts & Bolts of taxonomy, 231
Nuttal, G.H.F., 295
Nybakken, O.E., 212
Nylon, 153
Nymphs, 155

O

Ocean, 32, 40, 42, 44–46, 144, 151, 157
Octopus californicus bimaculatus, 204
Oenological, 30
Official Register, 318
Offprints, 254
O'Hara, J.R., 295
Ohno, S., 295
Oldroyd, H., 295
Olinguito, 43
Omaniundu, 43, 57
Oman, P.W., 295
Online identification programme, 296
Online Key, 183
Onomatophore, 151
Ontogeny, 289
Onychophora, 217
Oocyte, ovarian, 75
Oosterbroek, P., 295
Opah, 43
Operation taxonomic unit (OTU), 97
Operators, 129
Opinions, 224, 268
Optical discs, 281
Opuntia lindheimeri, 79
Order, 2, 20, 21, 44, 64, 86, 100, 114, 115, 120, 131, 136, 148, 150, 174, 176, 179, 182, 183, 213, 219, 268, 280, 288
Organdy cloth, 153
Orians, G.H., 295
Original descriptions, 196, 297

Origin of species, 137, 277, 279, 280, 286, 288, 290, 291
Origin of Species, 67, 135
Ornatype, 264
Ornithologists, 130
Ornstein, J., 280
O'Rourke, F.J., 295
Orr, H.A., 296
Orthodox taxonomists, 68, 98
Orthoptera, 154, 156, 170, 304
Orthotype, 264
Ostriker, J.P., 274
Oxen, 95
Oxidation, 166
Oxygen, 10, 37, 44
Ozone depleting substances (ODS), 37
Ozone depletion, 25

P

Pacific Remote Islands Marine National Monument, 45
Painkiller, 30
Painter, R.H., 301
Palaeontologists, 9, 269
Palaeontology, 103, 133, 140, 210, 295
Palaeospecies, 133
Pamphlets, 197, 205, 209
Panama, 31
Panchen, A.L., 296
Pankhurst, R.J., 296
Paper, 10, 93–95, 162–164, 168, 172, 174, 211–213, 219, 221, 244, 254, 271, 272, 277, 278, 283, 288, 292, 306
Papp, C.S., 212, 296
Paralectotype, 261, 265
Parallelism, 100
Paranyms, 258
Paraphyletic, 112, 241
Paraphyly, 112, 282, 295
Parasites, 5, 7, 10, 31, 55, 155, 157, 158, 204, 299
Paraspecies, 133
Parataxa, 133
Paratype, 260, 261, 265

Parentheses, use of, 248
Paris Agreement 38
Park, O., 296
Parr, A.E., 296
Parthenogenesis, obligatory, 132
Partridge, T.R., 306
Pasteur, G., 296
Paterson, H.E.H., 296
Pathogen, 36
Patrimony, 89
Patterson, D.J., 296
Paul, M.R., 290
Paulav, G., 292
Pauley, L.K., 296
Pauropoda, 217
PBI, 17
PEET, 17, 298
Pemberton, C.E., 296
Pentastomida, 217
Peptides, 89, 92, 93
Perfluorotributylamin (PERTBA) 37
Periodicals, World list of Scientific, 225
Peripheral population, 5
Permian period 49
Permian, 42, 49
Perez-Higarda, G., 301
Pests, 4–8, 153–157, 166, 175, 209, 290
Pesticide, 6, 10
Peterson, AT., 284
Petrunkevitch, A., 297
Pettersen, B., 297
Phanerozoic, 42
Phenetic similarity, 100
Pheneticist, 105
Phenol, 168, 199, 201
Phenon (Pl. phena), 139
Phenotype, 138, 140
Pheromone, 5
Philapatry, 319
Philosophica, Botanica, 236
Photuris, 81, 82, 275
Phthiraptera, 155
Phyletic branching, 106
Phyllocode 151
Phyllogenetic, 75, 150, 300

Phylloscopus collybita, 79
 – sibilatrix,79
 – trochilus, 79
Phylocode, 240, 241, 275, 297, 298
Phylogenetic, 3, 12, 25, 67, 68, 83,
    86, 90, 91, 95, 105, 106, 108, 109,
    123–125, 128, 129, 131, 133, 136,
    148, 182, 233, 240–242, 248, 275,
    276, 278, 280–282, 288, 295, 302,
    304, 305
 – adaptations, 68
 – affinities, 90
 – branching, 106
 – relationship, 12, 86, 90, 131,
    242, 275, 278
 – sequence, 109
 – taxonomy, 108, 109, 280
 – tree, 67, 110, 123, 129, 241, 282,
    304, 305
Phylogeny, 20, 66, 67, 87, 90, 91,
    105, 106, 108, 109, 112, 241, 242,
    276, 291, 298, 303
Phylum, 42, 110, 114
Physics, 2, 118
Physiological characteristics, 79
Physiology, 13–15, 65, 68, 69, 179,
    278
Pictures, 72, 191, 195, 232
Piechocki, R., 297
Pigeons, 93, 289
Pigs, 13, 95
Pin, Entomological, 169
Pinning block, 170, 172
Pisces, 218
Placozoa, 217
Planarian, 75
Planet, 21, 24–26, 37, 38, 40, 45,
    107, 144, 152
Planet, P.J., 299
Plankton net, 153, 158
Plant Protection, 222
Plant taxonomy, 226
Plant-eating dinosaurs 107
Plaster-of-paris, 163
Plastic, bags, vials, 198, 201
Plastotype, 261, 265
Platnick, N., 294
Plato, 63, 78, 119

Platyhelminthes, 217
Platypus 107
Plecoptera, 154
Plenary powers, 259, 268
Plesiotype, 260, 261, 265
Poaching, 36, 48, 49
Pogonophora, 217
Pointing, 172, 173
Points, 23, 48, 100, 140, 155, 169,
    172, 173, 182, 187, 198, 241, 256,
    272, 302
Polaszek, A., 297
Pollen, 28, 50
Pollination, 28
Pollinators 28, 54
Pollutant, 10
Pollution, 10, 11, 36–39, 44, 45, 49,
    279
Polyclave Identification, 182
Polymeric, 90
Polymorphic, 76, 89, 140, 264
 – morph, 89
 – stages, 76
Polymorphism, 119
Polynucleotides, 90
Polyphyletic, 109, 112, 113, 124,
    241
 – origin, 109
Polyphyly, 109, 112, 295
Polyploidy, 86
Polyporous pith, 172
Polysaccharides, 90
Polytene chromosomes, 86
Polythene bags, 200
Polythetic, 134
Polytopic, concept, 138
Polytypic, 68, 126, 134–136, 140
Polyzoa, 217
Population, 5, 7, 10, 11, 20, 25, 35,
    36, 40–42, 46, 49, 54, 68, 71, 83,
    86, 87, 100, 108, 109, 114, 118,
    120–129, 134–141, 143–145, 203,
    205, 231, 254, 282, 286, 287, 289,
    305
 – allopatric, 141
 – apomictic, 123
 – natural, 68, 120, 128
 – taxonomy, 68, 135

Porifera, 217
Porous, 93, 168
Ports, 8
Potassium cyanide, 163
Poultry, 13
Praying mantids, 81
Preamble, 242
Precipitate, 92, 93
Precipitin, 92
Predator, 5, 6, 7, 104, 107, 157
Pre-Linnaean name, 255
Preoccupied name, 254
Preparation of specimens, 167
Preservation, 9, 50, 143, 154, 166,
    167, 274, 279, 295, 304
Preservatives, 168
Priapulida, 217
Primary types, 146, 260, 261
Primates, 93
Priority, Law of, 251
Procecidochares utilis, 7
Proparte, 250, 251
Prospecting, 9
Prosser, C.L., 297
Protein, 29–31, 81, 83, 84, 89–96,
    166, 276–279, 291, 300–302, 306
    – taxonomy, 90
Proterotypes, 260
Protochordata, 147, 218
Protozoa, 217
Protozoological, 221
Protura, 154
Pseudotaxa, 135
Pseudotype, 265
Psocoptera, 155
Psychae, 63
Psychic, 118
Psychidae, 81
Pterygote, 112
Public health, 8, 21
Publication, 2, 24, 68, 100, 135, 143,
    205, 210, 211, 214–216, 218, 219,
    221–224, 228, 229, 243, 245, 250,
    252, 254, 264, 268, 269, 271, 272,
    274, 276, 281, 287, 306
Puig, H., 283
Punctuation marks, use of, 249
Pupae, 4, 29, 78, 163, 201

Pupal cases, 81
Purine, 129
Purvis, A., 273
Pyle, R., 284
Pyrgomorphidae, 88
Pyrimidine, 93, 129

Q

Quarantine, 8
Quardrupeda, 64
Quasi-three dimensional, 72
Queensland coast 32
Queiroz, K., 280
Quicke, D.L.J., 297

R

Races, geographical, 89, 136, 140
Radiates, 66
Radiation, 37, 145
Radioactive, 9, 37, 83
Rainstorms, 28
Rainforests, 21, 28, 54, 103
Ramsen, D., 296
Ran of Kutch, 35
Rana tigrina, 114, 258
Ranaivosolo, R., 293
Randall, H.M., 297
Ranidae, 114
Rank, 111, 113, 114, 139, 188, 241,
    245, 248, 251, 252, 257, 275
RANWA, 27
Rassenkreis, 140
Raven, P.H., 305
Raw, F., 297
Raymond, S., 297
Recapitulation, 67
Rechnitzer, A.B., 277
Redfield, A.C., 297
Register, 149, 217, 266, 297, 298
Rehmet, A., 293
Relaxing, 168, 169
Remane, A., 297
Rensch, B., 297
Replacement name, 254
Repressors, 129
Reprints, 214
Reproductive community, 120

Reproductively isolated, 89, 120, 128, 130
Reptiles, 35, 42, 48, 86, 144, 147, 157, 246, 281, 294, 295
Residues, 5, 10
Resistant plant variety, 5
Resistant strains, 5
Reticulate evolution, 109
Retrieval system, 65, 67, 180
Review of Agricultural Entomology, 221
Review of Applied Entomology, 221, 222
Review of Medical and Veterinary Entomology, 222
Reyment, R.A., 276
Rhino, 144
Ribonucleic acid (RNA), 321
Ribs, 104
Richardson, K.S., 283
Ride, W.D.L., 297
Ridgway, R., 297
Rieppel, O., 297
Rivas, L.R., 298
Robins, 114
Robinson, W.H., 298
Robotic Ants 33
Robson, G.C., 298
Rocks, 9, 32
Rockwood, E.S., 287
Rodman, J.E., 298
Rodrigues, D.P., 298
Rogers, D.J., 298
Rohlf, F.J., 298
Rohmberger, J.A., 298
Rollings, R.C., 298
Ronald, S., 298
Rosen, D., 298
Rosenbers, G. 298
Rosenkerg, M.S., 298
Rose-Ringed Parakeet, 9
Ross, E.S., 208, 298
Ross, H.H., 298
Rossetto, C.J., 298
Rothanaler, W., 298
Rotifera, 217
Royal, S. 284
Rubinoff, D., 299

Ruggiero, M., 276
Rundle, H.D., 299
Ruse, M., 299
Ryle, R. 298

S

Sabrosky, C.W., 299
Sachtlebev, H., 299
SAFE, 27
Sailor, R.I., 299
Salamander, 43, 107
Salivary gland, 89
Salt, G., 299
Sampling, 177, 292
Sanctuary, 46, 50
Sand degradation, 51
Sande, M.K., 299
Sarkar, I.N., 299
Savory, T., 299
Sawhney, V., 293
Saxena, K.N., 299
Scala Naturae, 66
Schaffer, H.E., 287
Schenk, E.T., 299
Schildkrant, C.L., 299
Schindewolf, O.H., 299
Schmidt, R.S., 300
Schmidt, W.L., 300
Schram, R.F., 298
Scolytidae, 72
Scorpion, 30
Scott, R., 300
Scudder, G.G.E., 300
Secondary types, 261
Sedimentary rocks, 9
Seed, 50, 152
Selander, R.K., 300
Semen Facial 31
Semispecies, 122, 123, 141
Senn, H.A., 300
Sensu, lato, 228, 250
Separates, 214, 219, 254
Septicemia, 44
Sequential branching, 111
Sereno, P.C., 300
Serum (PI. Sera), 92, 94
Seshachar, B.R., 300

Settler, R., 300
Sexual dimorphism, 119
Shark, 35, 43, 144
Sharov, A.G., 300
Sheep, 13, 35, 95
Shells, 10, 78, 81
Shellac gel, 172
Shrimp, 103
Sibley, C.G., 300
Sibling species, 8, 78–81, 96, 119,
   128, 131, 204, 290, 303
Sibson, R., 287
Sign, of congruence, 256
Silk, 11, 43
Silvestri, F., 300
Simonetti, J.A., 300
Simpson, G.G., 301
Siphonaptera, 157
Sipuncula, 217
Sister groups, 113
Slicker, 235
Slide box, 201
Smirnov, E.S., 301
Smith, D.W., 297
Smith, H.M., 301
Smith, K.B., 301
Smith, N.A., 279
Smith, R.C., 301
Smith, R.I., 301
Smith, S.G., 301
Smith, W.J., 301
Smithies, O., 301
Snails, 30, 94, 135, 288
Snakes, 98, 210, 300
Sneath, P.H.A., 301
Social insects, 162
Society for conservation of
   Biodiversity, 52
Software, biodiversity, 233
Soil-micro-biota, 54
Sokal, R.R., 301, 302
Solbrig, O.T., 302
Song pattern, 79, 80
Sonographs, 79
Sound, 11, 16, 52, 68, 80, 125, 150,
   166, 304
South Delhi Municipal
   Corporation 56

Specialist, 143, 178, 179, 196, 197,
   207, 267
Speciation, 4, 41, 117, 122, 124, 127,
   141, 147, 226, 279, 282, 284, 287,
   288, 290, 296, 300, 303, 305, 306
Species, 3–17, 19–28, 30, 32–36,
   41–44, 46–51, 53–55, 63–69, 71–
   73, 75, 78–90, 92, 93, 95–97, 101,
   103–115, 117–141, 143–150, 167,
   175, 176, 178–182, 188, 191, 193,
   196, 202–207, 209–211, 226–229,
   231–233, 235, 236, 239–242,
   244–249, 251–253, 255–267, 273,
   275, 277, 279–286, 288–306
– agamo, 132
– allopatric, 131, 134, 141
– aphanic, 130
– apomictic, 132
– biological, 12, 34, 62, 118–128,
   133
– collective, 121
– complexes, 15, 68, 72, 86
– contemporaneous, 133
– continental, 131
– cosmopolitan, 132
– form, 133
– fossil, 103, 116, 266
 – inquirenda, 255
– insular, 131, 132
– invasive, 19, 21, 36, 49
– linnaean, 67
– monotypic, 133, 134, 135, 140
– montane, 132
– morpho-geographical, 132
– nascent, 141
– non-dimensional, 133
– panmictic, 132
– pantropical, 132
– parapatric, 132
– paraspecies, 133
– polytypic, 126, 133, 134, 135,
   136, 140
– registry, 148
– successional, 133
– transient, 133
– tropicopolitan, 132
Species concept, 64, 68, 118–125,
   127–129, 135, 146, 273, 275, 277,

279–282, 284, 285, 290, 291, 295, 297, 300–303, 305
- Biological, 64, 118, 120, 121, 122, 123, 124, 125, 127, 128
- Cohesion, 126
- Differential fitness, 127
- Ecological, 125
- Evolutionary, 119, 123
- Genotypic cluster, 125
- Genic, 126, 127
- Phylogenetic, 124, 125
- Recognition, 119, 124
- Typological, 118, 119
Species-group name, compound, 249
Species group, 112, 258
Species Programme, 34
Specific name, 244, 245, 254
Spectrophotometry, 92, 96
Sperm, 32, 50
Sphagnum, 202
Spices, 29
Spider rain 43
Spider, 31, 43, 46, 252
Splayfoot, 43, 57
Sponges, 78
Spraying, 6
Spreading board, 174, 175
Square brackets, use of, 248
Staging, 169, 172
Staniland, L.N., 302
Stanley, J.T., 302
Starting-point date, 252
Statistical method, 100, 101
Statistician, 15
Stebbins, G.L., 302
Stephen, W.P., 302
Stereomicroscope, 72
Sterile, 123
Stevens, P.F., 302
Stevenson, H.J.R., 302
Steyskal, G.C., 302, 303
STI, 149
Stimulus,
Stomach ulcer, 30
Stone, N.D., 303
Stone, W.S., 296
Storey, M., 303

Stratiographic, 166, 266
Stratosphere, 37
Strepsiptera, 156
Strickland, H.E., 303
- code, 237
Stricto, 228, 250
Stuffed animal, 33
Subclass, 115
Subcohort, 115
Subfamily, 244
Subgenus, 141, 220, 244, 250, 251, 288, 289
Subkingdom, 115
Suborder, 115
Sub-Saharan Africa 30
Subspecies, 16, 50, 71, 114, 122, 130, 133–141, 206, 207, 229, 244, 245, 248, 255, 257, 273, 277, 281, 282, 285, 287, 288, 291, 292, 303, 304, 306
- annual, 138
- ecological, 138
- geographical, 137
- geological, 138
- polytopic, 138
- seasonal, 137
- temporal, 137
Substitute name, 254
Subtribe, 115
Successional cline, 140
Suction, 160, 161,162
Suffixes, use of, 252
Sugars, 89
Sulawesi Island, 43
Sunscreen, 32, 37
super salamander, 107
Superclass, 115
Supercohort, 115
Superfamily, 115, 244
Supergenus, 115
Superorder, 115
Superphylum, 115
Superspecies, 115, 141
- chronological, 141
- geographical, 141
Supertribe, 115
Supra-Specific, 3
Sussdorf, D.H., 277

Sustainable agriculture 54
Sustainable development, 20, 21,
    27, 51
Sweeping, 153, 154–157
Swift, H.H., 286
Swimming worms, 103
Sylvester Bradley, P.C., 303
Symbiont, 204
Symmorphic, 130
Sympatric, 79, 124, 131, 137, 138,
    140, 141, 298
Sympatry, 131
Symphylida, 217
Symplesiomorphic, 106
Synapomorphic, 106
Synchronic, 137, 138
Syngameon, 123
Synonym, 3, 67, 145, 208, 229, 246,
    249, 255–257, 259
    – homotypic, 256
    – nomenclatural, 256
    – objective, 256
Synonymy, 151, 238, 255, 256, 301,
    302
    – objective, 256
    – subjective, 256
Synthesys, 17
Syntopy, 131
Syntypes, 264, 266, 267
Syphilis, 30
Systema Naturae, 1
Systematic Association, 226, 231
    – Biology, Society,
Sytematic zoology, 219, 225, 226

T

T. rex, 108
Tardigrada, 217
Tautonym, 255
Taxa (s. taxon), 2, 3, 14, 21, 63, 66,
    67, 85, 86, 89, 95, 97, 98, 105,
    109, 112–114, 128, 150, 151, 178,
    181–184, 202–206, 208, 210, 213,
    216, 220, 226–228, 235, 238, 241,
    242, 244, 245, 250, 254, 258–260,
    281, 285
    – richness, 21

– search, 230
Taxeonomie, 1
Taximeters, 97
Taximetrics, 97
Taxinomie, 1
Taxinomy, 1
Taxometry, 97
Taxonometrics, 97
Taxonomic data working group,
    232
Taxonomic Entomology, 223
Taxonomic explosion, 68
Taxonomic Indexing, 218, 233
Taxonomic inflation, 16
Taxonomie, 1
Taxonomy, 1–5, 8, 10, 12–17,
    19–23, 27, 58, 59, 63–69, 71, 83,
    85, 86, 89, 90, 93, 95, 97, 98, 100,
    101, 105, 106, 108–110, 114, 124,
    128, 130, 132, 134, 135, 137, 139,
    143, 146, 147, 149, 151, 205, 210,
    213, 215, 217, 221–223, 226, 227,
    229–233, 235, 236, 238, 259, 262,
    272–306
Taylor, C., 280
Tchistovitch, T., 303
Telechow, A., 303
Templeton, A.R., 303
Temporal, 117, 122, 137, 138
Tenebrio molitor (yellow
    mealworms), 33
Tephritidae, 78, 86, 87, 206, 207,
    284
Termites, 29, 154, 200, 204
Terpenes, 90
Tertiary types, 261
Tertiary, 42, 184, 186–188, 261
Tetrachloroethane, 165
Tettigoniidae, 80
The Majesty of Life, 19
The Marine Trench 45
Thielcke, G., 303
Thompson, W.R., 303
Thorpe, W.H., 303
Throckmorton, L.H., 303
Thysanoptera, 156
Thysanura, 111, 154
Ticks, 72, 157, 190

Tikataalik rosae, 103
Tilden, J.W., 303
Tillage, 35
Tinbergen, N., 303
Tindall, B.J., 284
Ting, C.T., 306
Tiselius, A., 304
Tissues, 91–93, 96
Toad, 29, 125, 145, 146, 209
Tobias, M.A., 282
Topotype, 260, 261, 264, 265
Towards stability in names of
   animals, 225
Townes, H.K., 286
Toxicology, 69
Toxin, anti fungal, 31
Trabecular layers, 78
Tracks, 263
Traditional taxonomists, 85
Traditional Taxonomy, 3, 85, 101
Trails, 263
Trait, 41, 55, 203, 231
Transient, 133
Trawling, 151
Tree of Life, 14, 103, 110, 148, 232
Trehane, P., 284
Trematodes, 78
Triassic period, 42, 107
Tribe, 115, 243
Triceratops, 107
Trichoptera, 81, 86, 88, 156, 170,
   288
Trigynaspid mites, 97
Trilobita, 217
Trinomen, 244
Trinominal Nomenclature, 135,
   244
Trust of Zoological Nomenclature,
   238, 270
Tsunamis, 45
Tubbs, P., 284
Tundras, 26
Turbellaria, 75
Turner, B.L, 273, 304
Turtle, 46, 47, 96, 279, 306
Twins, 120
Type
   – Linnaean, 259

Type specimens, 175–177, 193,
   196, 259, 261, 264, 266
Typification, 238, 239, 250,
   259–261, 268
Typological, 98, 118, 119, 139–241
Typotype, 265

U

Ultrasonic, 80
Ultraviolet, 37, 75, 93, 145, 151
UNECC, 37
UNESCO World Heritage Panel,
   32
Uninominalism, 247
Uniparental, 123–125, 127, 132
Unisexual, 132
Unitary Taxonomy, 230, 233
United Nations Frame Work
   Convention, 56
United Nations, 29, 32, 51, 56, 222
Universal Web Taxonomy, 233
Universe, 25, 119
Uprety, D.R., 304
Urodeles, 86
US National Oceanic and
   Atmospheric Administration, 39
Usinger, R.L., 290, 292
Utahceratops, 107

V

Valdecasas, A.C., 305
Valentine, D.H., 304
Vampire Facial, 31
van der Kloot, W.G., 304
van Sande, M., 304
van Son, G., 304
van Tyne, J., 304
van Valen, L., 304
Valid name, 251–253, 257, 259
Variant, 122, 139, 140, 256, 286
Variation, 25, 81, 84, 86, 119, 139,
   140, 147, 204, 205, 231, 255, 256,
   260, 273, 276, 279, 280, 282, 290,
   297, 298, 302
Variety, 14, 24, 25, 34, 65, 114, 115,
   135, 139, 245
Vasopressin, 95

Vector (character state), 182
Vector, 7, 8, 182
Vernacular name, 235
Vertebrate, 4, 16, 28, 63, 66, 144,
   158, 176, 273, 275, 278, 281, 282,
   287, 290, 305
Vevers, H.G., 282
Vibrio, 44
Vienna Convention, 38
Vignes, R., 283
Virtual Museum of Natural
   History, 24
Vitelline membrane, 78
Vogel, H.J., 277
von Baer's Law, 66
Voss, E.G., 304
Voucher specimen, 265
Vultures, 43, 50

W

Wagele, J.W., 304
Wagner, R.P., 304
Wagner, W.H., 304
Wagstaffe, R., 304
Wainstein, B.A., 304
Walbank, B.E., 304
Walker, T.J., 304
Warburton, F.E., 304
Warren, B.C.S., 304
Wasp, 79, 155, 160
Wasraj Bait, 35
Wassef, H.Y., 289
Waterhouse, D.F., 304
Waterston, J., 304
Watertight, 69
Watson, L., 305
Watson, M., 289
Wax, 31
Web of Life, 34
Web, Taxonomy & Biodiversity,
   229
Websites, 229, 230, 271, 272
Weems, H.V., 280
Weighting, 105
Weintraub, L., 297
Weizenbaum, J., 305
Werren, J.H., 303

Western Ghats of India 35
Wetlands, 26, 46, 54
Wetmore, A., 305
Whales, 46, 63
Wheeler, Q.D., 295, 305
Whippey, 30
White, M.J.D., 305
Whiteflies, 78
White-Ringed Vulture, 35
Wikimedia Foundation, 232
Wikispecies, 232
Wild food, 54
Wild, R., 293
Wildlife, 9, 24, 26, 33, 35, 44, 48–51
Wiley, E.O., 305
Williams, C.A., 305
Williams, C.M., 304
Williams, W.T., 305
Willis, E.O., 305
Willmott, K., 290
Wilmoth, J.H., 305
Wilmowsky, N.J., 303
Wilson, A.C., 305
Wilson, E.O., 305, 306
Wilson, J.B., 306
Windbreaks, 28
Wing pigment, 92
Wire mesh, 162
Witworth, T.L., 306
Wolfe, S., 306
Woltereck, H., 306
Womble, W.H., 306
Wood, R., 285
Wooden boxes, 197
Woods, R.S., 306
Woodsford, F.P., 212
Woolley, J.B., 306
World Biosphere Reserve, 51
World Diversity Databases, 232
World Network of Biosphere
   Reserves, 49
World Wildlife Fund, 33, 44
Wright, C.A., 306
Wright, M.H., 274
Wrists, 104
Wu, C.L., 306
WWF, 33
Wyatt, G.R., 306

X

Xenophora, 81
XPER Programme, 183

Y

Yamuna Biodiversity Park, 44
Yeast, 79
Young ones, 64, 181
Young, F.N., 306

Z

Zhang, Z. 306

Zimmermann, W., 306
Zompro, O., 288
Zoo Outreach Organisation, 36
Zoo taxa, 232
Zoo, 36
Zoobank, 149, 150
Zoogeography, 68, 69, 208
Zoological Code, 243, 248, 249,
    252, 257, 264, 265, 269, 270
Zoological record, 215–221, 244
Zooplankton, 103
Zweig, G., 306
Zygote, 120